Dielectric Materials for Capacitive Energy Storage

Due to growing energy demands, the development of high-energy storage density dielectric materials for energy storage capacitors has become a top priority. *Dielectric Materials for Capacitive Energy Storage* focuses on the research and application of dielectric materials for energy storage capacitors. It provides a detailed summary of dielectric properties and polarization mechanism of dielectric materials and analyzes several international cases based on the latest research progress.

- Explains the advantages and development potential of dielectric capacitors.
- Discusses energy storage principles of dielectric materials as well as effects of polarization and breakdown mechanisms on energy storage performance.
- Summarizes achievements and progress of inorganic and organic dielectric materials as well as multidimensional composites.
- Details applications and features international case studies.
- Offers unique insights into existing issues and forecasts for future research priorities.

With its summary and large-scale analysis of the fields related to dielectric energy storage, this book will benefit scholars, researchers, and advanced students in materials, electrical, chemical, and other areas of engineering working on capacitors and energy storage.

Haibo Zhang is a full professor in the School of Materials Science and Engineering at the Huazhong University of Science and Technology (HUST), Wuhan China. Before joining HUST in 2014, he worked as a postdoctoral researcher at Kochi University Japan and Humboldt Research Fellow at Technique University of Darmstadt Germany from 2011 to 2014. His research interests include lead-free piezoelectric ceramics and piezoelectric devices, and dielectric materials for energy storage applications. He has published over 150 peer-reviewed SCI research papers and has 20 Chinese patents.

Hua Tan is currently an assistant professor in the School of Materials Science and Engineering at the Huazhong University of Science and Technology, Wuhan, Hubei, China. He received his PhD degree in Advanced Ceramic Materials from the Central European Institute of Technology, Brno University of Technology, Brno, Czech Republic, in 2020. He has published over 40 papers and received over 10 patents by now. His primary research interests include the rapid sintering of advanced ceramic materials and the preparation and properties of dielectric energy storage ceramics.

Emerging Materials and Technologies

Series Editor: Boris I. Kharissov

The *Emerging Materials and Technologies* series is devoted to highlighting publications centered on emerging advanced materials and novel technologies. Attention is paid to those newly discovered or applied materials with potential to solve pressing societal problems and improve quality of life, corresponding to environmental protection, medicine, communications, energy, transportation, advanced manufacturing, and related areas.

The series takes into account that, under present strong demands for energy, material, and cost savings, as well as heavy contamination problems and worldwide pandemic conditions, the area of emerging materials and related scalable technologies is a highly interdisciplinary field, with the need for researchers, professionals, and academics across the spectrum of engineering and technological disciplines. The main objective of this book series is to attract more attention to these materials and technologies and invite conversation among the international R&D community.

Nanofluids
Fundamentals, Applications, and Challenges
Shriram S. Sonawane and Parag P. Thakur

MXenes
From Research to Emerging Applications
Edited by Subhendu Chakroborty

Biodegradable Polymers, Blends and Biocomposites
Trends and Applications
Edited by A. Arun, Kunyu Zhang, Sudhakar Muniyasamy and Rathinam Raja

Bioinspired Materials and Metamaterials
A New Look at the Materials Science
Edward Bormashenko

Computational Studies
From Molecules to Materials
Edited by Ambrish Kumar Srivastava

2D Semiconductors for Environmental Remediation
Edited by Honey John, Nisha T Padmanabhan, Sona Stanly and Jith C Janardhanan

Materials from Natural Sources
Edited by Ramesh Gardas, Neha Patni and Amita Chaudhary

Dielectric Materials for Capacitive Energy Storage
Edited By Haibo Zhang and Hua Tan

For more information about this series, please visit: www.routledge.com/Emerging-Materials-and-Technologies/book-series/CRCEMT

Dielectric Materials for Capacitive Energy Storage

Edited by Haibo Zhang and Hua Tan

CRC Press
Taylor & Francis Group
Boca Raton London New York

CRC Press is an imprint of the
Taylor & Francis Group, an **informa** business

Designed cover image: Shutterstock

First edition published 2025
by CRC Press
2385 NW Executive Center Drive, Suite 320, Boca Raton FL 33431

and by CRC Press
4 Park Square, Milton Park, Abingdon, Oxon, OX14 4RN

CRC Press is an imprint of Taylor & Francis Group, LLC

ISBN: 978-1-032-58249-8 (hbk)
ISBN: 978-1-032-59396-8 (pbk)
ISBN: 978-1-003-45449-6 (ebk)

DOI: 10.1201/9781003454496

Typeset in Times
by codeMantra

Contents

Preface

This book delves into the intricate realm of dielectric materials and their pivotal role in advancing the field of energy storage. As the demand for efficient and sustainable energy solutions escalates, the exploration and optimization of dielectric materials become increasingly crucial.

This scholarly endeavor is crafted to serve as a definitive resource for researchers, engineers, and students engaged in the interdisciplinary pursuit of energy storage technologies. This book synthesizes a wealth of knowledge encompassing the fundamentals of dielectric materials, the diverse landscape of inorganic and organic dielectric materials, the innovative realm of composite dielectric materials, and the intricate processes involved in the fabrication of film and multilayer ceramic capacitors. Furthermore, it comprehensively covers the wide-ranging applications of these materials in various industries.

The opening section of this volume meticulously explores the fundamentals of dielectric materials for energy storage. Understanding the underlying principles governing the behavior of dielectric materials is paramount for developing advanced capacitive energy storage systems. Through an in-depth examination of the theoretical foundations, this section aims to provide readers with a solid grasp of the essential concepts that underpin subsequent discussions.

The journey into dielectric materials continues with a meticulous exploration of both inorganic and organic dielectric materials. Inorganic materials have long been stalwarts in the field, providing stability and reliability, while organic materials present a frontier of innovation with their unique properties. This section delves into the distinct characteristics, advantages, and challenges associated with each class of dielectric materials, offering a comparative analysis that aids researchers in selecting the most suitable materials for their specific applications.

A paradigm shift in dielectric materials research involves the development of composite materials. This section scrutinizes the synergistic effects achieved by combining different dielectric components, resulting in enhanced performance characteristics. The exploration of composite dielectric materials represents a pivotal step toward tailoring materials with superior energy storage capabilities, paving the way for the next generation of capacitors.

The practical realization of dielectric materials in capacitors necessitates a deep understanding of the processing techniques involved. This section elucidates the intricate processes involved in the fabrication of film capacitors and multilayer ceramic capacitors. From material preparation to the assembly of capacitive devices, this segment serves as a practical guide for engineers and researchers engaged in the manufacturing of these essential components.

The concluding part of this volume explores the diverse applications of dielectric materials across various industries. From consumer electronics to renewable energy systems, the impact of dielectric materials on modern technology is profound. This section provides insights into the practical implementations of dielectric materials, showcasing their versatility and adaptability in addressing the evolving needs of different sectors.

Dielectric Materials for Capacitive Energy Storage aspires to be a definitive reference for the scientific community, fostering a deeper understanding of dielectric materials and their role in advancing energy storage technologies. The contributors to this volume have brought forth their expertise to create a resource that not only surveys the current state of the field but also sparks inspiration for future innovations. We hope this work serves as a catalyst for further exploration and breakthroughs in the exciting and dynamic realm of capacitive energy storage.

Contributors

Shuaikang Huang
Huazhong University of Science and
Technology, Wuhan, P.R. China

Shenglin Jiang
Huazhong University of Science and
Technology, Wuhan, P.R. China

Mohsin Ali Marwat
Ghulam Ishaq Khan (GIK) Institute
of Engineering Sciences and
Technology, Topi, Pakistan

Bing Xie
Nanchang Hangkong University,
Nanchang, P.R. China

Zuo-Guang Ye
Simon Fraser University, Burnaby, BC,
V5A 1S6, Canada

1 Fundamentals of Dielectric Materials for Capacitive Energy Storage

Haibo Zhang, Hua Tan, and Shenglin Jiang

1 INTRODUCTION

The problem of energy consumption caused by the increasing population needs to be solved urgently. The contradiction between finite non-renewable energy and the escalating demands of human energy engenders a confluence of challenges such as the depletion of fossil energy, global warming and environmental pollution. With the vigorous development of renewable energy sources such as solar and wind energy, the problem of energy mismatch in time and space has become more and more prominent. The conversion of renewable energy into electrical energy and their storage through energy storage devices can effectively solve this problem.[1–8] It places higher demands on the performance of energy storage devices and the innovative development of the energy storage fields.

Figure 1.1 shows the main energy storage performance parameters of current efficient energy storage devices. The battery has a high-energy density, but its inherent characteristic that the energy inside it is stored by chemical reactions leads to its low charge-discharge rates and power density. Achieving higher power density, better stability and safety in batteries requires more innovations in related fields.[10–13] With lower energy density but higher power density than batteries, electrochemical capacitors are another highly efficient energy storage devices, which are often used in aircraft emergency door opening devices, brake assists and starter generators in automobiles. However, the shortcomings of electrochemical capacitors, such as electrolyte leakage, low-rated voltage and poor temperature resistance, limit their further development.[14–19]

Compared with batteries and electrochemical capacitors, dielectric capacitors have lower cost, faster charge-discharge rates (<0.01 s), higher power density (>10^5 W/kg), lower loss and higher operating voltages. Thus, they are the preferred choice if energy storage applications with high voltage, low cost and large-scale fabrication are considered.[20–33] Dielectric capacitors are the core components of electric vehicle inverters, medical pacemakers, HVDC transmission systems and radars.[34–36] They are widely used in various fields encompassing but not limited to medical, automotive, military, communications, electronics, oil and gas exploration.[37]

DOI: 10.1201/9781003454496-1

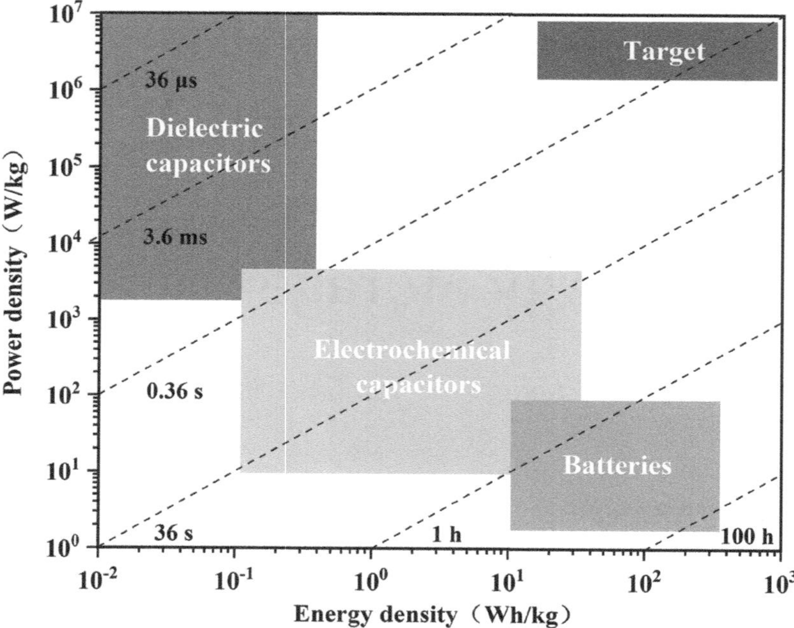

FIGURE 1.1 Power density/energy density comparison chart of different energy storage devices. The dotted line represents the energy release time.[9]

The dielectric materials of conventional dielectric capacitors can be classified into polymers and ceramics. Ceramic dielectric materials have ultra-high permittivity, high modulus and good high-temperature consistency. Polymer dielectric materials have remarkable properties such as low cost, high-rated voltage, high-energy storage efficiency and easy processing. In addition, the polymer dielectric has a unique self-healing behavior in the case of local dielectric breakdown over a large dielectric film. When a breakdown or failure occurs, the film forms an open circuit with other parts instead of a short circuit, thus avoiding overall damage. Dielectric capacitors can generally be divided into two main categories according to the materials used, ceramic capacitors and film capacitors. Ceramic capacitors, also known as multilayer ceramic capacitors, consist of ceramic dielectric foils with printed electrodes (inner electrodes) laminated in a staggered manner, sintered at high temperature at one time to form a ceramic chip, and then sealed with metal layers (outer electrodes) at both ends of the chip to form a monolithic-like structure. With the rapid development of the world electronics industry, multilayer chip ceramic capacitors, a basic component of the electronics industry, are also moving forward at an incredible rate, increasing by 10%–15% every year. The global demand for multilayer ceramic capacitors is over 200 billion pieces, 70% of which are from Japan, followed by Europe, America and Southeast Asia (including China). With the improvement of the reliability and integration of multilayer ceramic capacitors, the scope of their use is getting wider and wider. Specifically, they are widely used in various electronic machines and electronic devices for military and civilian use, such as computers, telephones, program-controlled switches, precision test instruments and radar communications. Film capacitors are constructed by using metal foils as electrodes, which are wounded with plastic films such as polyethylene, polypropylene, polystyrene or polycarbonate from both ends to form a cylinder. Film capacitors are mainly used in many industries such as electronics, home appliances, communication, electric power, electrified railroad, hybrid cars, wind power, solar power, etc. The steady development of these industries has driven the growth of the film capacitor market. With the development of technology level, the renewal cycle of multiple industries such as electronics, home appliances and communication is getting shorter and shorter. Moreover, film capacitors, with their good electrical performance and high reliability, become indispensable electronic components to promote the renewal of the above industries. In the coming years, with the further development of digitalization, informatization, and network construction; along with the augmented national investments in power grid construction, electrified railroad construction, energy-saving lighting and hybrid cars; as well as the upgrading of consumer electronic products, the market demand for film capacitors will further show a rapid growth trend.

The cores of these capacitors are dielectric ceramics and dielectric organic materials. Based on energy storage dielectric materials, this book first introduces the basics of energy storage dielectric materials, mainly including key parameters, polarization mechanism and breakdown mechanism. Then, the current problems of dielectric material systems and their solution strategies are introduced, and the research progress of dielectric energy storage materials is further presented from three aspects: inorganic, organic and composite materials. It also introduces how to design and prepare high-performance dielectric capacitors briefly from two aspects: multilayer ceramic capacitors and organic film capacitors. Finally, we summarize the achievements and problems of energy storage capacitors in recent years and offer a comprehensive overview of the practical applications and further prospects across different energy storage and dielectric fields.

2 PRINCIPLES AND CHARACTERISTIC PARAMETERS OF ENERGY STORAGE

2.1 ENERGY STORAGE PARAMETERS

A dielectric capacitor is an energy storage device composed of two parallel electrodes and a dielectric between the electrodes.[38, 39] When the capacitor is charged, the two electrodes are filled with equal amounts of opposite charges, forming an external electric field. Dipoles and free charges in the dielectric migrate locally in the electric field, which leads to polarization while generating an internal electric field.[40, 41] The direction of the internal electric field is opposite to the external electric field and increases with the accumulation of the charge. Once the internal electric field is consistent with the external electric field, the charging ends. At this time, the maximum charge accumulation is Q_{max}. The stored energy (W) in the capacitor can be calculated from Equation (1.1).

$$W = \int_0^{Q_{max}} V dq \tag{1.1}$$

where V is the voltage across the electrode and dq is the charge increment. The formula for calculating the energy storage density (U_s) of a dielectric is as follows:[42]

$$U_s = \frac{W}{Ad} = \frac{\int_0^{Q_{max}} V dq}{Ad} = \int_0^{D_{max}} E \, dD \tag{1.2}$$

E is the external electric field strength; dD is the electric displacement increment; D_{max} is the maximum electric displacement under the maximum applied electric field.

The discharge energy density is the amount of energy that can be released in a unit volume of dielectric after charging is complete, expressed as U_d, which is obtained by integrating the hysteresis loop:[43, 44]

$$U_d = \int_{P_r}^{D_{max}} E \, dD \tag{1.3}$$

where P_r is the remnant polarization at $E = 0$ MV/m, which is one of the important parameters indicating the ferroelectric properties of the dielectrics. The electric displacement (D) is related to the relative permittivity (ε_r) of the material and the polarization (P) of the dielectric[45, 46]

$$D = P + \varepsilon_0 E = \varepsilon_r \varepsilon_0 E \tag{1.4}$$

where ε_0 is a constant (8.85×10^{-12} F/m) indicating the vacuum permittivity. In linear dielectric materials, the ε_r is independent of the magnitude of the electric field.[47, 48] Therefore, the U_d calculation formula (Equation 1.3) can be expressed as follows in a linear dielectric:

$$U_d = \int_0^{E_{max}} \varepsilon_0 \varepsilon_r E \, dE = \frac{1}{2} \varepsilon_0 \varepsilon_r E^2 \tag{1.5}$$

In the actual charging and discharging process, the strong coupling between dipoles leads to the occurrence of hysteresis, which inevitably results in energy loss (U_l), and the U_l increases with increasing temperature and electric field. The energy storage density (U_s) consists of U_d and U_l. The charge-discharge efficiency (η) of dielectric, also known as energy efficiency, is calculated from the ratio of U_d to U_s.[44]

$$\eta = \frac{U_d}{U_s} = \frac{U_d}{U_d + U_l} \tag{1.6}$$

The D–E hysteresis loop is a hysteresis phenomenon of displacement with electric field change, which produces a loop where the charge and discharge curves do not coincide. The ferroelectric domains in a ferroelectric dielectric with the same polarization direction are oriented under the electric field to overcome the interdomain

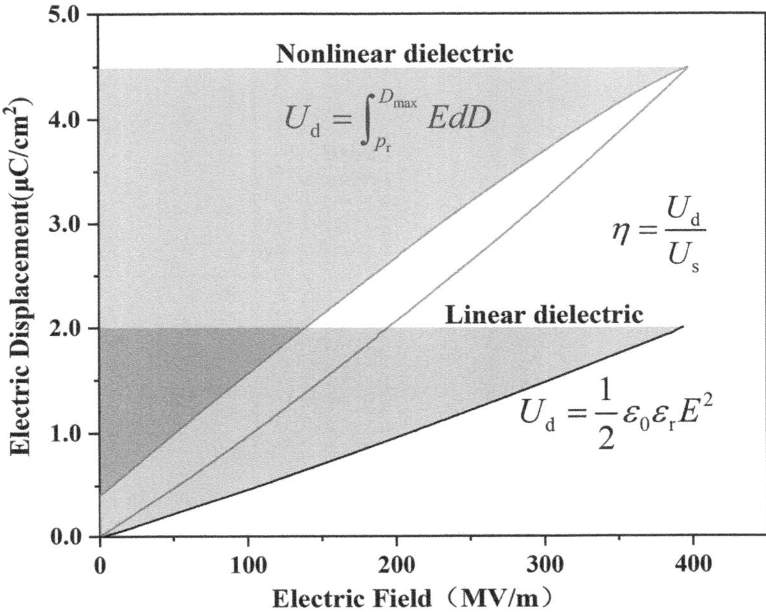

FIGURE 1.2 Characteristic D–E hysteresis loops of linear and nonlinear dielectrics under the same electric field. The shaded area is the U_{d} of the dielectric.[9]

forces and exhibit a nonlinear D–E relationship. Linear dielectrics do not have ferroelectric domains but still have other losses and exhibit a narrow D–E loop. The D–E loops can characterize the energy storage properties of the material. As shown in Figure 1.2, integrating the hysteresis loop of nonlinear dielectric, the integral of the orange shaded area is the U_{d}, and the integral of the white area is the U_{l}. In practical applications, the energy density we need is actually the dielectric material in the integral of the discharge curve over the electric displacement axis.

Energy storage performances of dielectric materials can be reflected by P–E loops measured by a modified Sawyer–Tower circuit. Meanwhile, the energy storage characteristics of capacitors, including effective discharging time ($t_{0.9}$) and power density (P), are more accurately reflected by the charging-discharging curve recorded at a specific RLC circuit. The simple equivalent circuit model is exhibited in Figure 1.3(a). In the charging process ("1" connected to "3"), the potential difference (Φ) between two surfaces of capacitor equals the applied voltage (V), representing the charging process is finished. And then, the vacuum switches automatically and quickly rotate ("2" connected to "3") to achieve the discharging process. Current flows to the oscilloscope via a load (R_0), which is recorded as wave function as a function of time. The discharging density is given by

$$J(t) = \frac{\int_0^t V(t)I(t)\,dt}{\varphi} = \frac{\tau V_0^2}{2\varphi R_0}\left(1 - e^{-\frac{2t}{\tau}}\right) \tag{1.7}$$

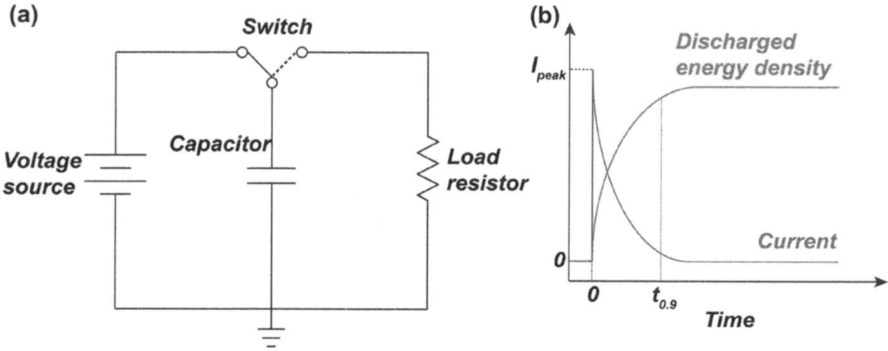

FIGURE 1.3 (a) Circuit of charging-discharging. (b) Discharging current (I) and energy density (J) versus time.[49]

where φ: the effective volume; τ: the relaxation time; $V(t)$: voltage as function of time; $I(t)$: current as function of time.

Discharging current (I) and energy density (J) versus time (t) are shown in Figure 1.3(b). It is reported that the value of J calculated by Equation (1.7) is generally smaller than that of U_d. For these reasons, it may be closely related to two factors. First, the domains cannot switch and orientate promptly due to fast discharge speed, and thus the energy is not completely released. Second, the circuit has equivalent series resistance that generates Joule energy. In order to acquire higher value of J and U_d, reducing domain size (RFEs) may be an effective method.

Power density (P) is also an important parameter of dielectric ceramic capacitors, which is determined as follows:

$$P = \frac{\pi f \varepsilon_0 \varepsilon_\mathrm{r} E_\mathrm{b}^2}{2 \tan \delta} \tag{1.8}$$

where f: the testing frequency.

It can be seen that high ε_r, high E_b, and low $\tan\delta$ are the basic requirements of achieving high P. Meanwhile, P has an obvious dependence on frequency (f), which means high f corresponds to high P. In addition, $t_{0.9}$ is an important factor in influencing the practical application of ceramic capacitors. The smaller the value of $t_{0.9}$, the stronger the pulse current can be generated in a short time. It is worth mentioning that the value of $t_{0.9}$ is dependent on not only the intrinsic properties of the dielectric materials but also some external factors.

Whether being with dielectric material or pulse capacitor, the reliability of work is a crucial factor to influence its application scenes. In general, the reliability of work requires maintaining the stability of electric properties of materials and devices under various external field stimuli such as thermal, electric, mechanical, and magnetic influences.

Temperature stability usually requires that dielectric properties and polarization of material have a gentle fluctuation as a function of temperature. For example, dielectric constant needs to maintain a ±15% variation over a temperature range from −55°C to 125°C, and even up to 300°C, especially in high-temperature working environments. Therefore, one should know the Joule heat categories: dielectric loss or leakage current, and to solve corresponding issues. Generally, different electric field conditions, such as cycle number, frequency, voltage category and magnitude, all require keeping good energy storage stability. Specially, fatigue endurance is a very important issue to influence capacitor working capability. In addition, mechanics and magnetic fields both influence the polarization behavior of the material, and their importance may play a crucial role in the future multi-functional coupling requirement. Therefore, the evaluation criterion of reliability of working should be a complicated project.

2.2 Relative Permittivity and Dielectric Loss

The relative permittivity (ε_r), also known as the dielectric constant, is critical to the dielectric properties of a material. The relative permittivity is a parameter that characterizes the ability of a dielectric material to bind charges and polarization. It can be calculated by the ratio of the dielectric capacitance (C_x) measured by adding a dielectric layer between the two electrode plates of the capacitor and the vacuum capacitance (C_0) measured under a vacuum.

$$\varepsilon_r = \frac{C_x}{C_0} \tag{1.9}$$

The permittivity can also be calculated by coating electrodes on two parallel surfaces of the dielectric and measuring the impedance.[50] The relative permittivity is expressed as follows:

$$\varepsilon_r = \varepsilon_r' - i \cdot \varepsilon_r'' \tag{1.10}$$

ε_r' is the real part of the relative permittivity, which determines the capacitance; ε_r'' is the imaginary part and represents the dielectric loss. The degree of loss is usually expressed by the dielectric loss factor (tan δ).[51]

$$\tan \delta = \frac{\varepsilon_r''}{\varepsilon_r'} \tag{1.11}$$

Dielectric loss is mainly composed of conduction loss and polarization loss (also known as relaxation loss), which is inversely proportional to the η. As the dielectric is not completely insulated, there is current leakage in the dielectric during the charging and discharging process. The conduction loss originates from the energy loss due to the leakage current generated by the free carriers. The polarization loss is related to the rotational loss of the bound charge (or dipole) of the dielectric under polarized and depolarized electric fields. Meanwhile, the dielectric/conductivity mismatch between filler and matrix or at the multilayer interface also contributes to polarization loss. Dielectric loss is frequency-dependent. Under a low-frequency electric field, the conduction loss is the main form of loss, and the loss value is almost constant. At a high-frequency electric field, the dipole orientation cannot immediately follow the change of a high-frequency external electric field, and the polarization loss gradually increases and dominates.[52–54] The ambient temperature and humidity in the measurement conditions also have a large effect on the dielectric loss. Under high temperatures and humidity, the leakage current and conduction loss of the dielectric increases rapidly, which is related to the breakdown mechanism and the conduction mechanism.[55, 56]

2.3 POLARIZATION MECHANISMS

Polarization is an important reason for the energy storage properties of dielectrics. [57, 58] Polarization is the vector sum of the dipole moments produced by an electric dipole in a unit volume of the dielectric. The polarization is correlated with ε_r and the equation is as follows:

$$P = (\varepsilon_r - 1)\varepsilon_0 E \tag{1.12}$$

The total polarization of dielectric is the sum of different polarization mechanisms including electronic polarization, ionic polarization, dipolar polarization and interfacial polarization.[59] Figure 1.4 shows the frequency dependence of four polarizations with different mechanisms and their corresponding dielectric loss peaks.

The mechanism of electronic polarization is shown in Figure 1.5(a). Atoms have a symmetrical structure with overlapping positive and negative charge centers, which exhibit electrical neutrality in the absence of an electric field. When an electric field is applied, the lighter electrons move more readily than the nucleus. The center of the negative charge deviates from the center of the positive charge, which creates an induced dipole moment called electronic polarization.[41, 59] Electronic polarization exists in all dielectrics, but the dipole moment generated by electronic polarization is very small and will disappear after removing the external electric field.[60]

The principle of ionic polarization is shown in Figure 1.5(b). In the initial state, anions and cations are connected end-to-end. The dipole moments created by anions and cations are neutralized by each other, and the whole does not exhibit polarization characteristics. When an electric field is applied in a certain direction, cations move in the positive direction of the electric field, while anions move in the opposite direction. The positive dipole moment increases and the negative dipole moment decreases in the dielectric, resulting in ionic polarization.[38, 61, 62] The enhancement of ionic polarization can lead to a large improvement in polarization properties, which is one of the sources of the high ε_r of ceramics such as $BaTiO_3$.[63] Electronic polarization and ionic polarization are intrinsic within the molecule and are independent of temperature. In a pulsed electric field, the dipoles of electronic polarization and ionic polarization can rapidly follow the electric field changes and resonate in the ultraviolet and infrared range frequencies, respectively.[64]

Dipolar polarization, also known as orientation polarization, is the orientation of the intrinsic dipoles of molecules or molecular chains in the direction of the electric field. As shown in Figure 1.5(c), under an electric field, the internal dipole is subjected

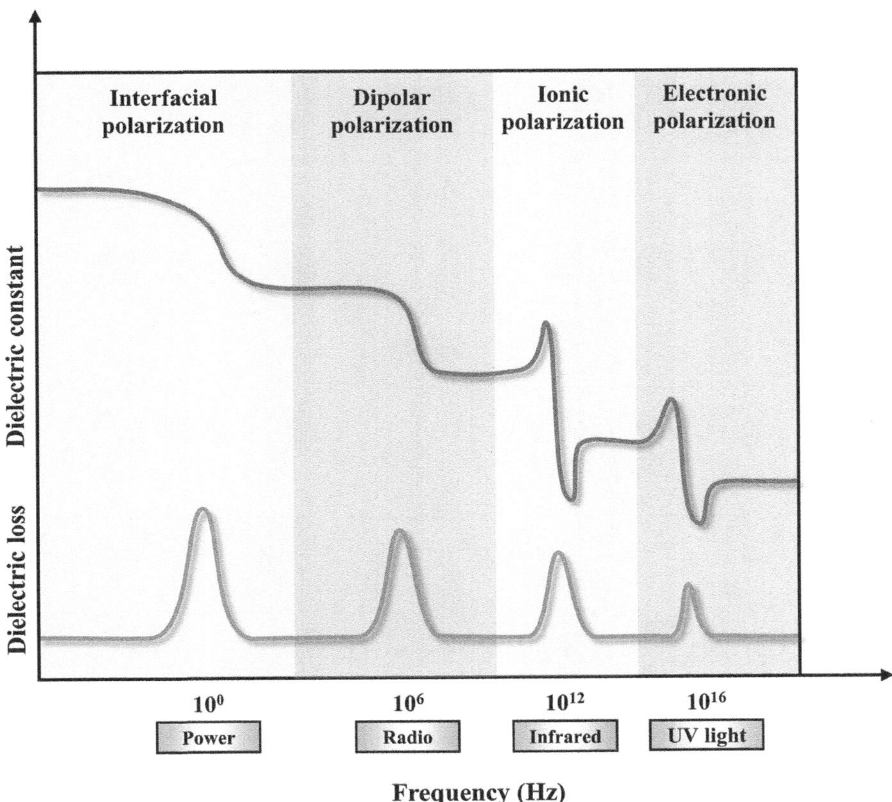

FIGURE 1.4 Different types of polarizations and their frequency dependences. Dielectric constant and corresponding loss depicted with red and bluelines, respectively.

to electric field forces and has a tendency to form an ordered arrangement in the electric field direction, thus exhibiting orientation polarization.[65, 66] Temperature and frequency affect the dipolar polarization of dielectrics. Excessively high or low temperatures will block the ordered alignment of polymer chain segments along the electric field direction, which reduces the dipolar polarization of dielectrics. As the frequency of the electric field increases, the change in dipole orientation cannot follow the change in the electric field, resulting in a weak dipole polarization. In polymer systems, dipolar polarization is mainly influenced by the geometry of the polar groups and polymer chain segments. Introducing polar groups into polymer chains is a viable modification method for achieving excellent dielectric properties.[41, 66, 67]

Interfacial polarization, also called Maxwell–Wagner–Sillars polarization, occurs mainly at the two-phase interface of dielectrics. As shown in Figure 1.5(d), interfacial polarization is generated by the accumulation of interfacial charges, which is related to the difference in the dielectric properties of the materials on both sides of the interface.[59, 66, 67] The interfacial polarization is the main reason for the high ε_r of polymer-inorganic composites. In polymer systems, interfacial polarization may exist

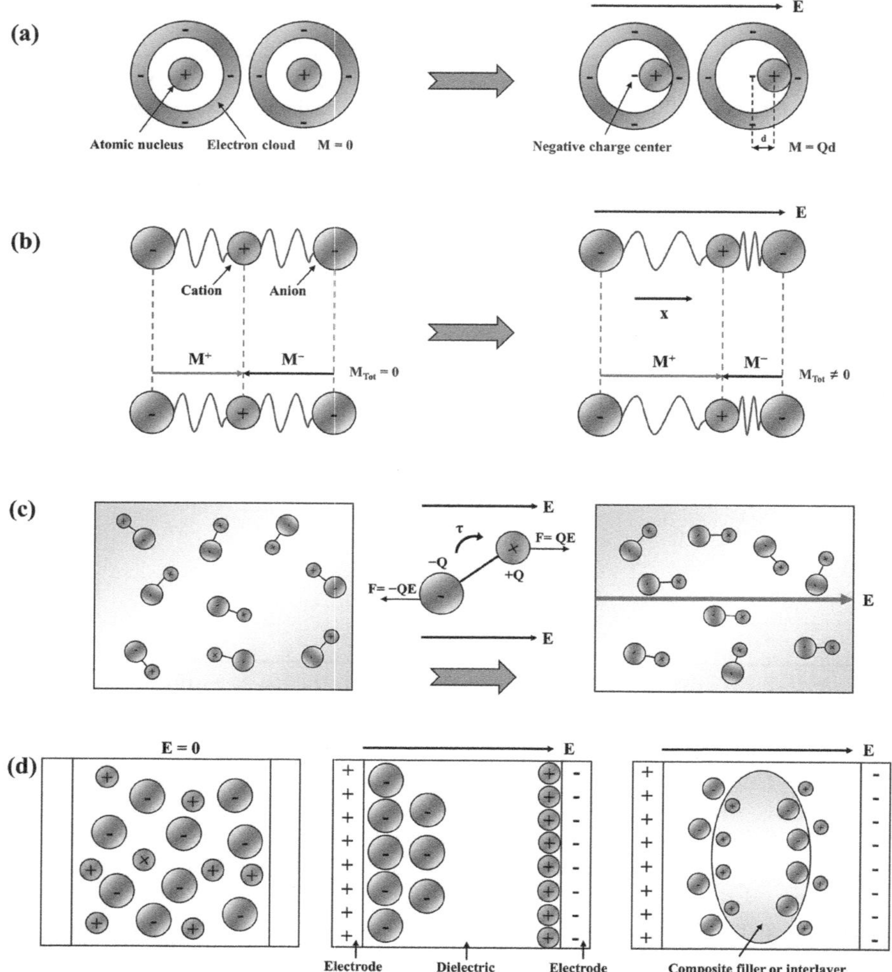

FIGURE 1.5 Schematic diagrams of various polarizations: (a) electronic polarization, (b) ionic polarization, (c) dipolar polarization and (d) interfacial polarization.[9]

at the two-phase interface of the polymer, the impurities and defects in the polymer, the crystal-amorphous interface, and the interface between the dielectric layer and the electrodes. Numerous studies have demonstrated that designing polymers into multilayer structures can enhance interfacial polarization and significantly improve the dielectric properties of materials. Dipolar polarization and interfacial polarization are greatly affected by temperature, and the ε_r will change accordingly.[65] During the discharge process, the discharge of dipolar polarization and interfacial polarization is slow. It takes a certain time for the dipole to return to the equilibrium state, which can only occur in the frequency range shown in the relaxation region of Figure 1.4.[64]

3 BREAKDOWN MECHANISMS

Dielectric breakdown means that the dielectric will lose its dielectric properties and become a conductor under an excessive electric field. The maximum electric field of the dielectric before this transition occurs is called the breakdown strength (E_b). The U_d of dielectrics is positively related to the square of the E_b.[68–70] Therefore, the E_b is crucial to the energy storage properties of the dielectric. The E_b is generally measured by applying a slowly increasing DC voltage to both sides of the sample until the breakdown. In order to accurately measure the E_b of materials, the Weibull distribution is usually used to analyze the measurement data, and the function is expressed as follows:[70]

$$P(E) = 1 - \exp\left[-\left(\frac{E_i}{E_w}\right)^{\beta}\right] \tag{1.13}$$

The Weibull distribution is a continuous probability distribution. To facilitate the calculation, the logarithmic form of the Weibull distribution is commonly used to calculate the Weibull characteristic breakdown strength (E_w).

$$\ln(-\ln(1 - P(E))) = \beta \ln(E_i) - \beta \ln(E_w) \tag{1.14}$$

$P(E)$ denotes the cumulative breakdown probability, which is estimated by different estimators;[71] E_i is the experimentally measured breakdown value; β is the shape factor, which indicates the stability of the measured value. By a linear fit of $P(E)$ and E_i, the β can be obtained and the breakdown strength at $P(E) = 63.2\%$ can be estimated and defined as E_w.

In particular, when conductive polymers are composite with polymers as organic fillers, the ε_r and E_b of the composite system follow the percolation theory.[72–74]

$$\varepsilon_r \propto (p_c - p)^{-s} \tag{1.15}$$

$$E_b \approx \frac{(p_c - p)^{v}}{\ln(L)} \tag{1.16}$$

Where s is the critical exponent, which is related to the dimensionality of the fillers; v is the applied voltage; L denotes the linear dimension of the polymer matrix; when the volume fraction of the conductive polymer (p) is well below the percolation threshold (p_c), there is no significant change in the ε_r. As p increases close to p_c, ε_r increases significantly. However, an order of magnitude increase in conductivity leads to the formation of conductive pathways and a rapid decrease in the E_b of the dielectric. The change in electrical properties with filler volume fraction is known as the percolation phenomenon.[75, 76] The geometry of the filler is critical to achieve low p_c and high dispersion in the composite.[73] Zero-dimensional (0D) spherical fillers have the advantages of low cost, easy preparation and high dispersion, but p_c is generally high. The use of one-dimensional (1D) and two-dimensional (2D)

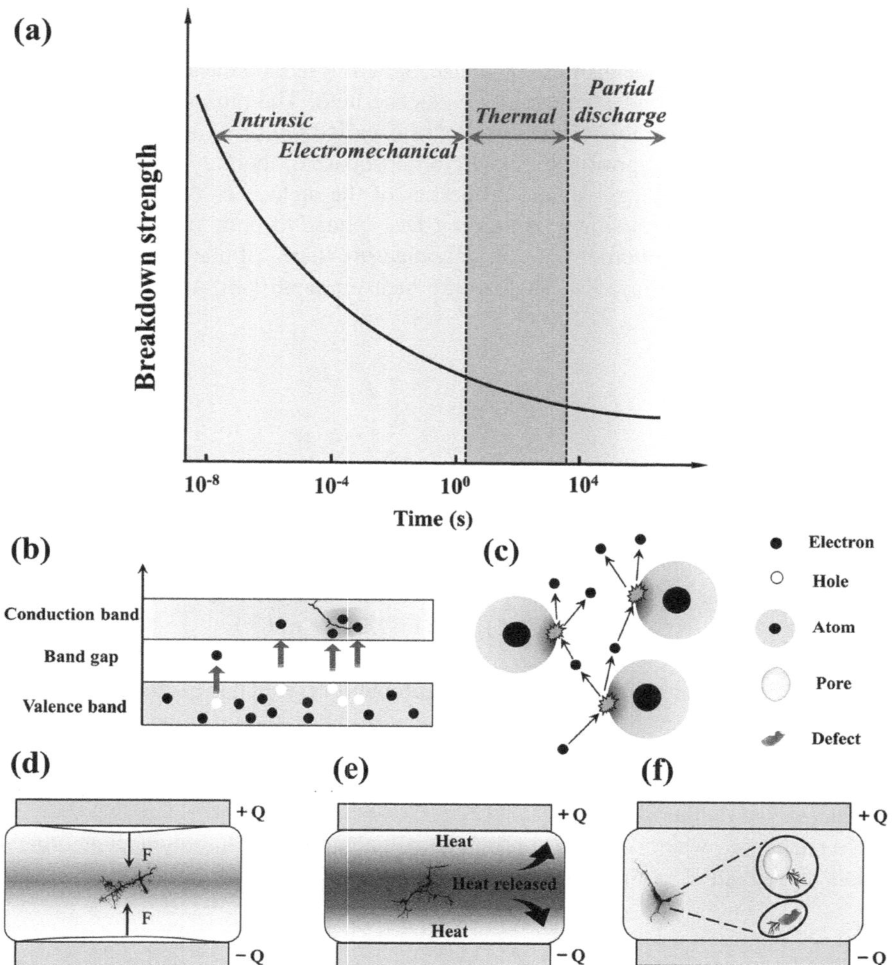

FIGURE 1.6 (a) Breakdown mechanism and the corresponding breakdown strength versus applied voltage time. Schematic diagram of different types of breakdown mechanisms: (b) electronic breakdown, (c) electron avalanche breakdown, (d) electromechanical breakdown, (e) thermal breakdown and (f) partial discharge breakdown.[80]

fillers can significantly reduce p_c. For example, fibrous conductive polymers facilitate the formation of conductive pathways and p_c decreases with increasing filler aspect ratio.[77]

The differentiated E_b divides the application scenarios of different materials. The defects, the width of the forbidden band, the thickness of the dielectric and the grain size of different materials will affect the E_b of materials. Furthermore, different measurement conditions, electrode state, sample thickness and voltage type (DC, AC, or pulse) of the same material will also cause dielectrics to exhibit different E_b.[70, 78, 79] According to the different breakdown mechanisms, the breakdown can be divided into intrinsic breakdown, electromechanical breakdown, thermal breakdown and partial discharge breakdown (Figure 1.6(a)).[80] These several generally accepted breakdown mechanisms are proposed to help explain the complex and unclear breakdown processes of dielectric materials.

3.1 INTRINSIC BREAKDOWN

The intrinsic breakdown describes the inherent breakdown strength of a uniform dielectric under experimental circumstances. The intrinsic breakdown can be classified into electronic breakdown and electron avalanche breakdown.[49] In an ideal dielectric, electrons can gain energy from the applied electric field to overcome the energy gap between the valence band and the conduction band and achieve conduction (Figure 1.6(b)). The larger the applied electric field and the smaller the energy gap, the more likely electron breakdown will occur.[81, 82] With the increase in the electrons of the conduction, it ultimately leads to electronic breakdown. In practical research, it is found that the defects in the dielectric can capture electrons, which form a defect energy level in the forbidden band. This reduces the width of the energy gap that needs to be overcome for electronic breakdown. The electronic breakdown strength (E_{ele}) can be expressed by the Equation (1.17).[49]

$$E_{ele} = \frac{A_{ele}}{\sqrt{\Delta V}} \exp\left(\frac{\Delta V}{2k_B T}\right) \tag{1.17}$$

where A_{ele} is a constant; ΔV is the energy required for an electron to transition from the defect energy level to the conduction band; k_B denotes the Boltzmann constant.

Electron avalanche breakdown occurs during electron conduction.[83, 84] Electrons move under the electric field and conduct energy to the lattice atoms through collisions. When the energy of a lattice atom surpasses its ionization energy, the atom releases electrons. The released electron continues to repeat the collision process, eventually producing an avalanche multiplication of electrons (Figure 1.6(c)).[85, 86] The electron avalanche breakdown is related to the dielectric thickness, and the electron avalanche breakdown strength (E_{ava}) of the material is calculated as follows:

$$E_{ava} = \frac{k}{ln\left(\frac{a_0 d}{40}\right)} \tag{1.18}$$

where a_0 and k are the constants determined by the material. From the above equation, the avalanche breakdown strength is proportional to the thickness (d) of the dielectric, which explains the phenomenon that thin-film dielectrics exhibit higher breakdown strength than thick films in the experiments. The increase in temperature and voltage will increase the energy of the electrons and accelerate the electron movement. High-energy electrons are more likely to cross the energy gap and exhibit stronger electron collisions, ultimately leading to a reduction in intrinsic breakdown strength.[87–90]

3.2 ELECTROMECHANICAL BREAKDOWN

When a high voltage is applied to the dielectric, different kinds of charges are generated on both sides of the dielectric in contact with the electrodes. These charges will attract each other to generate electrostatic pressure, which will deform the dielectric, as shown in Figure 1.6(d). Electromechanical breakdown occurs when the electrostatic pressure exceeds the mechanical compressive strength of the dielectric.[91–93] Electromechanical breakdown often occurs in solid dielectrics that are easily deformable and have low elastic modulus. The equation for calculating the electromechanical breakdown strength (E_{em}) is as follows:

$$E_{em} = \frac{d}{d_0} E_c \approx 0.6 \sqrt{\frac{Y}{\varepsilon_r \varepsilon_0}} \tag{1.19}$$

where d/d_0 is the residual ratio of the dielectric thickness. When the thickness of the dielectric becomes 60% of the original thickness, the imbalance of internal stress leads to electromechanical breakdown. E_c is the critical electrical stress; Y is the Young's modulus of the dielectric.

Electromechanical breakdown strength is positively correlated with Y and negatively correlated with ε_r of dielectrics. When the temperature is near the glass transition temperature (T_g) or even the melting point, the bond lengths and bond angles of the chains in the polymer move more freely.[94, 95] This results in a corresponding decrease in Young's modulus and causes the electromechanical breakdown.[96, 97] In addition, the electromechanical breakdown is associated with the electrostriction effect, which is a result of dielectric polarization and is prevalent in dielectrics. The electromechanical breakdown strength is inversely proportional to the electrostriction strain induced by the polarization.[98]

3.3 THERMAL BREAKDOWN

Thermal breakdown is a challenge for dielectric capacitors used in high-temperature applications, the schematic is shown in Figure 1.6(e). When an electric field is applied, there is leakage current and dielectric loss in the dielectric.[99–101] During charging and discharging, the dielectric loss causes the unreleased energy in the dielectric to be dissipated in the form of heat, thereby increasing the temperature of the dielectric. An increase in temperature causes a rapid increase in the conductivity of the dielectric, which leads to greater conduction loss and creates a positive feedback loop. Thermal breakdown happens as the accumulated heat exceeds the heat release.[101–103] The existence of thermal breakdown is the major reason for the high-temperature breakdown of many polymers, which reduces the dielectric performance and jeopardizes the stability and safety of the capacitor.[104, 105] Reducing the conductivity and dielectric loss of high-temperature dielectrics through modification, or improving the thermal conductivity of materials, will help expand the application of dielectric capacitors in high-temperature scenarios.

3.4 PARTIAL DISCHARGE BREAKDOWN

The above discussion of several different breakdown mechanisms for dielectric is based on the assumption that the dielectric is a pure homogeneous material, ignoring the effects of voids and pores in the dielectric. However, the practical preparation of continuous and uniform dielectrics is very difficult. There are voids and pores between the second phase and polymer matrix, between electrodes and matrix, and inside the matrix.[106–108] The voids and the low-ε_r medium inside voids will deteriorate the breakdown performance of the material, which generates an uneven electric field in the dielectric and leads to partial discharge breakdown (Figure 1.6(f)).[92, 108–110] In addition, the corona discharge generated by the gas medium in the void is also an important cause of partial discharge breakdown.[111] Partial discharge is related to long-term breakdown and has no significant impact on the short-term use of the dielectric. Long-term partial discharge will form dendritic breakdown pathways, corrosion and space charge accumulation, which gradually reduces the dielectric properties of the material, eventually causing the dielectric breakdown at normal operating voltages.[109]

4 CONDUCTION MECHANISMS

The leakage current present in the dielectric is one of the major reasons for dielectric loss and breakdown. As the electric field and temperature increase, the leakage current increases rapidly and the η decreases, which leads to more heat loss and affects the safety and stability of the dielectric.[112, 113] Depending on the source of carriers, the conduction mechanisms of dielectrics can be classified into bulk-limited conduction mechanisms and electrode-limited conduction mechanisms.[114–116] The bulk-limited conduction mechanisms are related to the intrinsic parameters of materials, such as trap energy level and density, carrier drift mobility and energy gap width. According to different mechanisms, the bulk-limited conduction mechanisms can be subdivided into Poole–Frenkel (P–F) emission, hopping conduction, ohmic conduction, space-charge-limited conduction and ionic conduction. The electrode-limited conduction mechanisms mainly depend on the electrical properties at the electrode/dielectric interface, which can be divided into Schottky emission, Fowler–Nordheim (F–N) tunneling, direct tunneling and thermionic field emission.[63] Among them, the P–F emission, hopping conduction and Schottky emission have the greatest influence on the leakage current at high temperatures.[116–118]

4.1 BULK-LIMITED CONDUCTION

Ohmic conduction is produced by the directional displacement of carriers in the valence and conduction bands in an electric field.[119] The current density of ohmic conduction follows Ohm's law which is proportional to the external electric field strength. However, ohmic conduction can only explain the conductance process under a low electric field. As the electric field increases rapidly, the change of carrier density and mobility in the polymer dielectric makes the current increase significantly, which no longer obeys Ohm's law.[120] The leakage current generated by ohmic conduction is small and can generally be ignored.

The Poole–Frenkel (P–F) emission and hopping conduction gradually increase in high-temperature environments, becoming the dominant conduction mechanisms for high-temperature polymer dielectrics.[112, 121] In the P–F emission mechanism, with the increase in electric field and temperature, the intrinsic carriers in the traps obtain energy to transition from the trap energy level to the conduction band (Figure 1.7). The escaped carriers move directionally to form a leakage current.[122–124] Both P–F emission and Schottky emission are processes in which thermally excited carriers overcome energy barriers to achieve conduction. Therefore, P–F emission is also known as internal Schottky emission. The current density of P–F emission is calculated as follows:

$$J = \sigma_0 E \exp\left[\frac{-q\left(\phi_B - \sqrt{qE/\pi\varepsilon_r\varepsilon_0}\right)}{k_B T}\right] \tag{1.20}$$

where σ_0 is the conductivity; $q\Phi_B$ is the trap energy level; k_B represents the Boltzmann constant. By increasing the barrier height, the leakage current of the P–F emission mechanism can be effectively reduced.

The hopping conduction is related to the tunnel effect, which is determined by the quantum characteristics of electrons. As shown in the energy band diagram (Figure 1.7), although the energy of the carriers in the trap state is lower than the potential barrier, the carriers can still engage in "hopping" between the trap sites to form conduction through the tunnel effect.[122, 125–127] The current density of hopping conduction is expressed as:

$$J = qanv \exp\left(\frac{qaE - E_a}{k_B T}\right) \tag{1.21}$$

where a is the average distance between traps; n is the concentration of carriers in the conduction band; v is the thermal vibrational frequency of carriers in the traps; E_a is the activation energy to overcome the potential barrier.[128, 129] Moreover, space charge-limited conduction and ionic conduction exist in the dielectric, but contribute less to the leakage current due to harsh conditions and complex mechanisms.[130–133]

4.2 Electrode-Limited Conduction

Schottky emission is the dominant conduction mechanism for electrode-limited conduction at high temperatures. Due to the difference in band structure between the electrode and the dielectric, there is a Schottky barrier between the electrode/dielectric interface, as shown in Figure 1.7. As the temperature increases, the carriers in the Fermi energy level at the electrode gain energy from the environment, cross the interfacial barrier by thermal excitation and enter the dielectric to form a leakage current, a process known as Schottky emission.[112, 115, 134] In high temperature

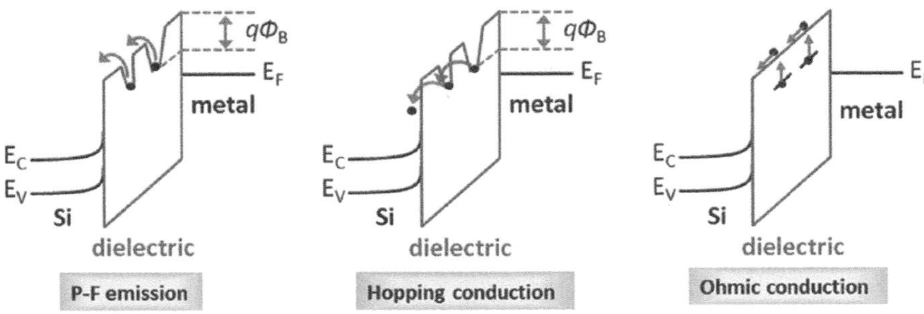

FIGURE 1.7 Energy band diagrams of P–F emission, hopping conduction and Schottky emission. E_V is the valence band, E_C is the conduction band, E_F is the Fermi level and $q\Phi_B$ is the barrier height.[115]

and high electric field environments, electrode carrier emission increases rapidly and becomes the main source of leakage current.[115, 135] The expression for Schottky emission is as follows:

$$J = AT^2 \exp\left[\frac{-q\left(\phi_B - \sqrt{qE / 4\pi\varepsilon_r\varepsilon_0}\right)}{k_B T}\right] \tag{1.22}$$

where A is the Richardson constant; $q\Phi_B$ is the Schottky barrier height.

The mechanism of both Fowler–Nordheim (F–N) tunneling and direct tunneling is related to the tunnel effect.[136] If the interfacial barrier of the electrode/dielectric is small, carriers can enter the conduction band directly via the tunneling phenomenon. The thermionic field emission is more difficult to occur, which elaborates that carriers with energy above the Fermi energy level can conduct through tunneling into the dielectrics.[137, 138]

Various conduction mechanisms lead to increased leakage currents and impair energy storage performance, especially in high-temperature environments.[115] The conduction can be effectively suppressed by modification methods such as surface treatment, introduction of deep traps and multiphase design, which we elaborate in subsequent chapters.[139, 140]

REFERENCES

1. Chen, H.; Cong, T. N.; Yang, W.; Tan, C.; Li, Y.; Ding, Y. Progress in electrical energy storage system: A critical review, *Prog. Nat. Sci.* **2009**, 19, 291–312. DOI: 10.1016/j.pnsc.2008.07.014.
2. Dincer, I. Renewable energy and sustainable development: A crucial review, *Renewable Sustainable Energy Rev.* **2000**, 4, 157–175. DOI: 10.1016/S1364-0321(99)00011-8.
3. Gross, R.; Leach, M.; Bauen, A. Progress in renewable energy, *Environ. Int.* **2003**, 29, 105–122. DOI: 10.1016/S0160-4120(02)00130-7.
4. Hall, P. J.; Bain, E. J. Energy-storage technologies and electricity generation, *Energy Policy* **2008**, 36, 4352–4355. DOI: 10.1016/j.enpol.2008.09.037.

5. Jurasz, J.; Canales, F. A.; Kies, A.; Guezgouz, M.; Beluco, A. A review on the complementarity of renewable energy sources: Concept, metrics, application and future research directions, *Sol. Energy* **2020**, 195, 703–724. DOI: 10.1016/j.solener.2019.11.087.

6. Kousksou, T.; Bruel, P.; Jamil, A.; El Rhafiki, T.; Zeraouli, Y. Energy storage: Applications and challenges, *Sol. Energy Mater. Sol. Cells* **2014**, 120, 59–80. DOI: 10.1016/j.solmat.2013.08.015.

7. Panwar, N. L.; Kaushik, S. C.; Kothari, S. Role of renewable energy sources in environmental protection: A review, *Renewable Sustainable Energy Rev.* **2011**, 15, 1513–1524. DOI: 10.1016/j.rser.2010.11.037.

8. Tai, Q.; Tang, K. C.; Yan, F. Recent progress of inorganic perovskite solar cells, *Energy Environ. Sci.* **2019**, 12, 2375–2405. DOI: 10.1039/c9ee01479a.

9. Huang, S.; Liu, K.; Zhang, W.; Xie, B.; Dou, Z.; Yan, Z.; Tan, H.; Samart, C.; Kongparakul, S.; Takesue, N.; Zhang, H. All-organic Polymer dielectric materials for advanced dielectric capacitors: Theory, property, modified design and future prospects, *Polym. Rev.* **2022**, 1–59. DOI: 10.1080/15583724.2022.2129680.

10. Li, M.; Lu, J.; Chen, Z.; Amine, K. 30 years of lithium-ion batteries, *Adv. Mater.* **2018**, 30, 1800561. DOI: 10.1002/adma.201800561.

11. Rajagopalan, R.; Tang, Y.; Ji, X.; Jia, C.; Wang, H. Advancements and challenges in potassium ion batteries: A comprehensive review, *Adv. Funct. Mater.* **2020**, 30, 1909486. DOI: 10.1002/adfm.201909486.

12. Tang, X.; Zhou, D.; Li, P.; Guo, X.; Sun, B.; Liu, H.; Yan, K.; Gogotsi, Y.; Wang, G. MXene-based dendrite-free potassium metal batteries, *Adv. Mater.* **2020**, 32, 1906739. DOI: 10.1002/adma.201906739.

13. Tarascon, J. M.; Armand, M. Issues and challenges facing rechargeable lithium batteries, In *Materials for Sustainable Energy*; Co-Published with Macmillan Publishers Ltd, London, **2010** pp. 171–179.

14. Kötz, R.; Carlen, M. Principles and applications of electrochemical capacitors, *Electrochim. Acta* **2000**, 45, 2483–2498. DOI: 10.1016/S0013-4686(00)00354-6.

15. Kusko, A.; Dedad, J. Stored energy - Short-term and long-term energy storage methods, *IEEE Ind. Appl. Mag.* **2007**, 13, 66–72. DOI: 10.1109/MIA.2007.4283511.

16. Liu, H.; Liu, X.; Wang, S.; Liu, H.; Li, L. Transition metal based battery-type electrodes in hybrid supercapacitors: A review, *Energy Storage Mater.* **2020**, 28, 122–145. DOI: 10.1016/j.ensm.2020.03.003.

17. Simon, P.; Gogotsi, Y. Materials for electrochemical capacitors, In *Nanoscience and Technology*; Co-Published with Macmillan Publishers Ltd, London, **2009** pp. 320–329.

18. Simon, P.; Gogotsi, Y. Perspectives for electrochemical capacitors and related devices, *Nat. Mater* **2020**, 19, 1151–1163. DOI: 10.1038/s41563-020-0747-z.

19. Yao, K.; Chen, S.; Rahimabady, M.; Mirshekarloo, M. S.; Yu, S.; Tay, F. E. H.; Sritharan, T.; Lu, L. Nonlinear dielectric thin films for high-power electric storage with energy density comparable with electrochemical supercapacitors, *IEEE Trans. Ultrason. Ferroelectr. Freq. Control* **2011**, 58, 1968–1974. DOI: 10.1109/TUFFC.2011.2039.

20. Mannodi-Kanakkithodi, A.; Treich, G. M.; Huan, T. D.; Ma, R.; Tefferi, M.; Cao, Y.; Sotzing, G. A.; Ramprasad, R. Rational co-design of polymer dielectrics for energy storage, *Adv. Mater.* **2016**, 28, 6277–6291. DOI: 10.1002/adma.201600377.

21. Marwat, M. A.; Ma, W.; Fan, P.; Elahi, H.; Samart, C.; Nan, B.; Tan, H.; Salamon, D.; Ye, B.; Zhang, H. Ultrahigh energy density and thermal stability in sandwich-structured nanocomposites with dopamine@Ag@BaTiO3, *Energy Storage Mater.* **2020**, 31, 492–504. DOI: 10.1016/j.ensm.2020.06.030.

22. Marwat, M. A.; Xie, B.; Zhu, Y.; Fan, P.; Ma, W.; Liu, H.; Ashtar, M.; Xiao, J.; Salamon, D.; Samart, C.; Zhang, H. Largely enhanced discharge energy density in linear polymer nanocomposites by designing a sandwich structure, *Compos. Part A Appl. Sci. Manuf.* **2019**, 121, 115–122. DOI: 10.1016/j.compositesa.2019.03.016.

23. Shehzad, K.; Ul-Haq, A.; Ahmad, S.; Mumtaz, M.; Hussain, T.; Mujahid, A.; Shah, A. T.; Choudhry, M. Y.; Khokhar, I.; Ul-Hassan, S.; Nawaz, F.; Rahman, F. u.; Butt, Y.; Pervaiz, M. All-organic PANI–DBSA/PVDF dielectric composites with unique electrical properties, *J. Mater. Sci.* **2013**, 48, 3737–3744. DOI: 10.1007/s10853-013-7172-5.

24. Sherrill, S. A.; Banerjee, P.; Rubloff, G. W.; Lee, S. B. High to ultra-high power electrical energy storage, *Phys. Chem. Chem. Phys.* **2011**, 13, 20714–20723. DOI: 10.1039/c1cp22659b.

25. Wang, Y.; Chen, J.; Li, Y.; Niu, Y.; Wang, Q.; Wang, H. Multilayered hierarchical polymer composites for high energydensity capacitors, *J. Mater. Chem. A* **2019**, 7, 2965–2980. DOI: 10.1039/c8ta11392k.

26. Wu, X.; Gandla, D.; Lei, L.; Chen, C.; Tan, D. Q. Superior discharged energy density in polyetherimide composites enabled by ultra-low ZnO@BN core-shell fillers, *Mater. Lett.* **2021**, 290. DOI: 10.1016/j.matlet.2021.129434.

27. Xie, B.; Wang, Q.; Zhang, Q.; Liu, Z.; Lu, J.; Zhang, H.; Jiang, S. High energy storage performance of PMMA nanocomposites utilizing hierarchically structured nanowires based on interface engineering, *ACS Appl. Mater. Interfaces* **2021**, 13, 27382–27391. DOI: 10.1021/acsami.1c03835.

28. Xie, B.; Zhang, H.; Zhang, Q.; Zang, J.; Yang, C.; Wang, Q.; Li, M. Y.; Jiang, S. Enhanced energy density of polymer nanocomposites at a low electric field through aligned BaTiO3 nanowires, *J. Mater. Chem. A* **2017**, 5, 6070–6078. DOI: 10.1039/c7ta00513j.

29. Yu, L.; Ke, S.; Zhang, Y.; Shen, B.; Zhang, A.; Huang, H. Dielectric relaxations of high-k poly(butylene succinate) based all-organic nanocomposite films for capacitor applications, *J. Mater. Res.* **2011**, 26, 2493–2502. DOI: 10.1557/jmr.2011.271.

30. Zhang, C.; Yu, Y.; Ding, Y.; Yang, T.; Duan, G.; Hou, H. β-Cyclodextrin toughened polyimide composites toward all-organic dielectric materials, *J. Mater. Sci. Mater. Electron.* **2017**, 29, 1182–1188. DOI: 10.1007/s10854-017-8020-1.

31. Zhang, H.; Marwat, M. A.; Xie, B.; Ashtar, M.; Liu, K.; Zhu, Y.; Zhang, L.; Fan, P.; Samart, C.; Ye, Z. G. Polymer matrix nanocomposites with 1D ceramic nanofillers for energy storage capacitor applications, *ACS Appl. Mater. Interfaces* **2020**, 12, 1–37. DOI: 10.1021/acsami.9b15005.

32. Zhang, H.; Zhu, Y.; Li, Z.; Fan, P.; Ma, W.; Xie, B. High discharged energy density of polymer nanocomposites containing paraelectric SrTiO3 nanowires for flexible energy storage device, *J. Alloys Compd.* **2018**, 744, 116–123. DOI: 10.1016/j.jallcom.2018.02.052.

33. Zhang, X.; Zhao, Y.; Wu, Y.; Zhang, Z. Poly(tetrafluoroethylene-hexafluoropropylene) films with high energy density and low loss for high-temperature pulse capacitors, *Polymer* **2017**, 114, 311–318. DOI: 10.1016/j.polymer.2017.03.022.

34. Castro, M.; Kumar, B.; Feller, J. F.; Haddi, Z.; Amari, A.; Bouchikhi, B. Novel e-nose for the discrimination of volatile organic biomarkers with an array of carbon nanotubes (CNT) conductive polymer nanocomposites (CPC) sensors, *Sens. Actuators, B* **2011**, 159, 213–219. DOI: 10.1016/j.snb.2011.06.073.

35. Hao, X. A review on the dielectric materials for high energy-storage application, *J. Adv. Dielectr.* **2013**, 03, 1330001. DOI: 10.1142/S2010135X13300016.

36. Johnson, R. W.; Evans, J. L.; Jacobsen, P.; Thompson, J. R.; Christopher, M. The changing automotive environment: High-temperature electronics, *IEEE Trans. Electron. Packag. Manuf.* **2004**, 27, 164–176. DOI: 10.1109/TEPM.2004.843109.

37. Barber, P.; Balasubramanian, S.; Anguchamy, Y.; Gong, S.; Wibowo, A.; Gao, H.; Ploehn, H. J.; Zur Loye, H. C. Polymer composite and nanocomposite dielectric materials for pulse power energy storage, *Materials* **2009**, 2. DOI: 10.3390/ma2041697.

38. Huan, T. D.; Boggs, S.; Teyssedre, G.; Laurent, C.; Cakmak, M.; Kumar, S.; Ramprasad, R. Advanced polymeric dielectrics for high energy density applications, *Prog. Mater. Sci.* **2016**, 83, 236–269. DOI: 10.1016/j.pmatsci.2016.05.001.

39. Yao, Z.; Song, Z.; Hao, H.; Yu, Z.; Cao, M.; Zhang, S.; Lanagan, M. T.; Liu, H. Homogeneous/inhomogeneous-structured dielectrics and their energy-storage performances, *Adv. Mater.* **2017**, 29, 1601727. DOI: 10.1002/adma.201601727.

40. Guo, M.; Jiang, J.; Shen, Z.; Lin, Y.; Nan, C. W.; Shen, Y. High-energy-density ferroelectric polymer nanocomposites for capacitive energy storage: Enhanced breakdown strength and improved discharge efficiency, *Mater. Today* **2019**, 29, 49–67. DOI: 10.1016/j.mattod.2019.04.015.

41. Feng, Q. K.; Zhong, S. L.; Pei, J. Y.; Zhao, Y.; Zhang, D. L.; Liu, D. F.; Zhang, Y. X.; Dang, Z. M. Recent progress and future prospects on all-organic polymer dielectrics for energy storage capacitors, *Chem. Rev.* **2022**, 122, 3820–3878. DOI: 10.1021/acs.chemrev.1c00793.

42. Dang, Z. M.; Yuan, J. K.; Yao, S. H.; Liao, R. J. Flexible nanodielectric materials with high permittivity for power energy storage, *Adv. Mater.* **2013**, 25, 6334–6365. DOI: 10.1002/adma.201301752.

43. Cheng, L.; Liu, K.; Gao, H.; Fan, Z.; Takesue, N.; Deng, H.; Zhang, H.; Hu, Y.; Tan, H.; Yan, Z.; Liu, Y. Energy storage performance of sandwich structure composites with strawberry-like Ag@SrTiO3 nanofillers, *Chem. Eng. J.* **2022**, 435. DOI: 10.1016/j.cej.2022.135064.

44. Sun, L.; Shi, Z.; Wang, H.; Zhang, K.; Dastan, D.; Sun, K.; Fan, R. Ultrahigh discharge efficiency and improved energy density in rationally designed bilayer polyetherimide-BaTiO3/P(VDF-HFP) composites, *J. Mater. Chem. A* **2020**, 8, 5750–5757. DOI: 10.1039/d0ta00903b.

45. Cheng, L.; Gao, H.; Liu, K.; Tan, H.; Fan, P.; Liu, Y.; Hu, Y.; Yan, Z.; Zhang, H. Research progress on multilayer-structured polymer-based dielectric nanocomposites for energy storage, *Macromol. Mater. Eng.* **2022**. DOI: 10.1002/mame.202100822.

46. Li, X.; Yang, Y.; Wang, Y.; Pang, S.; Shi, J.; Ma, X.; Zhu, K. Enhanced energy storage density of all-organic fluoropolymer composite dielectric via introducing crosslinked structure, *RSC Adv.* **2021**, 11, 15177–15183. DOI: 10.1039/d1ra01423d.

47. Dang, Z. M. 9 - Polymer nanocomposites with high permittivity, In *Nanocrystalline Materials* (Second Edition); Tjong, S. C. Ed.; Elsevier, Oxford, **2014** pp. 305–333.

48. Guo, N.; DiBenedetto, S. A.; Tewari, P.; Lanagan, M. T.; Ratner, M. A.; Marks, T. J. Nanoparticle, size, shape, and interfacial effects on leakage current density, permittivity, and breakdown strength of metal oxide–polyolefin nanocomposites: Experiment and theory, *Chem. Mater.* **2010**, 22, 1567–1578. DOI: 10.1021/cm902852h.

49. Yang, L.; Kong, X.; Li, F.; Hao, H.; Cheng, Z.; Liu, H.; Li, J. F.; Zhang, S. Perovskite lead-free dielectrics for energy storage applications, *Prog. Mater. Sci.* **2019**, 102, 72–108. DOI: 10.1016/j.pmatsci.2018.12.005.

50. Liu, S.; Wang, J.; Shen, B.; Zhai, J.; Hao, H.; Zhao, L. Poly(vinylidene fluoride) nanocomposites with a small loading of core-shell structured BaTiO3@Al2O3 nanofibers exhibiting high discharged energy density and efficiency, *J. Alloys Compd.* **2017**, 696, 136–142. DOI: 10.1016/j.jallcom.2016.11.186.

51. Dang, Z. M.; Yuan, J. K.; Zha, J. W.; Zhou, T.; Li, S. T.; Hu, G. H. Fundamentals, processes and applications of high-permittivity polymer–matrix composites, *Prog. Mater. Sci.* **2012**, 57, 660–723. DOI: 10.1016/j.pmatsci.2011.08.001.

52. Maharolkar, A. P.; Murugkar, A. G.; Khirade, P. W.; Mehrotra, S. C. Microwave dielectric relaxation & polarization study of binary mixture of methylethylketone with nitrobenzene, *Bull. Chem. Soc. Ethiop.* **2019**, 33. DOI: 10.4314/bcse.v33i2.15.

53. Prateek; Bhunia, R.; Siddiqui, S.; Garg, A.; Gupta, R. K. Significantly enhanced energy density by tailoring the interface in hierarchically structured TiO2–BaTiO3–TiO2 nanofillers in PVDF-based thin-film polymer nanocomposites, *ACS Appl. Mater. Interfaces* **2019**, 11, 14329–14339. DOI: 10.1021/acsami.9b01359.

54. Tanaka, T. Dielectric nanocomposites with insulating properties, *IEEE Trans. Dielectr. Electr. Insul.* **2005**, 12, 914–928. DOI: 10.1109/TDEI.2005.1522186.

55. Jiang, J.; Zhang, X.; Dan, Z.; Ma, J.; Lin, Y.; Li, M.; Nan, C. W.; Shen, Y. Tuning phase composition of polymer nanocomposites toward high energy density and high discharge efficiency by nonequilibrium processing, *ACS Appl. Mater. Interfaces* **2017**, 9, 29717–29731. DOI: 10.1021/acsami.7b07963.

56. Li, Q.; Yao, F. Z.; Liu, Y.; Zhang, G.; Wang, H.; Wang, Q. High-temperature dielectric materials for electrical energy storage, *Annu. Rev. Mater. Res.* **2018**, 48, 219–243. DOI: 10.1146/annurev-matsci-070317-124435.

57. Kim, M. P.; Um, D. S.; Shin, Y. E.; Ko, H. High-performance triboelectric devices via dielectric polarization: A review, *Nanoscale Res. Lett.* **2021**, 16, 35. DOI: 10.1186/s11671-021-03492-4.

58. Nan, C. W.; Shen, Y.; Ma, J. Physical properties of composites near percolation, *Annu. Rev. Mater. Res.* **2010**, 40, 131–151. DOI: 10.1146/annurev-matsci-070909-104529.

59. Fan, B.; Zhou, M.; Zhang, C.; He, D.; Bai, J. Polymer-based materials for achieving high energy density film capacitors, *Prog. Polym. Sci.* **2019**, 97. DOI: 10.1016/j.progpolymsci.2019.06.003.

60. Qiu, J.; Gu, Q.; Sha, Y.; Huang, Y.; Zhang, M.; Luo, Z. Preparation and application of dielectric polymers with high permittivity and low energy loss: A mini review, *J. Appl. Polym. Sci.* **2022**, 52367. DOI: 10.1002/app.52367.

61. Dakin, T. W. Conduction and polarization mechanisms and trends in dielectric, *IEEE Electr. Insul. Mag.* **2006**, 22, 11–28. DOI: 10.1109/MEI.2006.1705854.

62. O'Hara, A.; Balke, N.; Pantelides, S. T. Unique features of polarization in ferro-electric ionic conductors, *Adv. Electron. Mater.* **2022**, 8, 2100810. DOI: 10.1002/aelm.202100810.

63. Dong, J. F.; Deng, X. L.; Niu, Y. J.; Pan, Z. Z.; Wang, H. Research progress of polymer based dielectrics for high-temperature capacitor energy storage, *ACTA PHYS SIN-CH ED* **2020**, 69. DOI: 10.7498/aps.69.20201006.

64. Prateek; Thakur, V. K.; Gupta, R. K. Recent progress on ferroelectric polymer-based nanocomposites for high energy density capacitors: Synthesis, dielectric properties, and future aspects, *Chem. Rev.* **2016**, 116, 4260–317. DOI: 10.1021/acs.chemrev.5b00495.

65. Baldwin, A. F.; Ma, R.; Huan, T. D.; Cao, Y.; Ramprasad, R.; Sotzing, G. A. Effect of incorporating aromatic and chiral groups on the dielectric properties of poly(dimethyltin esters), *Macromol. Rapid Commun.* **2014**, 35, 2082–2088. DOI: 10.1002/marc.201400507.

66. Chen, Y.; Du, Y. K.; Yue, Y. C.; Liu, K. H.; Chen, Y.; Qin, L.; Jiang, C. B.; Zhang, L.; Cao, W. Q.; Pan, R. K. Correlation between energy storage density and differential dielectric constant in ferroelectrics, *J. Electron. Mater.* **2020**, 49, 659–667. DOI: 10.1007/s11664-019-07786-3.

67. Wei, J.; Zhu, L. Intrinsic polymer dielectrics for high energy density and low loss electric energy storage, *Prog. Polym. Sci.* **2020**, 106, 101254. DOI: 10.1016/j.progpolymsci.2020.101254.

68. Laghari, J. R.; Sarjeant, W. J. Energy-storage pulsed-power capacitor technology, *IEEE Trans. Power Electron.* **1992**, 7, 251–257. DOI: 10.1109/63.124597.

69. Nehm, F.; Klumbies, H.; Richter, C.; Singh, A.; Schroeder, U.; Mikolajick, T.; Mönch, T.; Hoßbach, C.; Albert, M.; Bartha, J. W.; Leo, K.; Müller-Meskamp, L. Breakdown and protection of ALD moisture barrier thin films, *ACS Appl. Mater. Interfaces* **2015**, 7, 22121–22127. DOI: 10.1021/acsami.5b06891.

70. Shen, Z. H.; Wang, J. J.; Jiang, J. Y.; Huang, S. X.; Lin, Y. H.; Nan, C. W.; Chen, L. Q.; Shen, Y. Phase-field modeling and machine learning of electric-thermal-mechanical breakdown of polymer-based dielectrics, *Nat. Commun.* **2019**, 10, 1843. DOI: 10.1038/s41467-019-09874-8.

71. Fothergill, J. C. Estimating the cumulative probability of failure data points to be plotted on Weibull and other probability paper, *IEEE Trans. Dielectr. Electr. Insul.* **1990**, 25, 489–492. DOI: 10.1109/14.55721.

72. Beale, P. D.; Duxbury, P. M. Theory of dielectric breakdown in metal-loaded dielectrics, *Phys. Rev. B* **1988**, 37, 2785-2791. DOI: 10.1103/PhysRevB.37.2785.

73. Shah, Z. M.; Khanday, F. A.; Malik, G. F. A.; Jhat, Z. A. Fabrication of polymer nanocomposite-based fractional-order capacitor: A guide, In *Fractional-Order Design*, **2022** pp. 437–483. DOI: 10.1016/B978-0-32-390090-4.00020-2

74. Palsaniya, S.; Mukherji, S. Enhanced dielectric and electrostatic energy density of electronic conductive organic-metal oxide frameworks at ultra-high frequency, *Carbon* **2022**, 196, 749–762. DOI: 10.1016/j.carbon.2022.05.042.

75. Yuan, J. K.; Dang, Z. M.; Yao, S. H.; Zha, J. W.; Zhou, T.; Li, S. T.; Bai, J. Fabrication and dielectric properties of advanced high permittivity polyaniline/poly(vinylidene fluoride) nanohybrid films with high energy storage density, *J. Mater. Chem.* **2010**, 20. DOI: 10.1039/b923590f.

76. Palsaniya, S.; Nemade, H. B.; Dasmahapatra, A. K. Graphene based PANI/MnO2 nanocomposites with enhanced dielectric properties for high energy density materials, *Carbon* **2019**, 150, 179–190. DOI: 10.1016/j.carbon.2019.05.006.

77. Zhang, L.; Lu, X.; Zhang, X.; Jin, L.; Xu, Z.; Cheng, Z. Y. All-organic dielectric nanocomposites using conducting polypyrrole nanoclips as filler, *Compos. Sci. Technol.* **2018**, 167, 285–293. DOI: 10.1016/j.compscitech.2018.08.017.

78. Rodríguez, A. E.; Morgan, W. L.; Touryan, K. J.; Moeny, W. M.; Martin, T. H. An air breakdown kinetic model, *J. Appl. Phys.* **1991**, 70, 2015–2022. DOI: 10.1063/1.349487.

79. Tan, D. Q. The search for enhanced dielectric strength of polymer-based dielectrics: A focused review on polymer nanocomposites, *J. Appl. Polym. Sci.* **2020**, 137. DOI: 10.1002/app.49379.

80. Liu, X. J.; Zheng, M. S.; Chen, G.; Dang, Z. M.; Zha, J. W. High-temperature polyimide dielectric materials for energy storage: Theory, design, preparation and properties, *Energy Environ. Sci.* **2022**, 15, 56–81. DOI: 10.1039/d1ee03186d.

81. Fröhlich, H.; Mott, N. F. On the theory of dielectric breakdown in solids, *Proceedings of the Royal Society of London. Series A. Mathematical and Physical Sciences* **1947**, 188, 521–532. DOI: 10.1098/rspa.1947.0023.

82. Gerson, R.; Marshall, T. C. Dielectric breakdown of porous ceramics, *J. Appl. Phys.* **1959**, 30, 1650–1653. DOI: 10.1063/1.1735030.

83. Owate, I. O.; Freer, R. ac breakdown characteristics of ceramic materials, *J. Appl. Phys.* **1992**, 72, 2418–2422. DOI: 10.1063/1.351586.

84. Seitz, F. On the theory of electron multiplication in crystals, *Phys. Rev.* **1949**, 76, 1376–1393. DOI: 10.1103/PhysRev.76.1376.

85. Guo, H. J.; Duan, B. X.; Wu, H.; Yang, Y. T. Breakdown mechanisms of power semiconductor devices, *IETE Tech. Rev.* **2019**, 36, 243–252. DOI: 10.1080/02564602.2018.1450652.

86. Ling, Y. M. Study on breakdown characteristics of a low-pressure dielectric barrier discharge, *J. Plasma Phys.* **2007**, 73, 417–426. DOI: 10.1017/S0022377806004715.

87. Shen, Z. H.; Wang, J. J.; Lin, Y.; Nan, C. W.; Chen, L. Q.; Shen, Y. High-throughput phase-field design of high-energy-density polymer nanocomposites, *Adv. Mater.* **2018**, 30, 1704380. DOI: 10.1002/adma.201704380.

88. Sun, Y.; Bealing, C.; Boggs, S.; Ramprasad, R. 50+ years of intrinsic breakdown, *IEEE Electr. Insul. Mag.* **2013**, 29, 8–15. DOI: 10.1109/MEI.2013.6457595.

89. Sun, Y.; Boggs, S. A.; Ramprasad, R. The intrinsic electrical breakdown strength of insulators from first principles, *Appl. Phys. Lett.* **2012**, 101, 132906. DOI: 10.1063/1.4755841.

90. Xu, A.; Jin, D.; Wang, Y.; Tan, X. Study on breakdown phenomenon of electrode induced by high-energy protons bombarding, *IEEE Trans. Plasma Sci.* **2020**, 48, 888–893. DOI: 10.1109/TPS.2020.2975634.

91. Cai, Z.; Wang, X.; Li, L. Phase-field modeling of electromechanical breakdown in multilayer ceramic capacitors, *Adv. Theor. Simul.* **2019**, 2, 1800179. DOI: 10.1002/adts.201800179.

92. Li, J.; Zhang, Y.; Xia, Z.; Qin, X.; Peng, Z. Action of space charge on aging and breakdown of polymers, *Chin. Sci. Bull.* **2001**, 46, 796–800. DOI: 10.1007/BF02900426.

93. Liu, L.; Chen, H.; Li, B.; Wang, Y.; Li, D. Thermal and strain-stiffening effects on the electromechanical breakdown strength of dielectric elastomers, *Appl. Phys. Lett.* **2015**, 107, 062906. DOI: 10.1063/1.4928712.

94. Neusel, C.; Jelitto, H.; Schmidt, D.; Janssen, R.; Felten, F.; Schneider, G. A. Thickness-dependence of the breakdown strength: Analysis of the dielectric and mechanical failure, *J. Eur. Ceram. Soc.* **2015**, 35, 113–123. DOI: 10.1016/j.jeurceramsoc.2014.08.028.

95. Tan, D. Q. The search for enhanced dielectric strength of polymer-based dielectrics: A focused review on polymer nanocomposites, *J. Appl. Polym. Sci.* **2020**, 137, 49379. DOI: 10.1002/app.49379.

96. Zhang, L.; Wang, Q.; Zhao, X. Mechanical constraints enhance electrical energy densities of soft dielectrics, *Appl. Phys. Lett.* **2011**, 99, 171906. DOI: 10.1063/1.3655910.

97. Rahimabady, M.; Chen, S.; Yao, K.; Eng Hock Tay, F.; Lu, L. High electric breakdown strength and energy density in vinylidene fluoride oligomer/poly(vinylidene fluoride) blend thin films, *Appl. Phys. Lett.* **2011**, 99, 142901. DOI: 10.1063/1.3645619.

98. Cross, L. E.; Jang, S. J.; Newnham, R. E.; Nomura, S.; Uchino, K. Large electrostrictive effects in relaxor ferroelectrics, *Ferroelectrics* **2011**, 23, 187–191. DOI: 10.1080/00150198008018801.

99. Huang, X.; Xie, L.; Yang, K.; Wu, C.; Jiang, P.; Li, S.; Wu, S.; Tatsumi, K.; Tanaka, T. Role of interface in highly filled epoxy/$BaTiO_3$ nanocomposites. Part II- effect of nanoparticle surface chemistry on processing, thermal expansion, energy storage and breakdown strength of the nanocomposites, *IEEE Trans. Dielectr. Electr. Insul.* **2014**, 21, 480–487. DOI: 10.1109/TDEI.2013.004166.

100. Pukhov, A. A. Thermal mechanism of microwave breakdown in HTSC films: Theoretical arguments and experimental evidence, *Supercond. Sci. Technol.* **1999**, 12, 102–104. DOI: 10.1088/0953-2048/12/2/008.

101. Shen, Z. H.; Wang, J. J.; Jiang, J. Y.; Lin, Y. H.; Nan, C. W.; Chen, L. Q.; Shen, Y. Phase-field model of electrothermal breakdown in flexible high-temperature nanocomposites under extreme conditions, *Adv. Energy Mater.* **2018**, 8, 1800509. DOI: 10.1002/aenm.201800509.

102. Somogyi, K. Some critical conditions of the thermal breakdown in semiconductors, *Sens. Actuators, A* **1998**, 71, 58–62. DOI: 10.1016/S0924-4247(98)00176-9.

103. Xu, K. Q.; Zhao, K. Y.; Gu, Y.; Zeng, H. R.; Liu, Z. F.; Li, G. R.; Li, Y. X. Local elastic and thermal behaviors of dielectric breakdown regions in multilayer ceramic capacitors, *Phys Status Solidi Rapid Res Lett.* **2015**, 9, 745–748. DOI: 10.1002/pssr.201510343.

104. Li, Q.; Chen, L.; Gadinski, M. R.; Zhang, S.; Zhang, G.; Li, U.; Iagodkine, E.; Haque, A.; Chen, L. Q.; Jackson, N.; Wang, Q. Flexible high-temperature dielectric materials from polymer nanocomposites, *Nature* **2015**, 523, 576–579. DOI: 10.1038/nature14647.

105. Claude, J.; Lu, Y.; Wang, Q. Effect of molecular weight on the dielectric breakdown strength of ferroelectric poly(vinylidene fluoride-chlorotrifluoroethylene)s, *Appl. Phys. Lett.* **2007**, 91, 212904. DOI: 10.1063/1.2816327.

106. Feng, X.; Xiong, Q.; Gattozzi, A. L.; Hebner, R. E. Partial discharge experimental study for medium voltage DC cables, *IEEE Trans. Power Delivery* **2021**, 36, 1128–1136. DOI: 10.1109/TPWRD.2020.3002508.

107. Garner, A. L.; Loveless, A. M.; Dahal, J. N.; Venkattraman, A. A tutorial on theoretical and computational techniques for gas breakdown in microscale gaps, *IEEE Trans. Plasma Sci.* **2020**, 48, 808–824. DOI: 10.1109/TPS.2020.2979707.

108. Huang, X.; Li, Y.; Liu, F.; Jiang, P.; Iizuka, T.; Tatsumi, K.; Tanaka, T. Electrical properties of epoxy/POSS composites with homogeneous nanostructure, *IEEE Trans. Dielectr. Electr. Insul.* **2014**, 21, 1516–1528. DOI: 10.1109/TDEI.2014.004314.

109. Li, S. T.; Yin, G. L.; Chen, G.; Li, J. Y.; Bai, S. N.; Zhong, L. S.; Zhang, Y. X.; Lei, Q. Q. Short-term breakdown and long-term failure in nanodielectrics: A review, *IEEE Trans. Dielectr. Electr. Insul.* **2010**, 17, 1523–1535. DOI: 10.1109/TDEI.2010.5595554.

110. Li, Z.; Sheng, G.; Jiang, X.; Tanaka, T. Effects of inorganic fillers on withstanding short-time breakdown and long-time electrical aging of epoxy composites, *IEEJ Trans. Electr. Electron. Eng.* **2017**, 12, S10–S15. DOI: 10.1002/tee.22548.

111. Lu, B. X.; Feng, Q. K.; Sun, H. Y. The effect of environmental temperature on negative corona discharge under the action of photoionization, *IEEE Trans. Plasma Sci.* **2019**, 47, 149–154. DOI: 10.1109/TPS.2018.2872504.

112. Chiu, F. C. A review on conduction mechanisms in dielectric films, *Adv. Mater. Sci. Eng.* **2014**, 2014, 578168. DOI: 10.1155/2014/578168.

113. Kim, Y. S.; Lee, Y. H.; Lim, K. M.; Sung, M. Y. The effect of Al–Ta2O5 topographic interface roughness on the leakage current of Ta2O5 thin films, *Appl. Phys. Lett.* **1999**, 74, 2800–2802. DOI: 10.1063/1.124018.

114. Kojima, K.; Takai, Y.; Ieda, M. Electronic conduction in polyethylene naphthalate at high electric fields, *J. Appl. Phys.* **1986**, 59, 2655–2659. DOI: 10.1063/1.336970.

115. Liu, A.; Zhu, H.; Sun, H.; Xu, Y.; Noh, Y. Y. Solution processed metal oxide high-kappa dielectrics for emerging transistors and circuits, *Adv. Mater* **2018**, e1706364. DOI: 10.1002/adma.201706364.

116. Luo, H.; Zhou, X.; Ellingford, C.; Zhang, Y.; Chen, S.; Zhou, K.; Zhang, D.; Bowen, C. R.; Wan, C. Interface design for high energy density polymer nanocomposites, *Chem. Soc. Rev.* **2019**, 48, 4424–4465. DOI: 10.1039/c9cs00043g.

117. Ortiz, R. P.; Facchetti, A.; Marks, T. J. High-k organic, inorganic, and hybrid dielectrics for low-voltage organic field-effect transistors, *Chem. Rev.* **2010**, 110, 205–239. DOI: 10.1021/cr9001275.

118. Zhou, Y.; Wang, Q. Advanced polymer dielectrics for high temperature capacitive energy storage, *J. Appl. Phys.* **2020**, 127, 240902. DOI: 10.1063/5.0009650.

119. Hamadani, B. H.; Natelson, D. Gated nonlinear transport in organic polymer field effect transistors, *J. Appl. Phys.* **2004**, 95, 1227–1232. DOI: 10.1063/1.1635979.

120. Zhang, G.; Li, Y.; Tang, S.; Thompson, R. D.; Zhu, L. The role of field electron emission in polypropylene/aluminum nanodielectrics under high electric fields, *ACS Appl. Mater. Interfaces* **2017**, 9, 10106–10119. DOI: 10.1021/acsami.7b00095.

121. Angle, R. L.; Talley, H. E. Electrical and charge storage characteristics of the tantalum oxide-silicon dioxide device, *IEEE Trans. Electron Devices* **1978**, 25, 1277–1283. DOI: 10.1109/T-ED.1978.19266.

122. Chiu, F. C.; Lee, C. Y.; Pan, T. M. Current conduction mechanisms in Pr2O3/oxynitride laminated gate dielectrics, *J. Appl. Phys.* **2009**, 105. DOI: 10.1063/1.3103282.

123. Leitl, B.; Schmidt, G.; Pobegen, G.; Knoll, P.; Krenn, H. Conduction mechanisms in hydrogenated amorphous silicon carbide, *J. Non-Cryst. Solids* **2020**, 528, 119750. DOI: 10.1016/j.jnoncrysol.2019.119750.

124. Yıldız, D. E.; Karakuş, M.; Toppare, L.; Cirpan, A. Leakage current by Frenkel–Poole emission on benzotriazole and benzothiadiazole based organic devices, *Mater. Sci. Semicond. Process.* **2014**, 28, 84–88. DOI: 10.1016/j.mssp.2014.06.038.

125. Chiu, F. C. Conduction mechanisms in resistance switching memory devices using transparent boron doped zinc oxide films, *Materials* **2014**, 7. DOI: 10.3390/ma7117339.

126. Chiu, F. C.; Shih, W. C.; Feng, J. J. Conduction mechanism of resistive switching films in MgO memory devices, *J. Appl. Phys.* **2012**, 111, 094104. DOI: 10.1063/1.4712628.

127. Rubinger, C. P. L.; Leyva, M. E.; Soares, B. G.; Ribeiro, G. M.; Rubinger, R. M. Hopping conduction on carbon black/styrene–butadiene–styrene composites, *J. Mater. Sci.* **2012**, 47, 860–865. DOI: 10.1007/s10853-011-5864-2.

128. Santos, S. J.; Salazar, I. High electric field effect on hopping conduction in molecularly doped polymer systems, *Polymer* **1999**, 40, 4415–4418. DOI: 10.1016/S0032-3861(98)00761-7.

129. Vecchio, M. A.; Meddeb, A. B.; Ounaies, Z.; Lanagan, M. T. Conduction through plasma-treated polyimide: Analysis of high-field conduction by hopping and Schottky theory, *J. Mater. Sci.* **2019**, 54, 10548–10559. DOI: 10.1007/s10853-019-03574-w.

130. Akram, S.; Yang, Y.; Zhong, X.; Bhutta, S.; Wu, G.; Castellon, J.; Zhou, K. Influence of nano layer interface on space charge behavior and trap levels using polyimide sandwich structure, *IEEE Trans. Dielectr. Electr. Insul.* **2017**, 24, 3505–3514. DOI: 10.1109/TDEI.2017.006765.

131. Bodhane, S. P.; Shirodkar, V. S. Space–charge-limited conduction in vacuum-deposited PVDF films, *J. Appl. Polym. Sci.* **1999**, 74, 1347–1354. DOI: 10.1002/(SICI)1097-4628(19991107)74:6<1347::AID-APP4>3.0.CO;2-K.

132. Huang, H.; Chen, X.; Li, R.; Fukuto, M.; Schuele, D. E.; Ponting, M.; Langhe, D.; Baer, E.; Zhu, L. Flat-on secondary crystals as effective blocks to reduce ionic conduction loss in polysulfone/poly(vinylidene fluoride) multilayer dielectric films, *Macromolecules* **2018**, 51, 5019–5026. DOI: 10.1021/acs.macromol.8b01037.

133. Huang, H.; Chen, X.; Yin, K.; Treufeld, I.; Schuele, D. E.; Ponting, M.; Langhe, D.; Baer, E.; Zhu, L. Reduction of ionic conduction loss in multilayer dielectric films by immobilizing impurity ions in high glass transition temperature polymer layers, *ACS Appl. Energy Mater.* **2018**, 1, 775–782. DOI: 10.1021/acsaem.7b00211.

134. Calvet, L. E.; Wheeler, R. G.; Reed, M. A. Electron transport measurements of Schottky barrier inhomogeneities, *Appl. Phys. Lett.* **2002**, 80, 1761–1763. DOI: 10.1063/1.1456257.

135. Vecchio, M. A.; Meddeb, A. B.; Lanagan, M. T.; Ounaies, Z.; Shallenberger, J. R. Plasma surface modification of P(VDF-TrFE): Influence of surface chemistry and structure on electronic charge injection, *J. Appl. Phys.* **2018**, 124, 114102. DOI: 10.1063/1.5042751.

136. Chiu, F. C. Interface characterization and carrier transportation in metal/HfO2/silicon structure, *J. Appl. Phys.* **2006**, 100, 114102. DOI: 10.1063/1.2401657.

137. Baniecki, J. D.; Laibowitz, R. B.; Shaw, T. M.; Parks, C.; Lian, J.; Xu, H.; Ma, Q. Y. Hydrogen induced tunnel emission in Pt/(BaxSr1−x)Ti1+yO3+z/Pt thin film capacitors, *J. Appl. Phys.* **2001**, 89, 2873–2885. DOI: 10.1063/1.1339207.

138. Padovani, F. A.; Stratton, R. Field and thermionic-field emission in Schottky barriers, *Solid-State Electron.* **1966**, 9, 695–707. DOI: 10.1016/0038-1101(66)90097-9.

139. Pei, J. Y.; Zhong, S. L.; Zhao, Y.; Yin, L. J.; Feng, Q. K.; Huang, L.; Liu, D. F.; Zhang, Y. X.; Dang, Z. M. All-organic dielectric polymer films exhibiting superior electric breakdown strength and discharged energy density by adjusting the electrode–dielectric interface with an organic nano-interlayer, *Energy Environ. Sci.* **2021**, 14, 5513–5522. DOI: 10.1039/d1ee01960k.

140. Li, Y.; Cheng, S.; Wang, S.; Yuan, C.; Luo, Z.; Zhu, Y.; Hu, J.; He, J.; Li, Q. Multilayered ferroelectric polymer composites with high-energy density at elevated temperature, *Compos. Sci. Technol.* **2021**, 202. DOI: 10.1016/j.compscitech.2020.108594.

2 Inorganic Dielectric Materials for Capacitive Energy Storage

Haibo Zhang, Hua Tan, and Zuo-Guang Ye

1 INTRODUCTION

Dielectric energy storage materials can be generally classified into five categories according to the polarization – electric field relations, which are (1) linear dielectrics (LDs), (2) paraelectrics (PEs), (3) ferroelectrics (FEs), (4) relaxor ferroelectrics (RFEs), and (5) antiferroelectrics (AFEs). The PE characteristics for different kinds of dielectric materials are very well presented in the work of Palneedi et al.[1] as shown in Figure 2.1. For LD ceramics, such as TiO_2 ceramics, linear PE loops are obtained with a stable dielectric constant under an electric field excitation. They usually possess high breakdown field (E_b) values and ultralow dielectric loss (tan δ). However, their low dielectric constant (ε_r) and medium energy storage density (W) are obstacles to realizing practical applications. PEs have dipoles instead of ferroelectric domains, and can quickly return to their initial nonpolar state after the electric field is removed. Therefore, their PE curves are extremely fine in shape, have weak nonlinear characteristics, and do not have hysteresis phenomena under ideal conditions. Although the relative permittivity of PEs is higher than that of LDs, it is significantly lower than that of several other types of nonlinear dielectrics. For PEs, the ferroelectric Curie temperature is generally lower than room temperature, so they exhibit almost the same properties as linear dielectrics at room temperature and above. Since there is still some controversy in the classification, we classify PEs as linear dielectrics in this chapter. The typical paraelectric material strontium titanate, for example, has a relative permittivity of about 300. Normal FE ceramics, such as undoped $BaTiO_3$ and $PbTiO_3$ ceramics, usually possess large maximum polarization (P_{max}) and remnant polarization (P_r), and moderate E_b, their energy storage efficiency is extremely low due to the hysteretic switch of electric domains under the electric field, and it is different to increase energy storage density (U_d). In contrast, the RFE and AFE ceramics are very suitable for energy storage applications. They both have a large P_{max} and low P_r, and thereby a large energy storage density can be obtained. In particular, for RFE ceramics (e.g., $Bi_{0.5}Na_{0.5}TiO_3$-based ergodic RFE ceramics), the FE order is induced by the electric field and when the field is switched off, the FE domains relax back to the ergodic relaxor state with near-zero polarization, thus ensuring a large P_{max}, low P_r, and high energy storage density. For the AFEs, double hysteresis is obtained when the electric field surpasses the critical field, thus leading

DOI: 10.1201/9781003454496-2

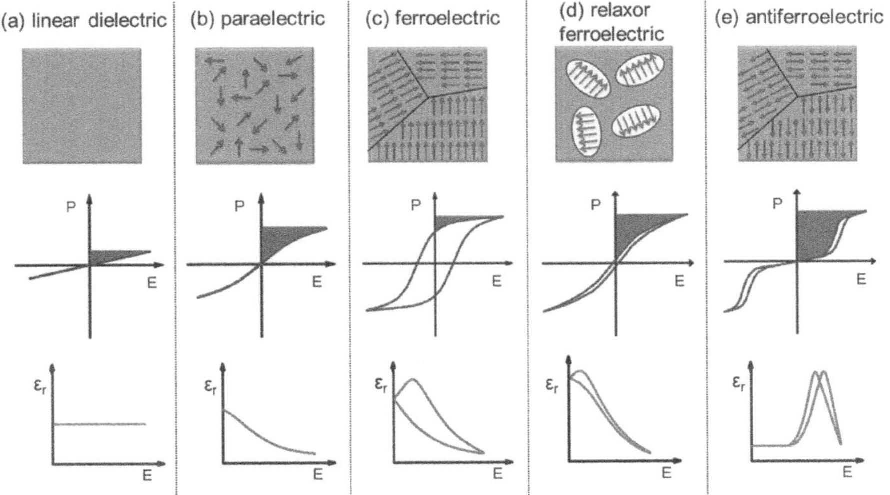

FIGURE 2.1 Schematic graphs of the PE loops for (a) linear dielectrics (LD); (b) paraelectrics, (c) ferroelectrics (FE); (d) relaxor ferroelectrics (RFE) and (e) antiferroelectrics (AFE).[1]

to a high energy density. Notably, the appearance of double hysteresis is unfavorable for achieving high efficiency. In addition, there have been a number of studies on the so-called relaxor-antiferroelectrics (R-AFE) in recent years, which present the same PE loops as RFEs, such as NN–SBT, NN–BNT, and so on.[2, 3] Since relaxor-antiferroelectrics are developed based on antiferroelectric materials, they are classified as antiferroelectrics in this work.

Inorganic dielectric energy storage materials can also be classified into linear dielectric, paraelectric, ferroelectric, relaxor ferroelectric, and antiferroelectric by the PE curves, as described above. Among them, PEs are usually compared with LDs due to their similar PE loops above room temperature. The PE curves of relaxor antiferroelectrics are essentially similar to those of relaxor ferroelectrics. Ferroelectrics are not suitable for energy storage applications because of their large remanent polarization, which leads to very low energy storage density and efficiency. However, ferroelectrics can be transformed into relaxor ferroelectrics by appropriate chemical modifications, thus improving the energy storage performance. In the following, the latest research progress in inorganic dielectric materials will be discussed in detail, starting from three different dielectric ceramics: linear dielectric ceramics, relaxor ferroelectric ceramics, and antiferroelectric ceramics. The dielectric glass-ceramics and ceramic films are also discussed in the respective subsections.

2 PROBLEMS AND STRATEGIES OF INORGANIC DIELECTRIC MATERIALS

As we mentioned in Chapter 1, the biggest problem of dielectric energy storage is the low energy storage density compared to other energy storage devices. If the energy

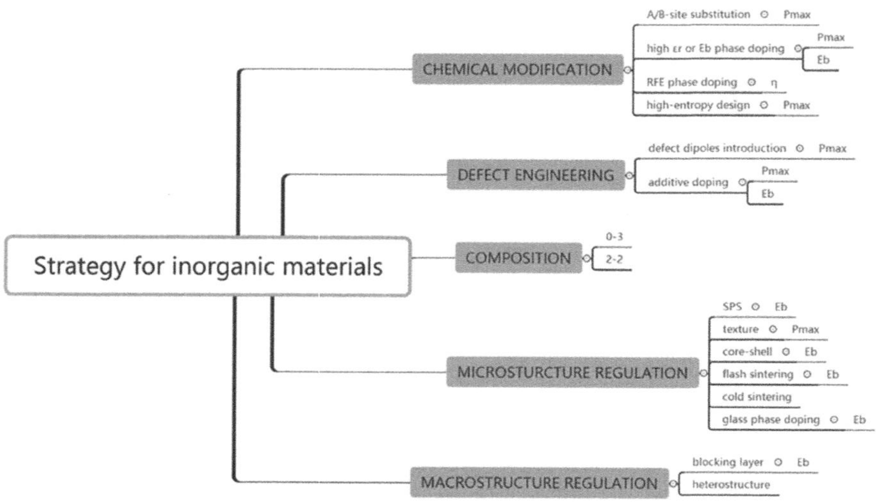

FIGURE 2.2   Strategic solutions to the problems of inorganic dielectric materials.

storage density of dielectric energy storage materials can be increased, dielectric capacitors will be useful in more fields. Therefore, improving the energy storage density of dielectric energy storage materials is an urgent problem to be solved. In addition, different material systems present different specific problems. For example, Pb-based and Bi-based inorganic dielectric materials suffer from easy volatilization of elements, leading to oxygen vacancies, antiferroelectric inorganic dielectric materials undergo strain-induced phase transitions, leading to microcracks, poly(vinylidene fluoride) (PVDF) organic dielectric material systems cannot be prepared and applied on a large scale due to low Young's modulus and most organic dielectric materials exhibit poor high-temperature performance. These problems will be specified in the subsequent chapters. In this chapter, a solution strategy for the problems common to inorganic dielectrics is proposed.

According to the PE curves, to improve the energy storage density of energy storage materials, it is mainly necessary to improve the polarization strength, breakdown field strength, and energy storage efficiency. Since the problems faced by inorganic and organic dielectric materials are different, we will discuss the problems and strategies separately. Figure 2.2 illustrates the various strategic solutions to the problems of inorganic dielectric materials.

2.1 CHEMICAL MODIFICATION

Chemical modification is the most commonly used strategy in improving dielectric materials. (1) A/B-site substitution, A-site substitution, B-site substitution, and simultaneous substitution of A- and B- sites are the approaches to increase ε_r and/or P_{max} in many ceramic systems. By replacing atoms in the crystal structure, the crystal structure is changed, causing lattice distortion and thus a series of effects.

A classic example is the increase of the dielectric constant of TiO_2 linear dielectrics from 10^2 to 10^4 by In and Nb co-doping, thus increasing the energy storage density. [4] Other examples will be described in detail in the following sections. (2) High ε_r or E_b phase doping: the addition or doping of high ε_r or E_b phase to the matrix is also a chemical modification strategy to improve the related properties. It is important to note that phase doping here is different from composition, in that phase doping implies the formation of a complete solid solution of the two phases, and the atoms in the lattice are replaced. The composition will be explained in Section 2.3. For example, $Ba_xSr_{1-x}TiO_3$ (BST) solid solutions combine the characteristics of high E_b of $SrTiO_3$ and high ε_r of $BaTiO_3$. With increasing BT content, the material changes from PE to FE. When x is from 0.3 to 0.4, the BST solid solutions present the best energy storage properties.[5] (3) RFE phase doping: RFE phase doping is similar to high ε_r or E_b phase doping, but here we classify them separately because this kind of doping has been shown to be an efficiency strategy for improving energy storage density in many systems, especially in NN-based antiferroelectric materials. In 2019, Qi et al.[3] reported NBT RFE phase doped NN AFE ceramics showing an ultrahigh U_d of 12.2 J/cm^3, which is the highest U_d among bulk ceramics at that time. The reason is attributed to the fine AFE domains and the large random fields in the R-AFE phase. (4) High-entropy design: the concept of high entropy first appeared in alloys, and the application of the concept to dielectric materials can be traced back to 2019. Pu et al.[6] synthesized a single-phase homogeneous A-site high entropy $(Na_{0.2}Bi_{0.2}Ba_{0.2}Sr_{0.2}Ca_{0.2})TiO_3$ ceramic which exhibited a relaxation behavior because of the disorder of the microscopic composition. We believe that research on high-entropy dielectric materials will definitely make further progress, especially with the applications of machine learning in materials science.

2.2 DEFECT ENGINEERING

Defect engineering generally involves two aspects. One is to improve the energy storage performance by artificially introducing defect dipoles through the substitution of different valence atoms, causing local structural distortion and enhancing the relaxation property. The other one is just the opposite, suppressing the generation of defects in the matrix by additives to improve the breakdown strength. These two approaches will be explained separately below. (1) Defect dipoles introduction: This approach is often confused with A- and B- site substitutions, but unlike A- and B- site substitutions, the introduction of defective dipoles is accomplished by doping with atoms of different valence. For example, the doping of A-site Sm^{3+} into the perovskite $Ca_{1-1.5x}Sm_x\square_{0.5x}TiO_3$ ceramics introduced A-site vacancies (VA). As a result of Sm^{3+} doping and VA, the grain size decreased while the ceramic density was improved.[7] (2) Additive doping: In the sintering of ceramics, some elements are prone to volatilization, such as Bi, Na, K, etc., and some elements are prone to chemical valence changes, such as Fe, etc., causing the generation of a large number of oxygen vacancies, reducing the resistivity of the material and decreasing the breakdown strength. For example, during the sintering of the BF–BT system, the Bi^{3+} is easy to evaporate and the Fe^{3+} is easy to be reduced to Fe^{2+}. Both are the reasons for the oxygen vacancies. With the addition of extra MnO_2, which has the multi-valence states (Mn^{2+},

Mn^{3+}, and Mn^{4+}), Mn^{3+} and Mn^{2+} most likely enter the B-sites to randomly substitute Fe^{3+} and Ti^{4+} ions due to their similar ionic radii. The adding of MnO_2 can restrain the conversion of Fe^{3+} into Fe^{2+} between Mn^{3+} and Fe^{2+}: $Mn^{3+} + Fe^{2+} - Mn^{2+} + Fe^{3+}$. Moreover, the Mn^{x+} ions can also compensate for cation deficiency due to the volatilization of Bi elements.[8]

2.3 COMPOSITION

Composition formation of composites in dielectric materials is a strategy to combine the advantages of two or more materials in energy storage performance. According to the spatial structure relationship, it can be divided into 0–3 composite and 2–2 composite. Simply put, they are particle doping and planar doping, respectively. (1) 0–3 composite: As mentioned earlier, the biggest difference between 0 and 3 composite and phase doping is the presence or absence of second-phase particles. Many studies based on the concept of composition end up as phase doping due to the diffusion during sintering. To solve the problem of diffusion, in 2020, Fan et al.[9] achieved the RE/AFE $0.94(Na_{0.82}K_{0.18})_{1/2}Bi_{1/2}TiO_3$–$0.06FeNbO_4$/$0.96NaNbO_3$–$0.04CaZrO_3$(NKBT–FN/NN–CZ) composite ceramic via the composition of fine and coarse particles, resulting in a narrow and oblique double hysteresis loops. Other methods to achieve 0–3 composite include core-shell structure, rapid sintering, and so on. (2) 2–2 composite: With the aid of the tape cast process, 2–2 composition can optimize the properties of two or more materials through stacked layer arrangements. This strategy is well illustrated in the work of Li et al.[10] They prepared 2–2 composite ceramics of $0.65Na_{0.5}Bi_{0.5}TiO_3$–$0.35SrTiO_3$ (NBST) with large polarization and the $0.45Na_{0.5}Bi_{0.5}TiO_3$–$0.55Sr_{0.7}Bi_{0.2}TiO_3$ (NBSBT) ceramics with high efficiency, which take the advantages of the respective two materials.

2.4 MICROSTRUCTURE REGULATION

The microstructure of ceramic materials has a large impact on the breakdown strength, and a dense and fine-grained structure is beneficial to a high breakdown field strength. In addition, some other methods of modifying the grain shape and orientation also have an effect on the dielectric properties of the materials. (1) Spark plasma sintering (SPS). This rapid sintering technique has already been applied in the fabrication of structure ceramics for decades due to the high densification and fine grains, which are very suitable for dielectric ceramics. Recently, Tan et al.[2] prepared NN-24SBT R-AFE ceramics by SPS and obtained an ultrahigh U_d of 12.2 J/cm^3 with a high η of 88%, which are the highest U_d and η at that time. The dense and fine-grained microstructure is the main reason for that. (2) Flash sintering (FS). The process of FS typically lasts less than 60 s, but it needs a complex pre-preparation. Moreover, one study showed that electrical breakdown cracks occurred inside the samples that had undergone FS because of the local charge accumulation during the flash firing process.[11] Therefore, the current research on FS of dielectric ceramics has not fully taken the advantage of the state of the art of this sintering technique. (3) Cold sintering. In 2017, Randall and colleagues developed a novel sintering method named cold sintering. In this process, the ceramic powders need to be uniformly moistened with a small amount of specific precursor solutions, followed by sintering at ultralow temperatures (generally $< 200°C$) under a pressure of MPa magnitude.[12] With this sintering technique, Ma et al.[13] prepared $BaTiO_3$ (BT) ceramics at an ultralow temperature of 180°C. The relative density is higher than 96% and the average size of grains in the annealed samples is as low as 122 nm. The BT ceramics exhibit high energy storage of 1.45 J/cm^3 with a high η of 85.6%. (4) Texture. There have been few studies on the texture of energy storage ceramics. The most interesting study is that of Li et al.[14] who achieved a high energy storage density by reducing the field strain and increasing the breakdown field strength of the material through the <111> directional texture in the NBT–SBT system. In contrast, the study of texture in other systems did not show its advantages. (5) Glass phase doping. Since glass phase doping changes the polycrystalline structure of ceramic materials into an amorphous structure, it is classified as microstructure modulation. The greatest benefit of this approach is the increase in material densities and the significant increase in material energy storage efficiency due to the linear dielectric nature of silica. However, the dielectric constant and polarization strength also decrease. This strategy has yielded some results in the study of thin film materials, but no significant studies on bulk materials have been reported for the time being.

2.5 MACROSTRUCTURE REGULATION

Macrostructure regulation is relative to microstructure regulation, which refers to the design and optimization of macrostructure on the basis of bulk or thin film materials. Currently, it mainly includes the formation of obstructive layers and heterogeneous structures. To some extent, it is similar to the concept of composition, especially the 2–2 composite. However, macrostructural regulation implies a higher level of design ideas and more sophisticated methods. (1) Blocking layer. The role of the blocking layer in ceramic materials is mainly to prevent electrical breakdown, and materials with high breakdown field strength are the first choice. The most commonly used materials are silica, alumina, and so on, which can be prepared by CVD, fusion, PVD, and other methods on the surface of bulk material or thin film material. Zhou et al.[15] deposited a thin alumina layer of 50 nm on the surface of $CaTiO_3$ ceramics and the U_d was increased by nearly two times, which showed great potential in the applications. (2) Heterogeneous structures. The concept in dielectric ceramics is very similar to composite, in that it consists of building a heterogeneous structure locally in the ceramic material, i.e., a 0–3 composite. It is also possible to build heterogeneous structures on a macroscopic scale, i.e., a 2–2 composite. In addition, heterostructures in thin film materials are more focused on device applications, i.e. multilayer ceramic capacitors, which we will explain in detail in a later section.

3 LINER DIELECTRIC CERAMICS

Linear dielectric ceramics usually possess characteristics of low ε_r and tan δ, as well as moderate E_b. It is thereby hard to obtain high U_d under a high electric field. In this regard, the researches on LDs are mainly concentrated on increasing ε_r or improving polarization behavior by maintaining high E_b. PEs, including the so-called superparaelectric relaxor ferroelectrics, have Curie temperatures below room temperature and exhibit slim nonlinear PE hysteresis. Therefore, at room temperature and above, PEs have practically the same properties as LDs. Since controversy still exists about the classification of these two, e.g. strontium titanate (ST) which is classified as a linear dielectric in some literature and as a PE in others, in this subsection, PEs are discussed with linear dielectrics. It should be noted that in the study of LDs, most researchers calculate the theoretical energy storage density only by measuring the breakdown field strength and dielectric constant. The authors feel that such a method is not comparable to the energy storage density calculated by PE curve measurements because the dielectric constant usually varies when chemical composition changes.

3.1 TiO$_2$-Based Ceramics

TiO$_2$ is a typical linear dielectric with characteristics of a low E_b (~4 kV/cm), low tan δ (<0.1%), moderate dielectric constant (~110), and wide band gap (~3.2 eV), which limit its applications.[16, 17] Generally, TiO$_2$ has three crystal structures including orthorhombic brookite, tetragonal rutile, and anatase. The rutile phase is more stable and easier to be synthesized than others, and hence gains more attention in ceramic bulks and films.[18, 19]

In order to enhance the energy storage density, one approach is to increase the ε_r. In 2013, Hu et al.[20] first reported that (In, Nb) co-doped TiO$_2$ ceramics displayed a giant dielectric constant (>10^4) as well as low tan δ (<5%), and possessed good temperature and frequency stabilities over a wide temperature range (80–450 K). They explained that 'triangular' In$_2^{3+}V_0^{\bullet\bullet}$ Ti^{3+} and 'diamond' shaped Nb$_2^{5+}$ Ti$_3^+A_{\mathrm{Ti}}$ (A = Ti$_3^+$/In$_3^+$/Ti$_4^+$) defect complexes are strongly correlated, giving rise to large defect-dipole clusters containing highly localized electrons that are together responsible for the excellent properties observed in the co-doped TiO$_2$. The temperature dependences of the dielectric properties and the structure of the In and Nb defect complexes are shown in Figure 2.3.

Since then, a large number of co-doped TiO$_2$-based energy storage ceramics have been investigated, such as: (Al, La) co-doped TiO$_2$, (Ta, Al) co-doped TiO$_2$, (Nb, La) co-doped TiO$_2$, (Ho, Ta) co-doped TiO$_2$ and so on.[21–23] A characteristic of these co-doped systems is their extremely high dielectric constant, typically greater than 10^4. However, the sharp reduction in E_b results in a low energy storage density. Currently, it is generally believed that the high dielectric constant is due to the localized defect polarization mechanism and the internal barrier layer capacitance mechanism,[24, 25] while the reduced E_b is related to the defects, such as the oxygen vacancy.

Alternatively, E_b can be improved via sintering process modifications of the TiO$_2$-based ceramics, such as refining grain size, adding glass phase, applying special sintering techniques (such as SPS), and so on.[26–28] Liu et al.[28] obtained a U_d of 1.15 J/cm^3 for alkali-free glass-modified TiO$_2$ ceramics at 501.7 kV/cm. Besides using sol-gel process, Wei et al.[29] obtained dense and uniform TiO$_2$-based ceramics with the DBS and storage density increased from 37.45 kV/mm and 0.76 J/cm^3 to 48.29 kV/mm and 1.05 J/cm^3, respectively. Unfortunately, the U_d of those ceramic bulks prepared using different modification ways is still low in the order of 1 J/cm^3, which is inferior to other energy storage dielectric materials.

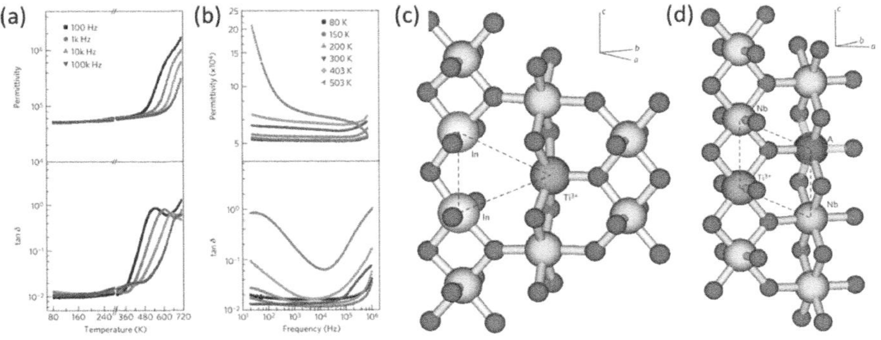

FIGURE 2.3 (a) Dielectric permittivity of 10% (Nb + In) co-doped TiO_2 measured from 80 to 700 K at given frequencies. A narrow data break occurs near the RT region (approximately 270–300 K), as two experimental set-ups were used for the low-temperature and high-temperature measurements, respectively. (b) Frequency dependences of dielectric permittivity and loss of 10% (Nb + In) co-doped TiO_2 measured at the given temperatures. Ball-and-stick illustration of the triangular $In_2^{3+}V_O^{\bullet\bullet}Ti^{3+}$ (c) and diamond $Nb_2^{5+}Ti^{3+}A_{Ti}$ (d) complexes common to stable (Nb + In) co-doped TiO_2. Unlabeled atoms are Ti^{4+} (yellow) and O^{2-} (red). The triangle- and diamond- shaped motifs are sketched with dashed lines.[20]

TABLE 2.1

Energy storage properties of some TiO_2-based linear dielectric materials

Composition	Type	E_b (kV/cm)	U_d (J/cm³)	Ref.
TiO_2	Film	1,800	15	[26]
TiO_2	Film	1,400	14	[30]
TiO_2	Bulk	350	0.76	[29]
TiO_2–CaO–MgO–Al_2O_3–SiO_2	Bulk	482.9	1.05	[29]
Mn-doped TiO_2–BaO–B_2O_3–Al_2O_3–SiO_2	Bulk	501.7	1.15	[28]

It should be noticed that high-quality TiO_2 films are a good direction owing to their high E_b. For instance, Chao and Dogan[30] fabricated 0.1-mm thick TiO_2 films using the tape-casting method, which achieve a U_d of 14 J/cm³ at 1,400 kV/cm. The U_d and E_b of some representative TiO_2-based dielectric materials are listed in Table 2.1.

3.2 SrTiO₃-BASED CERAMICS

$SrTiO_3$ (ST) is a cubic phase of ABO_3 type perovskite structure with a space group of $Pm\overline{3}m$ and a lattice constant of $a = 3.905$ Å.[31, 32] ST ceramics possess moderate ε_r of ~300 and E_b of ~100 kV/cm, and a low tan δ of ~10^{-3}, as well as good temperature-, frequency-independent dielectric properties, bias voltage stability, and thermoelectric energy conversion efficiency.[33–35] Therefore, it is considered to be a potential energy storage and conversion candidate. Compared with TiO_2 ceramics (~110) and polymer linear dielectrics (<10), ST has the advantage of relatively high ε_r, and thus is more suitable for pulse capacitor.

Firstly, it is necessary to increase ε_r, so as to increase U_d. Using a similar approach to the co-doped TiO_2 ceramics, Chen et al.[36–38] reported $(Bi,Sr)TiO_3$ ceramics with a giant dielectric constant (~5,000), discussed the related physical mechanisms, and proposed that the first and second ionizations of oxygen vacancies, and the corresponding thermal movement were the main reasons. To avoid the problem of Bi-containing oxide volatilization at high temperatures, Shen et al.[33, 39–41] used trivalent nonvolatile rare earth ions (Re^{3+} = La, Sm, Gd, Er, Nd, etc.) to replace Sr^{2+}, and designed three composition formulas based on three possible charge compensation mechanisms such as equimolar substitution, to introduce Sr or Ti vacancy in advance, and successfully fabricated ceramics with perovskite structure. Based on the experimentally verified feasibility of introducing ion vacancies in advance for charge compensation, the concept of "forced charge compensation mechanism" was proposed. For instance, Shen et al.[39] synthesized $Re_{0.02}Sr_{0.97}TiO_3$ ceramics by introducing Sr vacancy in advance, which displayed a high dielectric constant (2,750–4,530) and good bias voltage stability.

Besides utilizing Ca^{2+}, Ba^{2+}, and Pb^{2+} ions to replace Sr^{2+} on the A-site of ST could adjust the Curie temperature (T_c) to room temperature, thereby enhancing the dielectric constant.[42, 43] Especially, $Ba_xSr_{1-x}TiO_3$ (BST) solid solutions combine the characteristics of high E_b of $SrTiO_3$ and high ε_r of $BaTiO_3$ and have received a great deal of attention in recent years.[44] The structure and performance of BST can be adjusted over a wide range to meet the requirements of different applications. As the molar fraction of Ba increases from 0 to 1, the phase structure of BST varies from cubic paraelectric (ST) to tetragonal ferroelectric (BT), accompanied by a T_c increase from near 0 to ~393 K. According to the theoretical calculation of BST solid solutions by Fletcher et al.[45], it is easier to obtain an ideal energy storage property if the T_c of the ceramic composition is far away the working temperature. Therefore, the BST compositions with $x \leq 0.4$ would be more suitable pulse power capacitor candidates because their PE loops display a linear or weakly nonlinear characteristic at room temperature. In 2015, Wang et al.[42] investigated the energy storage performances of $Ba_xSr_{1-x}TiO_3$ ($x \leq 0.4$) ceramics and found that $Ba_{0.4}Sr_{0.6}TiO_3$ achieved the highest U_d of 0.33 J/cm³, while its relatively low and rapidly decreased η posed a serious problem hindering its application. By comparison, $Ba_{0.3}Sr_{0.7}TiO_3$ possessed a moderate U_d of 0.23 J/cm³, a high η ($\geq 95\%$), and a very low dielectric loss (tan $\delta = 7.6 \cdot 10^{-4}$ @ 1 kHz), making it more suitable for the fabrication of solid-state compact portable pulse power electronics.

The microstructure regulation of ST ceramics has been widely studied to increase the E_b. In 2014, Song et al.[46] prepared $Ba_{0.4}Sr_{0.6}TiO_3$ ceramics with various grain sizes (0.5–5.6 μm), and observed that the dielectric peak gradually depressed and broadened and E_b gradually increased with decreasing grain size, which should be closely related to the ratio of grain/grain boundary and polar nanoregions (PNRs). $Ba_{0.4}Sr_{0.6}TiO_3$ ceramic bulk with a grain size of 0.5 μm achieves a high $U_d = 1.28$ J/cm³ measured at the highest E_b of 243 kV/cm. Wu et al.[47] compared the microstructure and energy storage properties of SPS and conventionally sintered (CS) $Ba_{0.3}Sr_{0.7}TiO_3$ ceramics. The SPS sintered ceramics consist of tetragonal and cubic phases with an average grain size of 880 nm, while the CS ones are of the tetragonal phase only. The maximum U_d of the SPS samples is 1.13 J/cm³ at E_b = 230 kV/cm, which is approximately twice that of the CS samples (0.57 J/cm³). Moreover, the addition of suitable glass compositions is also an effective method to enhance E_b and reduce the sintering temperature of ST-based ceramics.[48, 49] In 2019, Shen et al.[50] used a homemade glass frit to modify BST so as to enhance the E_b (160 kV/cm) and reduce the high-temperature resistivity, which expanded the working temperature range for energy storage ceramic capacitor applications.

In terms of ST-based ceramic films, "Defect engineering" is an effective tool to enhance U_d by strengthening relaxor characteristics for ST-based ceramic films. Yang et al.[51] fabricated $(Sr_{1-1.5x}Bi_x)Ti_{0.99}Mn_{0.01}O_3$ (SBTM, x = 0.01, 0.05, 0.1) thin films with a thickness of 217 nm using a sol-gel method. As the x value increases, relaxor behaviors are gradually strengthened due to a slight rotation of the (TiO_6) octahedra induced by the formation of $Bi^{3+}-V_{sr}''$ defect complex. Under an electric field of 1,982 kV/cm, $(Sr_{0.85}Bi_{0.1})Ti_{0.99}Mn_{0.01}O_3$ displays a U_d of 24.4 J/cm³ accompanied by the largest ΔP ($P_{max}-P_r$ = 34.3 µC/cm²). The introductions of $Bi_{0.5}Na_{0.5}TiO_3$ (BNT) and $BiFeO_3$ (BF) have a similar effect.[52, 53] In the related mechanism studies, Hou et al.[54] investigated the influences of interface difference and thickness on the energy storage performances for ST thin films, and observed the existence of ionic diffusion layers and oxygen vacancies using high-resolution transmission electron microscope (HR-TEM). Moreover, they found that the E_b and P_{max} (up to 102 µC/cm²) along the positive direction were higher than the negative direction. Therefore, the maximum U_d of 307 J/cm³ was achieved in the ST thin films, which may be related to local electric field and redistribution of oxygen vacancies, as shown in Figure 2.4.

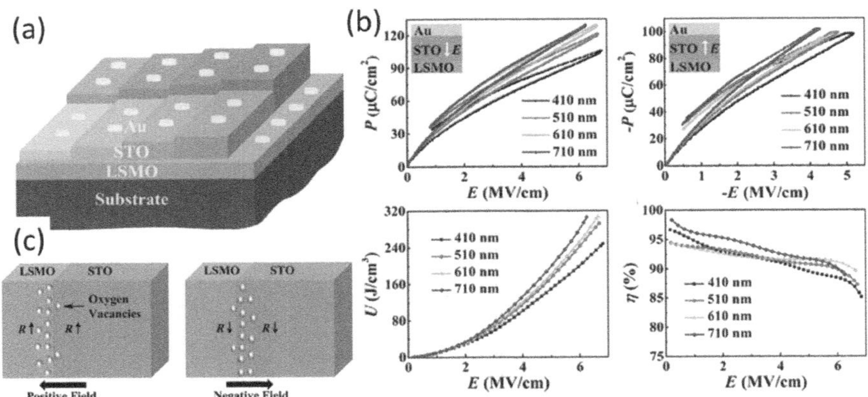

FIGURE 2.4 (a) Schematic illustration of capacitors with different thicknesses; (b) PE loops of Au/STO/LSMO capacitors under positive fields and negative fields; the insets show the orientation of E. Variations of recoverable energy density and energy storage efficiency under positive electric fields of the Au/STO/LSMO capacitors; (c) Schematic of the migration of oxygen vacancies at the STO/ LSMO interface.[54]

Various meaningful and interesting works have been carried out to optimize the macro/micro structures and to improve the energy storage properties of ST-based ceramic films. It is well known that an amorphous phase usually possesses higher E_b but lower ε_r than its crystalline counterparts. Gao et al.[55] studied the energy storage behaviors of amorphous ST thin films with different top electrodes, and proposed a "self-healing" mechanism based on the anodic oxidation reaction in aluminum electrolytic capacitors. At a relative humidity of 60%, amorphous ST films with Al top electrode achieved a U_d of 15.7 J/cm^3 at 3,500 kV/cm, which is nearly eight times that of the samples with an Au electrode. Since then, Gao et al.[56] inserted insulating Al$_2$O$_3$ as a blocking layer to form a heterostructure, and achieved the maximum U_d of 39.49 J/cm^3 at E_b = 7,542.3 kV/cm when the interface number equaled 4. Recently, Chen et al.[57] used Ca$_{0.2}$Zr$_{0.8}$O$_1$.8 (CSZ) as dead-layer to enhance E_b of Ba$_{0.3}$Sr$_{0.7}$Zr$_{0.18}$Ti$_{0.82}$O$_3$ (BSZT) thin films. Due to the formation of a high electron injection barrier, Schottky electron emission was suppressed, and thus E_b and U_d were enhanced from 5.4 to 6.3 MV/cm and 64.8 to 89.4 J/cm^3, respectively. The energy storage properties of typical ST-based ceramics are listed in Table 2.2.

TABLE 2.2

Energy storage properties of typical ST-based ceramics

Composition	Type	E_b (kV/cm)	U_d (J/cm^3)	Ref.
Ba$_{0.3}$Sr$_{0.7}$TiO$_3$	Bulk	90	0.23	[42]
Ba$_{0.4}$Sr$_{0.6}$TiO$_3$	Bulk	90	0.33	[42]
Ba$_{0.3}$Sr$_{0.7}$TiO$_3$	Bulk	140	0.57	[30]
Ba$_{0.3}$Sr$_{0.7}$TiO$_3$(SPS)	Bulk	230	1.13	[30]
Ba$_{0.3}$Sr$_{0.7}$TiO$_3$ + 2% BBSZ	Bulk	160	0.63	[50]
0.9(Sr$_{0.7}$Bi$_{0.2}$)TiO$_3$–0.1Bi(Mg$_{0.5}$Hf$_{0.5}$)O$_3$	Bulk	360	3.1	[35]
0.6(Ba$_{0.75}$Sr$_{0.25}$)TiO$_3$–0.4Bi(Mg$_{0.5}$Hf$_{0.5}$)O$_3$	Bulk	390	4.3	[44]
SrTiO$_3$ (STO) films grown on La$_{0.67}$Sr$_{0.33}$MnO$_3$ (LSMO)	Film	6,800	307	[54]
With Au and Al being deposited on SrTiO$_3$ thin films as top electrode	Film	–	15.7	[55]
Mn-doped 0.4BiFeO$_3$–0.6SrTiO$_3$ (BFSTO) thin film	Film	3,600	51	[53]
SrTiO$_3$ film with Al$_2$O$_3$ blocking layer	Film	75,423	39.49	[56]
(Sr$_{0.85}$Bi$_{0.1}$)Ti$_{0.99}$Mn$_{0.01}$O$_3$ thin film	Film	–	24.4	[51]
Sr$_{0.99}$5(Na$_{0.5}$Bi$_{0.5}$)0.005(Ti$_{0.99}$Mn$_{0.01}$)O$_3$ amorphous thin film	Film	–	65.3	[52]
Ba$_{0.3}$Sr$_{0.7}$Zr$_{0.18}$Ti$_{0.82}$O$_3$ (BSZT)	Film	–	89.4	[57]

3.3 CaTiO$_3$-Based Ceramics

CaTiO$_3$ ceramics are generally considered to be suitable for high-temperature energy storage due to their medium permittivity (~170) and high breakdown strength (~300 kV/cm), as well as large band gap (Eg of ~3.4 eV).[15, 58, 59] Undoped CaTiO$_3$ ceramics prepared by conventional sintering approach showed an energy storage density of 1.5 J/cm^3 (E_b = 435 kV/cm) and 0.25 J/cm^3 (E_b = 193 kV/cm), as reported by Zhou et al.[60] and Luo et al.[61] respectively. With the doping of BiScO$_3$, the U_d was enhanced to 1.55 J/cm^3 at E_b of 270 kV/cm, and the hybridization between O 2p and Sc 3d orbitals at high conduction band was found to enlarge the band gap, which was the essential reason for the enhancement of polarization and dielectric properties.[61] With the doping of Zr^{4+} (40%) on the B-site, U_d was further improved to 2.7 J/cm^3 (E_b = 756 kV/cm).[60] With the doping of SrTiO$_3$, Wang et al.[62] prepared Ca$_{0.5}$Sr$_{0.5}$TiO$_3$ ceramics with a moderate E_b of 270 kV/cm for probing high energy storage properties at high temperatures. Ouyang et al.[63] further modified Ca$_{0.5}$Sr$_{0.5}$TiO$_3$ ceramics with Mg^{2+} ion doping, and obtained an enhanced breakdown field of 460 kV/cm, a high recoverable energy density of 2.88 J/cm^3, and an energy efficiency of 90%. Pu et al.[59] presented a regulating strategy through Zr^{4+} doping and oxygen atmosphere treatment to reliably enhance the energy storage performances of Ca$_{0.5}$Sr$_{0.5}$TiO$_3$ ceramics. An energy density of 3.37 J/cm^3 and a 96% energy efficiency were obtained simultaneously in Ca$_{0.5}$Sr$_{0.5}$Ti$_{0.85}$Zr$_{0.15}$O$_3$ ceramics under a breakdown strength of 440 kV/cm following oxygen treatment, which are better results than other reported linear dielectric systems. Zhang et al.[64] reported A-site Sm^{3+}-doped perovskite Ca$_{1-1.5x}$Sm$_x\square_{0.5x}$TiO$_3$ ceramics with introduced A-site vacancies (VA). As a result, the grain size decreased while the ceramic density was improved. The dielectric breakdown strength was significantly improved from 429 kV/cm (x = 0) to 547 kV/cm (x = 0.1) while the dielectric linearity was maintained. The optimum energy storage density of 2 J/cm^3 (x = 0.02) with an ultrahigh energy efficiency of over 93% was achieved.

For the microstructure modification, the most remarkable result was obtained by Zhou et al. by SPS, and creating fine and uniform microstructures, as shown in Figure 2.5, which gave rise to greatly enhanced dielectric strength (E_b ~ 910 kV/cm) and energy storage density (U_d = 6.9 J/cm^3).[15]

For the CaTiO$_3$-based ceramic films, a typical composition is 0.8CaZrO$_3$–0.2CaTiO$_3$ (CZT), which yields a relative permittivity of ε_r ~ 34–75, a breakdown strength of 0.65–1.75 MV/cm, a thermal coefficient of capacitance (TCC) of 30 ppm/°C or 0.5% from −50°C to 125°C, and an energy density of up to 5 J/cm^3.[65, 66] Shay et al.[67] used tape-casting method to prepare the 0.8CaTiO$_3$–0.2CaHfO$_3$ single layer (~10 μm) capacitors, which exhibited an energy density as large as 9.0 J/cm^3 at 1.2 MV/cm. The energy storage properties of some CT-based ceramics are listed in Table 2.3.

3.4 Glass-Ceramics

Compared with ceramics or polymer-based dielectrics, glass-ceramics have high glass transition temperature and submicron grains. The microstructure of a

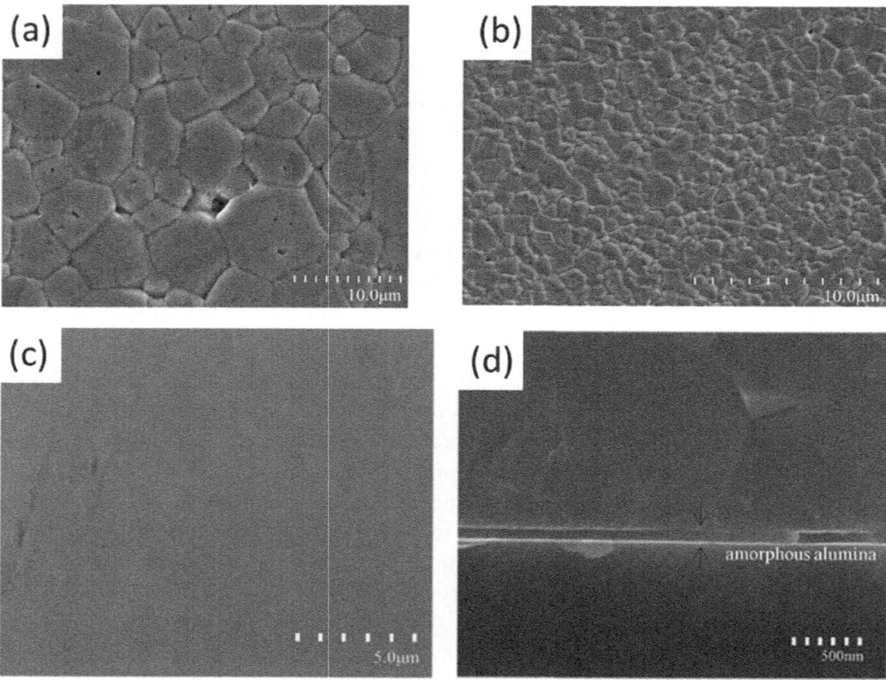

FIGURE 2.5 SEM images of polished and thermal etched surfaces of (a) CS and (b) SPS samples, SEM micrographs of the amorphous alumina coating deposited on the ceramic surface and (c) top view and (d) cross-section view.[15]

glass-ceramic is characterized by the presence of crystallites which are held together by an amorphous glass matrix. So it has the superiorities of high thermal stability and low dielectric loss, as well as low electric field dependence of the relative dielectric constant. Thus, it is recognized that glass-ceramics are considered to be one of the most promising materials for use in energy-storage devices.[68]

Over the past decade, niobate-based glass-ceramics with tetragonal tungsten bronze structure $(A_1)_4(A_2)_2C_4B_{10}O_{30}$ and cubic perovskite structure ABO_3 have attracted much attention owing to their superior E_b. Many efforts have been made to improve the dielectric properties and energy storage performances of these glass-ceramics. For instance, Xiu et al.[69] reported that the theoretical energy storage density was optimized when replacing a small amount of SiO_2 with B_2O_3 in niobate-based glass-ceramics. When the B_2O_3/SiO_2 proportion was 1.5–33.5, the optimal theoretical energy storage density reached 7.4 J cm^{-3}. Wang et al.[70] reported that the contents of SiO_2 and Nb_2O_5 have a great influence on the discharge properties of Na_2O–BaO–PbO–Nb_2O_5–SiO_2–Al_2O_3 glass-ceramics, and the maximum actual discharge energy density of 0.74 J cm^{-3} was achieved when the ratio of SiO_2/Nb_2O_5 was 30:32. Li et al.[71] found that the crystallization time was proportional to the grain size and indirectly affected the dielectric constant in BaO–SrO–Na_2O–Nb_2O_5–SiO_2 glass-ceramics. The highest dielectric constant of 240 was obtained

TABLE 2.3

Energy storage properties of some CT-based ceramics

Composition	Type	E_b (kV/cm)	U_d (J/cm³)	Ref.
$CaTiO_3$	Bulk	435	1.5	[60]
$CaTiO_3$	Bulk	193	0.25	[61]
$CaZr_{0.4}Ti_{0.6}O_3$	Bulk	756	2.7	[60]
$CaZr_{0.5}Ti_{0.5}O_3$	Bulk	756	2.6	[60]
$0.9CaTiO_3–0.1BiScO_3$	Bulk	270	1.55	[61]
$CaTiO_3$(SPS)	Bulk	910	6.9	[15]
$CaTiO_3$(SPS) with amorphous alumina thin films	Bulk	1,188	11.8	[15]
$Ca_{0.5}Sr_{0.5}Ti_{0.9}Zr_{0.1}O_3$	Bulk	390	2.05	[62]
$Ca_{0.5}Sr_{0.5}Ti_{0.85}Zr_{0.15}O_3$	Bulk	440	3.37	[59]
$Ca_{0.5}Sr_{0.5}Ti_{0.97}Sn_{0.03}O_3$	Bulk	330	2.06	[57]
$Ca_{0.5}Sr_{0.5}(Ta_xTi_{1-1.25x})$ $O_3–2wt\%SiO_2(x = 0.024)$	Bulk	360	2	[58]
$Ca_{1-1.5}xSm_x\square_{0.5x}TiO_3(x = 0.1)$	Bulk	547	2	[64]
$(Ca_{0.5}Sr_{0.5})_{0.99}Mg_{0.01}TiO_3$	Bulk	460	2.88	[63]
$0.8CaTiO_3–0.2CaHfO_3$	Film	1,100	9.0	[67]
0.5 mol% Mn-doped $0.8CaTiO_3–0.2CaHfO_3$	Film	1,200	9.5	[67]

in the sample with a crystallization time of 1,000 min. Zhou et al.[72] investigated $Na_2O–BaO–Nb_2O_5–SiO_2$ glass-ceramics and found that the grain size and crystallinity both raised with increasing crystalline temperature. Moreover, when this glass-ceramics was crystallized at 1,000°C, the charge energy density and discharge energy density reached 0.19 J/cm³ and 0.17 J/cm³, respectively, with the applied electric field of 10 kV/mm. Chen et al.[73] prepared $BaO–Na_2O–Bi_2O_3–Nb_2O_5–Al_2O_3–SiO_2$ (BNBN-AS) glass-ceramics by the technique of melt-quenching accompanied by the process of controlled crystallization. The theoretical energy storage density was markedly enhanced to the maximum value of 18.4 ± 1.3 J/cm³ at 950°C, which is approximately twice that at 850°C. Researches on various dopants like Gd_2O_3[74, 75], CeO_2[76–78], CaF_2[79], La_2O_3[80] and ZrO_2[81] have been carried out to improve the properties of ferroelectric glass-ceramics in the corresponding systems.

A large number of results have also been achieved by adding ceramic substrates to energy-storage amorphous films. Michael and Trolier-Mckinstry[82] explored composite thin films of nanocrystalline PbO particles in an amorphous $Pb_{1.1}TiO_{3.1}$ matrix in order to increase the stored energy density of dielectrics. Excess lead was added in the precursor solution to promote the formation of nanocrystallites in the amorphous matrix. The presence of PbO nanocrystals enhanced the moderately high permittivity of the film and the amorphous matrix facilitated the high breakdown field of 5 MV cm⁻¹. The lack of structural periodicity of the amorphous network and the isolated, randomly distributed lead oxide nanocrystals did not form a conduction pathway through the film, leading to an extremely low leakage current (in the order of 10^{-8} A/cm^2) and a good U_d of 28 J/cm^3. Nanocomposite films with nanocrystalline δ-Bi$_2$O$_3$ dispersed in an amorphous matrix of Bi$_{1.5}$Zn$_{0.9}$Nb$_{1.35}$Ta$_{0.15}$O$_{6.9}$ were deposited on Pt/Polyimide substrates for flexible energy storage applications, which exhibited a good U_d of 40 J/cm^3 at 3.8 MV/cm.[83] Gao et al.[55] explored the self-repairing behavior of amorphous STO film capacitors with Au and Al top electrodes in humid air. At 60% humidity, the U_d of Al/STO was estimated to be about 15.7 J/cm^3, which was higher than that of the sample with the Au electrode (13.1 J/cm^3). In a similar study on Bi-doped (2, 5, 10, and 15 mol%) amorphous STO films, Yao et al.[84] found that Bi-doping helped to suppress the formation of the oxygen vacancies and trap the charges, thus decreasing the leakage current and enhancing the breakdown field of the films. The energy storage properties of some glass-ceramics are listed in Table 2.4.

In summary, for linear dielectric ceramic bulks, giant dielectric constant can be observed in TiO$_2$-based ceramics with donor/acceptor co-doping on the B-site, and ST-based ceramics with donor/acceptor co-doping on the B-site or aliovalent doping on the A-site. Some controversies regarding the physical mechanisms, however, still exist. The energy storage properties of ceramic bulks are limited by the rapid decrease in E_b as a result of substitution. Adding a suitable glass phase and using special sintering techniques to refine grain size are both able to enhance E_b of ceramic

TABLE 2.4

Energy storage properties of some glass-ceramics

Composition	Type	E_b (kV/cm)	U_d (J/cm^3)	Ref.
SrO-BaO-Nb$_2$O$_5$-B$_2$O$_3$-SiO$_2$ with CeO$_2$ added	Bulk	1,250	3.39	[76]
BaO–SrO–Nb$_2$O$_5$–B$_2$O$_3$ + 0.5mol%Gd$_2$O$_3$	Bulk	1,075	6.94	[74]
SrO–BaO–Nb$_2$O$_5$ based glass-ceramics with B$_2$O$_3$/SiO$_2$	Bulk	–	7.4	[69]
Sr$_{0.5}$Ba$_{0.5}$Nb$_2$O$_6$ + La$_2$O$_3$	Bulk	1,127	7.1	[80]
Na$_2$O–BaO–Nb$_2$O$_5$–SiO$_2$	Bulk	100	0.17	[72]
Na$_2$O–BaO–PbO–Nb$_2$O$_5$–SiO$_2$–Al$_2$O$_3$	Bulk	300	17.9	[70]
BaO–Na$_2$O–Nb$_2$O$_5$ + CaF$_2$	Bulk	–	14.3	[79]
Bi$_2$O$_3$–Nb$_2$O$_5$–SiO$_2$–Al$_2$O$_3$ + CeO$_2$	Bulk	300	20.9	[77]
SrO–Na$_2$O–Nb$_2$O$_5$–SiO$_2$ + Ta$_2$O$_5$	Bulk	400	15.22	[85]
10PbTiO$_3$–10Fe$_2$O$_3$–30V$_2$O$_5$–50B$_2$O$_3$	Bulk	–	0.16	[86]
BaO–Na$_2$O–Bi$_2$O$_3$–Nb$_2$O$_5$–Al$_2$O$_3$–SiO$_2$	Bulk	1,878	18.4	[73]
26.5BaO-7.5K$_2$O–30Nb$_2$O$_5$–6Al$_2$O$_3$–30SiO$_2$-0.6PbO-xZrO$_2$ (x = 1.5)	Bulk	1,949	21.62	[87]
Pb$_{1.1}$TiO$_{3.1}$	Film	5,000	28	[82]
Bi$_{1.5}$Zn$_{0.9}$Nb$_{1.35}$Ta$_{0.15}$O$_{6.9}$	Film	3,800	40.2	[83]

bulks. For ceramic films, adjusting the suitable ratio of amorphous to crystalline phase or introducing a highly insulating layer would be a good way to improve their breakdown strength.

4 RELAXOR FERROELECTRIC CERAMICS

Ferroelectrics are a special class of dielectric materials that possess a spontaneous polarization (P_s) at a certain temperature range and the direction of P_s can be changed with an external electric field. Compared with linear dielectrics, ferroelectrics display an obvious nonlinear characteristic since the domains cannot fast respond to electric field stimulation. Generally, according to the characteristics of their PE loops, ferroelectrics can be classified into normal ferroelectrics and relaxor ferroelectrics. Normal ferroelectrics possess high P_{max} while their high P_r leads to the dissipation of most energy during the discharge process. By contrast, relaxor ferroelectrics exhibit a slim PE loop with a high P_{max} and a low P_r (i.e., a high $\triangle P$ = $P_{max} - P_r$), meaning that electric energy can be effectively released, and thus better energy storage performances. Note that strengthening the relaxor characteristics and enhancing E_b have become two important approaches for enhancing U_d. In this section, ferroelectrics are mainly referred to as relaxor ferroelectrics.

4.1 Pb-Based Relaxor Ferroelectric Ceramics

PbZr$_{1-x}$Ti$_x$O$_3$ (PZT, $0 \leq x \leq 1$) ceramics with compositions located at the morphotropic phase boundary (MPB) where Zr:Ti = 52:48, possess a high piezoelectric activity (d_{33} up to 300 pC/N) and good temperature stability, and have been an extremely popular dielectric material.[88, 89] In addition, other ceramic compositions, such as PZT 65/35, 70/30, have also received much attention for their good piezoelectric properties.[90, 91] For example, researches on the actuators based on PZT ceramics are accelerated due to a large electrostrain under a low electric field. And a small amount of La^{3+} (about 7–10 mol%) replacing Pb^{2+} can effectively strengthen the relaxor characteristics of PZT-based ceramics owing to a disrupted long-range ferroelectric order and diffuse phase transition.[92, 93] As such, (Pb,La)(Zr,Ti)O$_3$ (PLZT)-based relaxor ferroelectrics are considered to be a promising candidate for energy storage ceramic capacitors. Adjusting the La/Zr/Ti ratio can influence significantly the electric properties because of the turning of the complicated phase structures. In 2019, Kumar et al.[94] reported (Pb$_{0.89}$La$_{0.11}$)(Zr$_{0.70}$Ti$_{0.30}$)$_{0.9725}$O$_3$ (PLZT 11/70/30) ceramics with a U_d of only 0.85 J/cm^3 due to low electric field. Generally speaking, the energy density of Pb-based relaxor ferroelectric ceramic bulks is currently less than 3 J/cm^3.

In terms of Pb-based ceramic films, Hu et al.[95] investigated the effects of different Zr/Ti ratios on the energy storage properties of PLZT thin films at a fixed La content of 8 mol%. As the Ti/(Zr+Ti) ratio gradually increases, ε_r is enhanced while tan δ shows an opposite trend, indicating that the phase structure gradually transforms from relaxor ferroelectric into normal ferroelectric. At E_b = 2,180 kV/cm, the U_d and η of the PLZT 8/52/48 relaxor ferroelectric thin films are 30 J/cm^3 and 78%, respectively. Liu et al.[96] used Mn as a dopant to (Pb$_{0.91}$La$_{0.09}$)(Zr$_{0.65}$Ti$_{0.35}$)O$_3$ (PLZT 9/65/35) to enlarge the polarization difference of ($P_{max} - P_r$), and obtained a U_d of 30.8 J/cm^3 for the 1 mol% Mn-doped thick films. In addition, the addition of excess Pb is a common chemical compensation method to solve the problems of Pb volatilization during the annealing process and to suppress the formation of the pyrochlore phase.[97] In 2017, Zhang et al.[98] investigated the energy storage properties of PbZr$_{0.52}$Ti$_{0.48}$O$_3$-based thin films, and acquired a high U_d = 28.2 J/cm^3 for the PbZrO$_3$/PbZr$_{0.52}$Ti$_{0.48}$O$_3$ bilayer thin films at 2,410 kV/cm. In addition, they designed a sandwich structure of PbZr$_{0.52}$Ti$_{0.48}$O$_3$/Al$_2$O$_3$/PbZr$_{0.52}$Ti$_{0.48}$O$_3$ (PZT/AO/PZT) to enhance E_b.[99] Due to the formation of the so-called "built-in electric field" at the interface and the high insulating characteristic of AO, PZT/AO/PZT annealed at 550°C achieved a U_d of 63.7 J/cm^3 at 5.7 MV/cm. It should be mentioned that the PZT 52/48 system still possesses a relatively high P_r, restricting its energy storage properties.

Reasonable design and selection of heterostructure for Pb-based thin films is an important step in optimizing energy storage properties. In 2013, Zhang et al.[100] prepared a compositional gradient (Pb$_{1-x}$La$_x$)(Zr$_{0.65}$Ti$_{0.35}$)O$_3$ (PLZT, x = 0.08, 0.09, 0.1) thick films using a sol-gel method. The up-graded PLZT films possess a U_d of 12.4 J/cm^3 at 800 kV/cm, the down-graded ones of 8.9 J/cm^3, and the lowest single-composition ones of 7.1 J/cm^3. It is accepted that high texture quality and dense structure can both severely influence breakdown behaviors and energy storage performances. [101, 102] Nguyen et al.[101] deposited (Pb$_{0.9}$La$_{0.1}$)(Zr$_{0.52}$Ti$_{0.48}$)O$_3$ (PLZT 10/52/48) thin films using PLD on Si substrate choosing Ca$_2$Nb$_3$O$_{10}$ (CNOns) and Ti$_{0.87}$O$_2$ (TiOns) nanosheets as the template layer. Highly textured (001)-oriented PLZT 10/52/48 films grown on CNOns possess a high U_d of 58.4 J/cm^3, which exceeds the U_d of 44 J/cm^3 for (110)-oriented PLZT 10/52/48 films grown on TiOns. Generally speaking, the differences in lattice constants and thermal expansion coefficients in the heterointerfaces provide good conditions for stress. Ma et al.[103] utilized XRD to analyze the residual stress of PLZT 8/52/48 thick films, and found that compressive stress could to a certain extent improve the tunability of polarization, and enhance E_b and domain switch ability. In addition, Peng et al.[104] recently prepared Mn-doped Pb$_{0.97}$La$_{0.02}$(Zr$_{0.905}$Sn$_{0.015}$Ti$_{0.08}$)O$_3$ (PLZST) relaxor ferroelectric thin films, and proposed a "low-temperature poling" method to improve E_b and via a "wake-up" mechanism. The E_b and U_d at room temperature were greatly enhanced (nearly doubled) from 1,286 to 2,000 kV/cm, and from 16.6 to 31.2 J/cm^3, respectively. The energy storage performances of typical Pb-based relaxor ferroelectric materials are listed in Table 2.5.

4.2 Bi$_{0.5}$Na$_{0.5}$TiO$_3$-Based Ceramics

Bi$_{0.5}$Na$_{0.5}$TiO$_3$ (BNT) is a ferroelectric material first discovered by Smolenskii et al.[105] which possesses complicated phase structures and good dielectric, piezoelectric, and ferroelectric properties, especially a high P_{max} (~40 μC/cm^2).[106–109] It has been a popular research topic both for fundamental studies and practical applications in ferroelectric materials.[110–112] However, the characteristics of high P_r (~38 μC/cm^2), high E_c (~73 kV/cm), and poor sintering behavior of pure BNT ceramics hinder their energy storage applications.

With the substitution of BaTiO$_3$, the Bi$_{0.5}$Na$_{0.5}$TiO$_{3-x}$BaTiO$_3$ (x = 6%–7%) binary solid solution near MPB exhibits excellent electrical properties, and is one of the

TABLE 2.5

Energy storage performances of typical Pb-based relaxor ferroelectric materials

Composition	Type	E_b (kV/cm)	U_d (J/cm³)	η (%)	Ref.
$(1-x)Pb(Mg_{1/3}Nb_{2/3})O_3-xPbTiO_3$	Bulk	–	0.47	–	[91]
$(Pb_{0.89}La_{0.11})(Zr_{0.70}Ti_{0.30})_{0.9725}O_3$	Bulk	–	0.85	92.9	[94]
$(Pb_{1-x}La_x)(Zr_{0.65}Ti_{0.35})O_3$ (PLZT)	Film	800	12.4	–	[100]
$(Pb_{0.91}La_{0.09})(Zr_{0.65}Ti_{0.35})O_3 + Mn$	Film	–	30.8	–	[96]
4 mol% Nb-doped $PbZr_{0.4}Ti_{0.6}O_3$	Film	1,351	20	70	[97]
$(Pb_{1-x}La_x)(Zr_{0.65}Ti_{0.35})O_3$ (La/Zr/Ti = 8/52/48)	Film	25.9	30	78	[95]
PLZT film deposited on LNO/Ni	Film	2,600	85	65	[103]
$(PbZrO_3/PbZr_{0.52}Ti_{0.48}O_3)$ bilayer thin films	Film	2,410	28.2	–	[98]
$Pb_{0.9}La_{0.1}(Zr_{0.52}Ti_{0.48})O_3$ (PLZT)	Film	2,500	40.2	–	[102]
$Pb_{0.9}La_{0.1}(Zr_{0.52}Ti_{0.48})O_3$	Film	3,400	58.4	81.2	[101]
$PbZr_{0.52}Ti_{0.48}O_3/Al_2O_3/PbZr_{0.52}Ti_{0.48}O_3$ opposite double-heterojunction	Film	5,711	63.7	81.3	[99]
Mn-doped $Pb_{0.97}La_{0.02}(Zr_{0.905}Sn_{0.015}Ti_{0.08})O_3$	Film	2,000	31.2	–	[104]

most promising candidates for replacing Pb-based ceramics.[113–115] Thus, extensive energy storage studies have been carried out on this system. It is particularly important that strengthening the dynamic of polar nanoregions (PNRs) through disturbing long-range ferroelectric ordering or expanding nonergodic–ergodic phase transition range could optimize the polarization behavior to obtain a high U_d. In 2011, Gao et al.[116] firstly reported that $0.89Bi_{0.5}Na_{0.5}TiO_3–0.06BaTiO_3–0.05K_{0.5}Na_{0.5}NbO_3$ (0.89BNT–0.06BT–0.05KNN) ceramics possessed a U_d of 0.46 J/cm³ at 56 kV/cm. In 2016, Cao et al.[117] used Mn^{2+} to modify 0.7(0.94BNT–0.06BT)–0.3ST ceramics to reduce P_r by forming $Mn_{Ti}'' - V_{\ddot{O}}$ defect complex, which induced a local electric field influencing domain switch. Consequently, a U_d of 1.06 J/cm³ was obtained at 95 kV/cm with 1.1 mol% Mn, owing to a large ($P_{max} - P_r$) of up to 36.8 μC/cm². Note that, a similar phenomenon had been reported by Ren.[118] In 2017, Li et al.[119] incorporated $NaNbO_3$ into the $0.8Bi_{0.5}Na_{0.5}TiO_3–0.2SrTiO_3$ relaxor ferroelectric ceramics, and observed that the PE loops gradually became slimmer together with vanishing current peaks as the NN content increased, which was attributed to the nonergodic–ergodic phase transition. The 0.5 mol $NaNbO_3$ modified ceramics exhibited a high

U_d of 0.74 J/cm^3, accompanied by good high-temperature energy storage stability and charging-discharging capability. Indeed, preventing early saturation of the polarization is also an effective method to enhance U_d. Li et al.[120] selected Ba$_{0.3}$Sr$_{0.7}$TiO$_3$ to improve the energy storage performances of BNT-based dielectric ceramics. The PE loop of high-ε_r (Ba$_{0.3}$Sr$_{0.7}$)$_{0.35}$(Bi$_{0.5}$Na$_{0.5}$)$_{0.65}$TiO$_3$ (BS)$_{0.35}$(BNT)$_{0.65}$ relaxor ferroelectric ceramics originally presented an obvious clamped behavior with U_d = 1.04 J /cm^3, and η = 77% at 100 kV/cm, but its P_r was still high. They then chose antiferroelectric NaNbO$_3$ (NN) to optimize the polarization behavior of the 0.94BSBNT–0.06NN relaxor ferroelectric ceramics, which achieved a high U_d of 1.25 J/cm^3 at room temperature.[9] Besides, the system exhibited good high-temperature stability and fatigue endurance, which may be closely related to the reduction in domain size. To simultaneously increase U_d and η, Fan et al.[9] prepared a lead-free RE/AFE 0.94(Na$_{0.82}$K$_{0.18}$)$_{1/2}$Bi$_{1/2}$TiO$_3$–0.06FeNbO$_4$/0.96NaNbO$_3$–0.04CaZrO$_3$ (NKBT–FN/NN–CZ) composite ceramic with a multiphase structure, in which an antiferroelectric was embedded in a relaxor ferroelectric through a heat treatment of powder followed by the traditional solid phase reaction method. A high U_d of 2.85 J/cm^3 and a high η of 80% were simultaneously realized at the composition with a mass ratio of 70/30 at room temperature (RT), as shown in Figure 2.6. This study demonstrates that the formation of relaxor/antiferroelectric composite provides a highly effective method to improve the energy storage performance of lead-free ceramics.

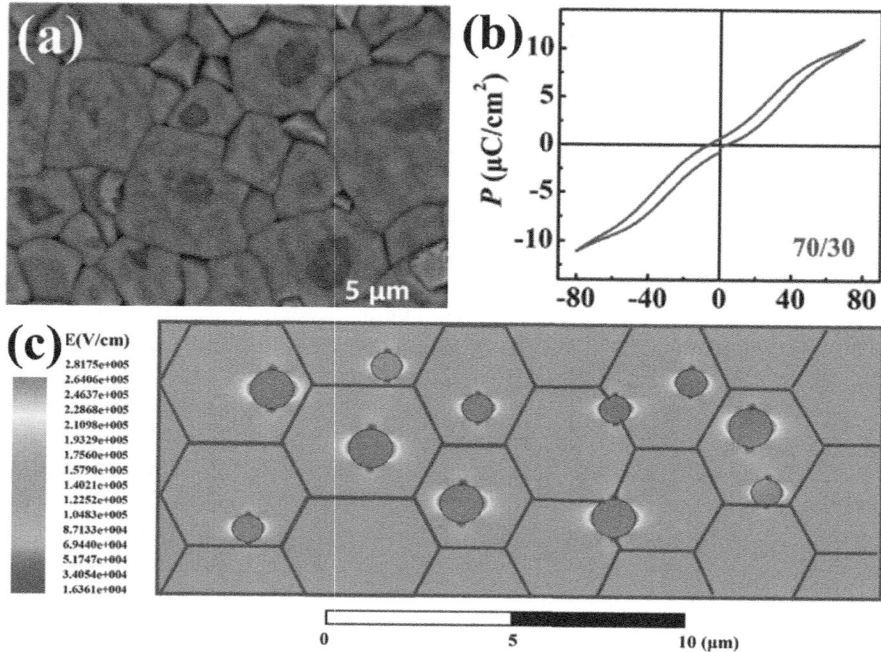

FIGURE 2.6　(a) The backscattered electron image (BEI) of 80(NKBT–FN)–20(NN–CZ) composite; (b) The PE loop of 70(NKBT–FN)–30(NN–CZ) composite; (c) The distribution of the electric field simulated for the 70/30 composite at 180 kV/cm.[9]

Besides chemical doping, the formation of multilayer structures and miniaturization of BNT-based ceramic capacitors constitute another significant research trend. In 2018, Li et al.[121] designed a series of $(1-x)Bi_{0.5}Na_{0.5}TiO_3-x(Sr_{0.7}Bi_{0.2})TiO_3$ (NBT–xSBT, x = 0.3–0.5) ceramics, and fabricated corresponding multilayer ceramic capacitors (MLCC) with a single layer thickness of 20 µm. A high U_d of 9.5 J/cm^3, together with η of 92%, was achieved in the NBT–0.45SBT MLCC. Furthermore, the energy storage properties of the MLCC displayed good temperature stability, fatigue endurance, and charging-discharging capability. Recently, Li et al.[14] improved the E_b and energy storage performances of NBT–0.35SBT MLCC under different stress states through controlling grain orientation. On the other hand, Yang et al.[122] studied the energy storage properties of gradient-structured $(SrTiO_3 + 0.5$ wt%$Li_2CO_3)/$ $(0.93Bi_{0.5}Na_{0.5}TiO_3-0.07Ba_{0.94}La_{0.04}Zr_{0.02}Ti_{0.98}O_3)$ (STL/(BNT–BLZT)) ceramics along the thickness direction, and a high U_d of 2.72 J/cm^3 at 294 kV/cm for the STL/(BNT–BLZT) ceramics. Due to the strict requirements of the harsh working environment such as high temperatures (>200°C or even >300°C), good insulation, and antioxidant capability, MLCC still faces many challenges.[123, 124] Meanwhile, material system selection, electrode design such as equivalent series resistance (ESR) and loss, cost control of fabrication, and so on, need further consideration.

The volatilization of Bi_2O_3 and Na_2O and the reduction of Ti at high temperatures could generate oxygen vacancies for the BNT-based thin films, resulting in a large leakage current. Single metal oxides $(MnO_2,$[125, 126] $Fe_2O_3,$[127] etc.) were used to modify BNT films, which led to different defect complexes to maintain charge balance and suppress oxygen vacancy formation and migration. For example, Mn-doped BNT thick films displayed a reduced leakage current due to the formation of $Mn_{Ti}'' - V_{\ddot{O}}$ defect complex, giving rise to a U_d of 30.2 J/cm^3.[126] In addition, controlling the annealing temperature could also reduce leakage current.[128] Peng et al.[129] deposited La/Zr modified 0.94BNT–0.06BT high epitaxial-quality thin films using PLD, and obtained P_{max} of 102 µC/cm^2 due to the complex phase composition and great relaxor dispersion. The (100)- and (111)-oriented $(Bi_{1/2}Na_{1/2})_{0.9118}La_{0.02}Ba_{0.0582}$ $(Ti_{0.97}Zr_{0.03})O_3$ (BNLBTZ) thin films exhibited large U_d of 137 and 154 J/cm^3, respectively, far exceeding other Pb-free, even Pb-based systems.

Different from single-composition MLCC, macrostructure modification of BNT-based films mainly focused on gradient compositions. It is widely accepted that BNT-based ceramic films are of p-type conduction due to many vacancies generating acceptor states in the band gap, and thus $p-n$ junctions and block layers can be applied to inhibit charge transportation. For instance, Guo et al.[130] studied the effect of the addition of $Bi_{3.25}La_{0.75}Ti_3O_{12}$ (BLT) and $Pb(Zr_{0.4}Ti_{0.6})O_3$ (PZT) dielectric layers on the pyroelectric and ferroelectric properties of 0.94BNT–0.06BT ceramic films, and observed a reduction of leakage current by 3 orders of magnitude. Besides, Chen et al.[131, 132] studied the effect of interface number on the energy storage properties of $0.94(Bi_{0.5}Na_{0.5})TiO_3-0.06BaTiO_3/BiFeO_3$ (abbreviated as BNBT/nBFO) multilayer film capacitors under a given total thickness. It was found that BNBT/2BF thin films exhibited a U_d of 31.96 J/cm^3 and a η of 61% at 2,400 kV/cm owing to enhanced insulating characteristics and high polarization. With the rapid development of flexible wearable devices in recent years, researches on the related materials and their energy storage performances have gradually increased. Qian et al.[132]

TABLE 2.6

Energy storage performances of the BNT-based relaxor ferroelectric materials

Composition	Type	E_b (kV/cm)	U_d (J/cm³)	η (%)	Ref.
$0.89Bi_{0.5}Na_{0.5}TiO_3$–$0.06BaTiO_3$–$0.05K_{0.5}Na_{0.5}NbO_3$	Bulk	56	0.59	–	[116]
$Na_{0.5}Bi_{0.5}TiO_3 + 11mol\%Mn$	Bulk	95	1.06	–	[117]
$(1-x)(0.8Bi_{0.5}Na_{0.5}TiO_3$–$0.2SrTiO_3)$–$xNaNbO_3$ $(x = 0.06)$	Bulk	70	0.74	–	[119]
$(SrTiO_3 + 0.5wt\%Li_2CO_3)/$ $(0.93Bi_{0.5}Na_{0.5}TiO_3$–$0.07Ba_{0.94}La_{0.04}Zr_{0.02}Ti_{0.98}O_3)$	Bulk	294	2.72	–	[122]
$(Na_{0.25}Bi_{0.25}Sr_{0.5})(Ti_{0.8}Sn_{0.2})O_3$	Bulk		3.4	90	[133]
$(1-x)$ $Bi_{0.5}Na_{0.5}TiO_3$–$xSr_{0.7}La_{0.2}TiO_3$	Bulk	315	4.14	92	[109]
$0.95Bi_{0.5}Na_{0.5}TiO_3$–$0.05SrZrO_3$	Bulk	230	3.14	79	[112]
$(Ba_{0.3}Sr_{0.7})_{0.35}(Bi_{0.5}Na_{0.5})_{0.65}TiO_3$	Bulk		1.25		[9]
$(Ba_{0.3}Sr_{0.7})_x(Bi_{0.5}Na_{0.5})_{1-x}TiO_3$ $(x = 0.5)$	Bulk	100	1.04	77	[120]
$Sr_{0.85}Bi_{0.1}TiO_3$ modified $Bi_{0.4465}Na_{0.4465}$ $Ba_{0.057}La_{0.05}TiO_3$	Bulk		4.55	90	[115]
$0.85[(1-x)Bi_{0.5}Na_{0.5}TiO_3$–$xBi_{0.1}Sr_{0.85}TiO_3]$–$0.15KNbO_3$ $(x = 0.3)$	Bulk	290	3.72	90.7	[111]
$(1-x)[0.8Bi_{0.5}Na_{0.5}TiO_3$–$0.2Ba(Zr_{0.3}Ti_{0.7})O_3]$–$xSr_{0.7}La_{0.2}TiO_3(x = 0.3)$	Bulk	210	2.6	92.2	[110]
$(Bi_{1/2}Na_{1/2})_{0.9118}La_{0.02}Ba_{0.0582}(Ti_{0.97}Zr_{0.03})O_3$	Film	–	154	–	[129]
$0.94(Na_{0.5}Bi_{0.5}TiO_3)$–$0.06BaTiO_3$	Film	3,310	17.2	74.3	[128]
$Na_{0.5}Bi_{0.5}Ti_{1-x}Mn_xO_3$ $(x = 0.01)$	Film	2,310	30.2	47.7	[126]
$0.94(Bi_{0.5}Na_{0.5})TiO_3$–$0.06BaTiO_3/BiFeO_3$	Film	2,400	31.96	61	[131]
$(Na_{0.5}Bi_{0.5})TiO_3$–$x(Sr_{0.7}Bi_{0.2})TiO_3$	MLCC		9.5	92	[121]
$[0.94(0.75Bi_{0.5}Na_{0.5}TiO_3$–$0.25NaNbO_3)$–$0.06BaTiO_3]$–$xCaZrO_3$ (CZ_100x)	MLCC	120	0.31–0.35	77	[123]
$(Na_{0.8}K_{0.2})_{0.5}Bi_{0.5}TiO_3/0.6(Na_{0.8}K_{0.2})_{0.5}Bi_{0.5}TiO_3$–$0.4SrTiO_3$	MLCC		73.7	68.1	[132]
$0.87BaTiO_3$–$0.13Bi(Zn_{2/3}(Nb_{0.85}Ta_{0.15})_{1/3})O_3$	MLCC	400	14.08		[124]
<111>-textured $Na_{0.5}Bi_{0.5}TiO_3$–$Sr_{0.7}Bi_{0.2}TiO_3$	MLCC	1,030	21.5		[14]

prepared the multilayer $(Na_{0.8}K_{0.2})_{0.5}Bi_{0.5}TiO_3/0.6(Na_{0.8}K_{0.2})_{0.5}Bi_{0.5}TiO_3-0.4SrTiO_3$ (NKBT/NKBT–ST)N (N = 2, 3, 6, 8) films on an F-Mica substrate, which displayed a U_d of 73.7 J/cm^3 at 3,077 kV/cm for N = 6. Under different testing conditions such as (–50)–200°C, 10^8 cycle numbers and 10^4 bending tests, the energy storage performances of the (NKBT/NKBT–ST)6 ceramic films still maintained good stability. The energy storage performances of the BNT-based relaxor ferroelectric materials are listed in Table 2.6.

4.3 BaTiO₃-Based Ceramics

$BaTiO_3$ (BT) with a simple perovskite structure possesses a moderate Curie temperature (T_c) of 120°C and a relatively high ε_r of 10^3 at RT, and has been a common dielectric material in passive capacitors.[134, 135] BT-based materials have also received a great deal of attention in memory and memristor for information storage and transfer. On the other hand, the structure and physical properties researches on BT-based materials have been continuously studied.[136–138] However, pure BT ceramics have some problems such as high P_r, and the reduction from Ti^{4+} to Ti^{3+} at high temperatures, which prevent them from achieving high U_d.[139]

In 2009, Ogihara et al.[140] synthesized a temperature-stable $0.7BaTiO_3-0.3BiScO_3$ (0.7BT–0.3BS) relaxor ferroelectric ceramics with a thickness of 0.2 mm, and obtained a U_d of 2.3 J/cm^3 at 225 kV/cm. When the thickness was reduced to 15 μm, U_d was enhanced to 6.1 J/cm^3 at 730 kV/cm. Yuan et al.[141] prepared $Bi(Mg_{1/2}Zr_{1/2})O_3$ (BMZ)-modified BT-based relaxor ferroelectric, and achieved a high U_d = 2.9 J/cm^3 for 0.85BT–0.15BMZ ceramics, much higher than that of pure BT ceramics (0.4 J/cm^3). The authors attributed the enhancement of U_d to PNRs, as evidenced by piezoelectric force microscopy (PFM) and transmission electron microscopy (TEM). It should be mentioned that other systems such as $BaTiO_3-Bi(Mg,Ti)O_3$ (BT–BMT) also displayed good energy storage capabilities due to the low tolerance factor of Bi-based compounds.[142, 143] In recent years, $Ba_{0.85}Ca_{0.15}Zr_{0.10}Ti_{0.90}O_3$ (BCZT)-based relaxor ferroelectrics gains much attention in their ferroelectric and pyroelectric properties because of the high ε_r at RT.[144–147] However, the U_d of BCZT-based ceramic bulks is generally less than 3 J/cm^3 limited by a low E_b. If the trade-off between ε_r and E_b could be optimized, BCZT-based ceramics would be a promising energy storage candidate.

"Core–shell" structure is a common way of modifying BT-based ceramics, especially in enhancing the temperature stability for ceramic capacitor applications. Wu et al.[148] utilized $BiScO_3$ (BS) as shell material to coat BT, and obtained a U_d of only 0.68 J/cm^3 at E_b = 120 kV/cm for the BT@3 mol% BS ceramics. In 2014, Su et al.[149] coated BT nanocrystals with $65PbO-20B_2O_3-15SiO_2$ and $65Bi_2O_3-20B_2O_3-15SiO_2$ glass phases, and reduced the sintering temperature to 900°C, as shown in Figure 2.7. Similarly, high-E_b materials such as SiO_2 and $SrTiO_3$ (ST) were both used as shell materials to improve the breakdown behavior.[150, 151] For example, Wu et al.[151] fabricated BT@ST relaxor ferroelectric ceramics using a sol-precipitation approach, where core-shell structure was illustrated by EDS. However, the U_d was only 0.22 J/cm^3 at 47 kV/cm with a η approaching 90%. It should be pointed out that high-E_b materials usually possess a low εr, and so the U_d of coated BT ceramics

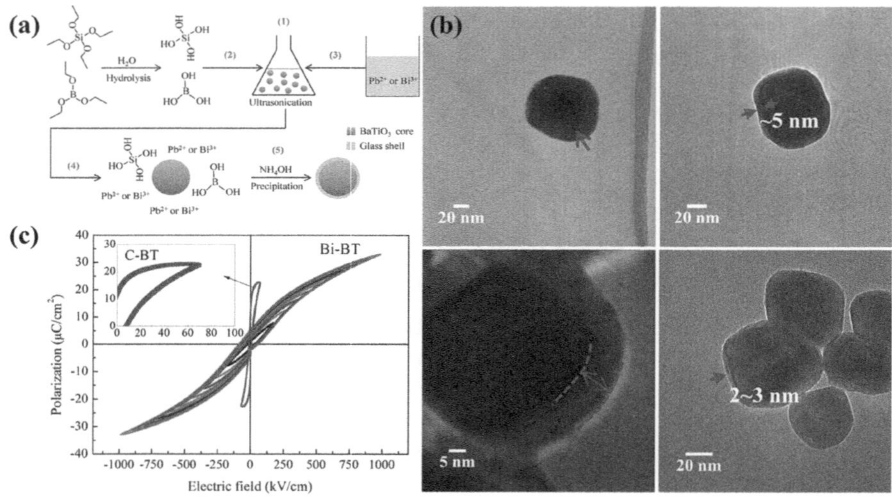

FIGURE 2.7   (a) Schematic illustration of the formation of BaTiO$_3$/low-melting glass core-shell nanoparticles; (b) TEM micrographs of as-prepared BaTiO$_3$/PbO–B$_2$O$_3$–SiO$_2$ core-shell nanoparticles with various shell thicknesses; (c) PE hysteresis loops of coarse-grained barium titanate (C-BT) and the Bi–BT nanocomposite with 16 mm thickness.[149]

can hardly be effectively enhanced at the sacrifice of ε_r. Bi-based materials such as BiScO$_3$ as shell material also suffer from the same problem of obtaining a high U_d. The reasons need to be further explored taking into account the complex composition gradient.

Ba(Zr,Ti)O$_3$ (BZT)-based ceramic films are considered to be a potential energy storage material since Zr^{4+} has a stable valence and a similar ion radius to Ti^{4+}, and BZT exhibits good relaxor characteristics. As the Zr^{4+} content increases, the relaxor degree gradually enhances, and a weak diffuse phase transition accompanied by a slim PE loop takes place when Zr^{4+} reaches 15 mol%, which is attributed to the inhomogeneous distribution of Zr and the related mechanical stress.[152, 153] Therefore, the energy storage performances of BZT-based ceramic systems have received much attention. Instan et al.[154] prepared 400 nm <100>-oriented Ba(Zr$_x$Ti$_{1-x}$)O$_3$ (x = 0.3, 0.4, 0.5) relaxor ferroelectric thin films using PLD on La$_{0.7}$Sr$_{0.3}$MnO$_3$/MgO substrates. Importantly, the U_d of all the ceramics can reach a scale of 10^2 J/cm^3 around $E_b \approx$ 3 MV/cm, and a maximum U_d of 156 J/cm^3 was obtained at the x = 0.3 composition. In 2017, Cheng et al.[155] reported the influence of thickness and substrate categories on the domain/phase of Ba(Zr$_{0.2}$Ti$_{0.8}$)O$_3$ thin films. With increasing thickness, the mismatch stress gradually released, the rhombohedral phase content increased, and a twined domain structure was formed. A high U_d of 166 J/cm^3 was achieved for Ba(Zr$_{0.2}$Ti$_{0.8}$)O$_3$ thin films at $E_b \approx$ 5.7 MV/cm. In addition, oxygen pressure was found to generate a positive effect on the energy storage properties of BT-based thin films.[156]

"Interface engineering", including interface compatibility, periodic number, and space charge, plays a critical role in enhancing the E_b and optimizing the polarization behavior of BT-based ceramic films. In 2012, Ortega et al.[157] deposited $BaTiO_3$/$Ba_{0.3}Sr_{0.7}TiO_3$ superlattice thin films on MgO single substrate using PLD, and acquired a U_d of 12.24 J/cm^3 measured from PE loop. It should be noted that a theoretical value of U_d = 46 J/cm^3 was calculated at E_b of 5.8–6 MV/cm. Sun et al.[158] studied the energy storage properties of laminated $Ba_{0.7}Ca_{0.3}TiO_3$/$BaZr_{0.2}Ti_{0.8}O_3$ (BCT/BZT) thin films with two layers as a period and the number is 2, 4, and 8. With increasing period, E_b was enhanced from 3 to 4.5 MV/cm, and a maximum U_d of 52.4 J/cm^3 was obtained for N = 8. Meanwhile, a series of multilayer structures consisting of two materials or a stack of different dielectric layers have become very promising model systems since some unique properties can be enhanced. However, due to the difference in electric properties of the stacked dielectric materials, some physical mechanisms regarding current leakage and charge distribution in heterostructure interface, still need to be further investigated.[159, 160] The energy storage performances of some BT-based relaxor ferroelectric materials are listed in Table 2.7.

4.4 BiFeO$_3$-Based Ceramics

As a multiferroic material with perovskite structure, $BiFeO_3$ (BF) exhibits a high T_c (~850 °C), a high Ps (~100 μC/cm^2), and a large S_{max} (~0.4%), and has been extensively studied in terms of its piezoelectric, magnetic, and quantum properties across

TABLE 2.7

Energy storage performances of the BT-based relaxor ferroelectric materials

Composition	Type	E_b (kV/cm)	U_d (J/cm^3)	η (%)	Ref.
0.7BaTiO$_3$–0.3BiScO$_3$	Bulk	730	6.1	–	[140]
SiO$_2$-coated BaTiO$_3$	Bulk		1.2	53.8	[150]
Bi(Mg$_{0.5}$Zr$_{0.5}$)O$_3$-modified BaTiO$_3$	Bulk	301.4	2.9	86.8	[141]
Ba$_{0.85}$Ca$_{0.15}$Zr$_{0.1}$Ti$_{0.9}$O$_3$ + (Bi$_2$O$_3$–B$_2$O$_3$–SiO$_2$)	Bulk	330	2.12	90.5	[145]
0.9(Sr$_{0.7}$Bi$_{0.2}$)TiO$_3$– 0.1Bi(Mg$_{0.5}$Zr$_{0.5}$)O$_3$ modified BaTiO$_3$	Bulk	370	4.03	96.2	[139]
BT@3 mol% BiScO$_3$	Bulk	120	0.68	–	[148]
BT@65PbO–20B$_2$O$_3$–15SiO$_2$ and 65Bi$_2$O$_3$–20B$_2$O$_3$–15SiO$_2$	Bulk	1,140	10	–	[149]
BT@ST	Bulk	47	0.22	90	[151]
BaTiO$_3$/Ba$_{0.3}$Sr$_{0.7}$TiO$_3$	Film	6,000	46		[157]
(100)-highly-textured BaZr$_x$Ti$_{1-x}$O$_3$	Film	3,000	214	72.8	[154]
Ba(Zr$_{0.2}$Ti$_{0.8}$)O$_3$	Film		166	96	[155]
Ba$_{0.85}$Ca$_{0.15}$Ti$_{0.90}$Zr$_{0.10}$O$_3$	Film	2,000	64.8	73	[156]
Laminated Ba$_{0.7}$Ca$_{0.3}$TiO$_3$/ BaZr$_{0.2}$Ti$_{0.8}$O$_3$	Film	4,500	52.4	–	[158]

various fields.[161, 162] The volatile nature of Bi_2O_3 and multiple valence states of Fe during sintering cause large dielectric loss and leakage current.[163] Nevertheless, BF relaxor ferroelectric is still considered a potential candidate for energy storage capacitors because of its high P_s.

$BiFeO_3$–$xBaTiO_3$ (BF–xBT, $0 \leq x \leq 0.5$) ceramics maintain a high T_c, and good ferroelectric and piezoelectric properties, resulting from complicated phase structure evolution. With increasing x value, the BF–xBT ceramics possess a high P_{max} near the MPB with BT ≈ 0.33 mol.[164–166] The best energy storage properties of the BF-based ceramics are basically found around BF–0.33BT system due to reduced P_r. For example, Liu et al.[167] added $Ba(Zn_{1/3}Ta_{2/3})O_3$ (BZT) into BT–0.34BT relaxor ferroelectric ceramics and obtained a U_d of 2.56 J/cm³ at 160 kV/cm. To enhance the breakdown strength of BF-based ceramics, Qi et al.[168] used $NaNbO_3$ (NN) combined with 0.1 wt% MnO_2 and 2 wt% $BaCu(B_2O_5)$ (BCB) as sintering aids to modify 0.67BF–0.33BT relaxor ferroelectric ceramics, and attained a maximum U_d of 8.12 J/cm³ accompanied to η of 90% with the $x = 0.1$ sample, which was related to the existence of nanodomains and a high density. In addition, Wang et al.[169] reported that the U_d of 0.62BF–0.3BT–0.08Nd$(Zr_{0.5}Zn_{0.5})O_3$ ceramic bulks was 2.45 J/cm³ at 240 kV/cm. When an MLCC device based on this system was made, U_d is enhanced to 10.5 J/cm³.

Similar to the BF–xBT systems, ST-modified BF-based energy storage ceramics have mainly been studied in the form of films due to thickness limitations. In 2013, Correia et al.[170] prepared 0.4BF–0.6ST thin films using PLD, and obtained a high U_d of 18.6 J/cm³ at 972 kV/cm. Recently, Pan et al.[171] reported (0.55–x)BF–xBT–0.45ST ($x = 0$–0.4) thin films with a polymorphic nanodomain structure grown on Nb-doped $SrTiO_3$ single crystal substrate. Under an electric field of 4.9 MV/cm, a maximum $U_d =112$ J/cm³ was achieved with the $x = 0.3$ composition, 53 J/cm³ higher than that of pure 0.55BF–0.45ST thin films, which is owing to the enhanced relaxor behaviors and breakdown strengthen. Moreover, the energy storage performances of this system maintained good stabilities in a wide temperature range of (–100)–150 °C and after 10^8 cycles, respectively. In addition, an important and interesting double P–E loop was observed in rare earth-substituted BF thin films due to the transition from orthorhombic to rhombohedral phase.[172]

In heterostructure design, McMillen et al.[173] first proposed to use a "artificial dead layer" to enhance E_b, and deposited an Al_2O_3 layer of 6 nm thick between 0.6BF–0.4ST film and ST substrate. The interface polarization behavior to some degree increased the hysteresis loss while the Al_2O_3 insulating layer enhanced E_b, giving rise to an enhancement of U_d from 13 to 17 J/cm³. Since then, a sandwich structure of a "soft layer" with high polarization and a "hard layer" with high E_b were constructed to prevent "electric tree" growth and to achieve high U_d.[174] It is particularly important that strong interface coupling factors should be considered in designing the structure and analyzing the properties.[175] The energy storage performances of some BF-based relaxor ferroelectric materials are listed in Table 2.8.

4.5 $K_{0.5}Na_{0.5}NbO_3$-Based Ceramics

$K_{0.5}Na_{0.5}NbO_3$ (KNN) is a binary solid solution of ferroelectric $KNbO_3$ and antiferroelectric $NaNbO_3$, and possesses a moderate d_{33} around 80 pC/N, a high T_c of

TABLE 2.8

Energy storage performances of the BF-based relaxor ferroelectric materials

Composition	Type	E_b (kV/cm)	U_d (J/cm³)	η (%)	Ref.
Ba(Zn$_{1/2}$Ta$_{2/3}$)O$_3$-modified BFO-based solution	Bulk	160	2.56	80	[167]
Bi$_{1-x}$Sm$_x$Fe$_{0.95}$Sc$_{0.05}$O$_3$	Bulk	230	2.21	76	[162]
BiFeO$_3$–BaTiO$_3$–NaNbO$_3$	Bulk		8.12	90	[168]
(1–x)Bi$_{0.83}$Sm$_{0.17}$Fe$_{0.95}$Sc$_{0.05}$O$_3$–x(0.85BaTiO$_3$–0.15Bi(Mg$_{0.5}$Zr$_{0.5}$)O$_3$)(x = 0.75)	Bulk	206	3.2	92	[163]
SrTiO$_3$-substituted BiFeO$_3$ thin films	Film	972	18.6	–	[170]
BaTiO$_3$/SrRuO$_3$ heterostructure	Film		51		[175]
BiFeO$_3$/BaTiO$_3$/BiFeO$_3$ sandwich-structured film	Film	2,320	18.5	82.3	[174]
BiFeO$_3$–BaTiO$_3$–SrTiO$_3$ solid-solution films	Film		112	80	[171]
0.3Bi(Fe$_{0.95}$Mn$_{0.05}$)O$_3$–0.7(Sr$_{0.7}$Bi$_{0.2}$)TiO$_3$	Film	3,000	61	75	[161]
(0.7–x)BiFeO$_3$–0.3BaTiO$_3$–xNd(Zn$_{0.5}$Zr$_{0.5}$)O$_3$(x = 0.8)	MLCC	700	10.5	87	[169]

420 °C, and complicated phase structure.[176, 177] Since the 1950s, the researches on KNN materials have been concentrated on the piezoelectric properties of the MPB compositions located at 47.5% KNbO$_3$ content.[177] The ratio of K/Na can be slightly varied around 0.5/0.5. However, the narrow sintering temperature range and the volatilization of K$_2$O and Na$_2$O at high temperatures both hinder its applications.

In 2016, Du et al. studied KNN-based energy storage ceramics and achieved a high U_d of about 4 J/cm³ for SrTiO$_3$ (ST)- and Bi(Mg$_{1/3}$Nb$_{2/3}$)O$_3$ (BMN)-modified KNN ceramics.[178, 179] And they used CuO,[180] ZnO,[181] etc., as adds to further improve the sintering behavior of KNN-based ceramics. It should be noted that the modified KNN-based energy storage ceramics with superfine grain size possess not only a high E_b but also good transparency. In addition, "phase boundary engineering" has proved to be a useful method to enhance electric properties of KNN ceramics. Recently, Yang et al.[182] proposed the term "morphotropic relaxor boundary (MRB)" in BT-modified KNN ceramics to enhance the electrostrain and dielectric permittivity. At the MRB the electrostrain increased by ~3 times and the permittivity increased by ~1.5 times over a wide temperature range of more than 100 K, as compared with the off-MRB compositions. In comparison, the studies of KNN-based ceramic films for energy storage have been relatively scarce, which might be related to the insolubility of Nb and volatilization of K and Na. In 2017, Won et al.[183] reported 6 mol% BiFeO$_3$-doped (K$_{0.5}$Na$_{0.5}$)(Mn$_{0.005}$Nb$_{0.995}$)O$_3$ (KNMN) thick film that possessed a slim PE loop, a U_d of 28 J/cm³, and a η of 90.3%. Recently, Huang

TABLE 2.9

Energy storage properties of some KNN-based relaxor ferroelectric ceramics and films

Composition	Type	E_b (kV/cm)	U_d (J/cm³)	η (%)	Ref.
$0.8(K_{0.5}Na_{0.5})NbO_3-0.2Sr(Sc_{0.5}Nb_{0.5})O_3 + ZnO$	Bulk	400	2.6	73.2	[181]
$(1-x)(K_{0.5}Na_{0.5})NbO_3-xSrTiO_3(x = 0.15)$	Bulk	400	4.03	–	[178]
$(1-x)(K_{0.5}Na_{0.5})NbO_3-xSrTiO_3(x = 0.2)$	Bulk	400	3.67	72.1	[178]
$0.9(K_{0.5}Na_{0.5})NbO_3-0.1Bi(Mg_{2/3}Nb_{1/3})O_3-1.0$ mol% CuO	Bulk	400	4.02		[180]
$(1-x)(K_{0.5}Na_{0.5})NbO_3-xBi(Mg_{2/3}Nb_{1/3})O_3(x = 0.1)$	Bulk	300	4.08		[179]
$(K_{0.5}Na_{0.5})(Mn_{0.005}Nb_{0.995})O_3 + BiFeO_3$	Film	2,000	28	90.3	[183]
$0.95(K_{0.49}Na_{0.49}Li_{0.02})(Nb_{0.8}Ta_{0.2})O_3-0.05CaZrO_3-x$mol% Mn	Film	3,080	64.6	84.6	[184]

et al.[184] used MnO_2 to reduce the leakage current of $0.95(K_{0.49}Na_{0.49}Li_{0.02})(Nb_{0.8}Ta_{0.2})O_3-0.05CaZrO_3-x$ mol% Mn (KNN–LT–CZ$_5$–x mol% Mn) thin films prepared by a sol-gel method, and achieved a high U_d = 64.6 J/cm³ under an electric field of 3,080 kV/cm at the x = 0.5 composition. The energy storage properties of some KNN-based relaxor ferroelectric ceramics and films are listed in Table 2.9.

4.6 High-Entropy Relaxor Ferroelectric Ceramics

The concept of high entropy was first introduced in 2003 by Ranganathan, and in 2004 by Ye and Cantor, respectively.[185, 186] The presence of different characteristic elements gives high-entropy materials the following four core effects that could benefit dielectric ceramics. (1) High-entropy effect. The high-entropy effect is the main concept of high-entropy materials, in which multiple components with nearly equal molar ratios increase the conformational entropy of the material and the stability of the single-phase structure at high temperatures. Thus, the high-entropy effect can be used to maintain the stability of the single-phase perovskite structure at high temperatures without segregation. (2) Lattice distortion. The lattice of a high-entropy material is composed of atoms of different sizes randomly occupying the lattice sites. These slightly unbalanced lattice sites and the random positions of the atoms lead to severe lattice distortions. The lattice distortion causes a decrease in the intensity of the X-ray diffraction peak, a decrease in the thermal and electrical conductivity, and an increase in the strength of the material. At the same time, the lattice stretching induced by the electric field becomes larger, and the polarization strength increases, which is beneficial to improve the energy storage properties of the material. (3)

Hysteresis diffusion. This effect generally occurs in high-entropy alloys. Traditional alloy materials in the formation process require high temperatures to provide energy to bring the atoms close together. When the temperature decreases, the material will be in a sub-stable state, so it is easy to precipitate the second phase. The complex elemental composition of high-entropy alloys and the highly chaotic internal state make the atomic diffusion very slow, and the internal stress from lattice distortion also hinders the atomic movement, leading to the formation of nanocrystals and amorphous phases when solidifying at lower temperatures, while keeping the microstructure stable. In the case of high-entropy ceramic materials, this property is manifested by the slow diffusion rate of grain boundaries within the material and the reduced grain growth drive, which can effectively refine the grains and improve E_b. (4) Cocktail effect. The "cocktail" effect will make the final performance of the material not equal to the sum of its component properties and unpredictable. The synergistic mixing of elements results in high-entropy materials with excellent properties. Therefore, high-entropy dielectric ceramics are also a family of promising ceramic materials for energy storage conversion.[187, 188]

In 2019, Pu et al.[189] synthesized a single-phase homogeneous A-site high-entropy $(Na_{0.2}Bi_{0.2}Ba_{0.2}Sr_{0.2}Ca_{0.2})TiO_3$ ceramic by using a solid-state method. It exhibited relaxation behavior because of the disorder of the microscopic composition. In addition, a discharge energy density of 1.02 J/cm^3 was displayed at an applied electric field of 145 kV/cm. In 2020, Liu et al.[190] reported the dielectric energy storage properties of $(Bi_{0.2}Na_{0.2}K_{0.2}Ba_{0.2}Ca_{0.2})TiO_3$ high-entropy ceramics with relaxor ferroelectric properties, but the energy storage density and efficiency are only 0.684 J/cm^3 and 87.5%. This rather poor energy storage performance was attributed to the volatilization of some elements, which resulted in deviation from the design composition, and the formation of a large number of oxygen vacancies, and the large difference in the sintering temperatures of the component that prevented the ceramics from well densifying. In 2021, Liu et al.[191] successfully synthesized pure-phase $(Bi_{1/6}Na_{1/6}Sr_{1/6}Ba_{1/6}Pb_{1/6}Ca_{1/6})TiO_3$ (BNSBPC) and $(Bi_{1/6}La_{1/6}Na_{1/6}K_{1/6}Sr_{1/6}Ba_{1/6})TiO_3$ (BLNKSB) high-entropy ceramics (HECs) by conventional solid-state method. The results confirmed the RFE behaviors. Xiong et al.[192] prepared a high-entropy relaxor-like ferroelectric ceramics with A-site disorder, $(Pb_{0.25}Ba_{0.25}Sr_{0.25}Ca_{0.25})TiO_3$, which exhibited a low loss (<0.015) from RT to 125°C. In 2022, Yang and Zheng prepared a HEC $(Bi_{0.2}Na_{0.2}K_{0.2}La_{0.2}Sr_{0.2})TiO_3$ (BNKLST) with a grain size as small as 45 nm, and, obtained a recoverable energy storage density of 0.959 J/cm^3 under an applied electric field of 180 kV/cm.[193] In 2021, Zhou et al.[194] introduced a bismuth-based HEC, $Bi(Zn_{0.2}Mg_{0.2}Al_{0.2}Sn_{0.2}Zr_{0.2})O_3$ (BZMASZ) into the $BaTiO_3$–$Na_{0.5}Bi_{0.5}TiO_3$ (BT-NBT) matrix, and obtained an excellent recoverable energy density of U_d ~ 3.74 J/cm^3, a high conversion efficiency of η ~ 82.2% and superior temperature stability of ± 4.5% from −30 to 200°C. Recently, a breakthrough in energy storage ceramics based on the high-entropy concept was achieved by Chen et al.[195] By introducing the (Li^+, Ba^{2+}, Bi^{3+}, Sc^{3+}, Hf^{4+}, Zr^{4+}, Ta^{5+} and Sb^{5+}) ions into the KNN lattice, a R–O–T–C multiphase coexistence was obtained on local scale, which refined the PNRs, leading to a huge improvement in energy storage performance and efficiency, with U_d = 10.06 J/cm^3 and η = 90.8%. Similarly, Yang et al.[196] designed the $(Bi_{3.25}La_{0.75})(Ti_{3-3x}Zr_xHf_xSn_x)O_{12}$ high-entropy system based on

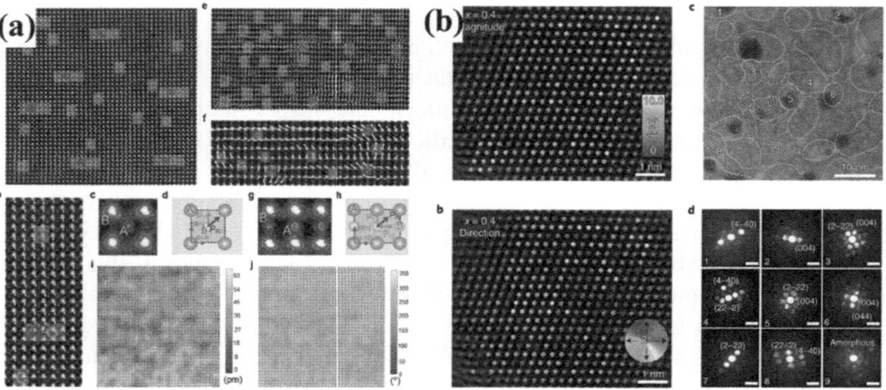

FIGURE 2.8 (a) R–O–T–C multiphase nanoclusters coexistence, atomic-resolution HAADF STEM polarization vector image along $[100]_c$ and polarization magnitude mapping;[195] (b) Lattice distortion, nanocrystalline grains and amorphous-like phase in the high-entropy films.[196]

TABLE 2.10

Energy storage properties of some high-entropy RFE ceramics

Composition	Type	E_b (kV/cm)	U_d (J/cm³)	η (%)	Ref.
$(Na_{0.2}Bi_{0.2}Ba_{0.2}Sr_{0.2}Ca_{0.2})TiO_3$	Bulk	145	1.02		[189]
$(Bi_{0.2}Na_{0.2}K_{0.2}Ba_{0.2}Ca_{0.2})TiO_3$	Bulk	129	0.684	87.5	[190]
$(Bi_{0.2}Na_{0.2}K_{0.2}La_{0.2}Sr_{0.2})TiO_3$	Bulk	180	0.959	67	[193]
$0.75BaTiO_3$–$0.25Na_{0.5}Bi_{0.5}TiO_3$ + $0.1\,mol\ Bi(Zn_{0.2}Mg_{0.2}Al_{0.2}Sn_{0.2}Zr_{0.2})O_3$	Bulk	273	3.74	82.2	[194]
KNN + (Li^+, Ba^{2+}, Bi^{3+}, Sc^{3+}, Hf^{4+}, Zr^{4+}, Ta^{5+}, Sb^{5+})	Bulk	740	10.06	90.8	[195]
$(Bi_{3.25}La_{0.75})(Ti_{3-3x}Zr_xHf_xSn_x)O_{12}$	Film	6,350	182	78	[196]

$Bi_4Ti_3O_{12}$ ceramic thin film and obtained a microstructure with the coexistence of nanograins and amorphous phases, resulting in a huge energy storage density of 182 J/cm³ under a 6,360 kV/cm electric field, and a 78% efficiency. The atomic scale TEM analysis is shown in Figure 2.8. These two works have significant effects on the research and applications of high-entropy energy storage ceramics.

The energy storage properties of some high-entropy RFE ceramics are summarized in Table 2.10.

In summary, for relaxor ferroelectric ceramics, the formation of PNRs due to disturbed long-range ferroelectric order or nonergodic–ergodic phase transition can strengthen relaxor characteristics with a slim PE loop. In addition, delaying the polarization saturation is also used to optimize polarization behavior. Designing an MLCC and "core–shell" structure with a non-ferroelectric phase as shell material is a commonly accepted method to improve electric breakdown behavior for ceramic bulks. By comparison, chemical modifications of ceramic films play two maim roles: to suppress the generation and transportation of vacancy defects to reduce leakage current; and to strengthen the relaxor characteristics and optimize polarization behavior. In the heterostructure of ceramic films, the matching degree of physical parameters such as lattice constant, thermal expansion coefficient, etc., the gradient sequence, and the template or new inert layers are all important factors influencing the energy storage performances

5 ANTIFERROELECTRIC CERAMICS

As a special group of ferroelectric materials, antiferroelectrics have many similarities with ferroelectrics but also still have obvious differences. In 1951, Kittel[197] originally proposed the concept of antiferroelectricity, predicted its existence, and gave some basic characteristics. Generally speaking, antiferroelectric materials possess P_s in sub-unit cells, but the direction of a P_s is antiparallel to that of neighboring sub-cells. Thus, antiferroelectrics do not exhibit a net polarization on a macroscopic scale. Antiferroelectrics have a unique feature: a double hysteresis loop under an external field, which is of particular importance. The P is linearly proportional to the E at a low electric field. When E exceeds the forward switching (AFE to FE) field E_{A-F}, antiferroelectric displays ferroelectric behavior, with P fastly increasing and gradually reaching P_{max}. Note that an antiferroelectric can undergo a ferroelectric–antiferroelectric phase transition after removing E (E_{F-A}), which is usually lower than E_{A-F}. Consequently, P sharply reduces and returns to the initial state (that is, $E = 0$, $P_r = 0$). A similar variation in P can be observed as E continuously increases in an opposite direction. It is hard to display a double hysteresis loop for most pure antiferroelectric materials at a low electric field, due to a high E_{A-F}, and conditions of high temperature and strong electric field are generally needed to induce the AFE to FE transition. However, for energy storage applications, it is crucial for antiferroelectrics to display double PE loop, especially for ceramic bulks. On the contrary, a typical high squareness PE loop, and the corresponding internal strain induced by antiferroelectric–ferroelectric phase transition both result in a low U_d and reduced device life.[198] In this regard, enhancing E_b and E_{A-F} (corresponding to stabilizing AFE phase) on the one hand, and allowing the induced FE state to easily relax back to the AFE phase (corresponding to reducing the difference of $\Delta E = E_{A-F}-E_{F-A}$) are two particularly important issues to acquire good energy storage performances.

5.1 PbZrO$_3$-Based Antiferroelectric Ceramics

PbZrO$_3$ (PZ) is a prototype antiferroelectric and serves as a model system, which has been extensively studied so far. The origin of the phase transition mechanism, however, is not well understood.[199] Despite this, the characteristic double PE loop with a high P_{max} and a low P_r makes it suitable for energy storage capacitors. Ye et al.[200] prepared a new solid solution of complex perovskite structure, $(1-x)$PbZrO$_3$–xPb(Mn$_{1/2}$W$_{1/2}$)O$_3$ (PZ-xPMnW, with $x = 0$–0.1), they found that the induced ferroelectric polarization (PInd) was enhanced, whereas the critical field (ECr) was decreased, with increasing PMnW concentration. Although the U_d is poor, the softening of antiferroelectric order and the improvement in induced polarization make the

PZ-PMnW ceramics an interesting material system for such applications as energy storage devices. A new binary AFE solid solution of $(1-x)PbZrO_3-xPb(Mg_{1/2}Mo_{1/2})O_3$ (PZ-PMM, $x = 0.00-0.10$) was investigated by Ye et al.[201] At 160°C, a typical double hysteresis loop can be displayed for all the compositions, and a U_d of 0.58 J/cm^3 was obtained at $x = 0.08$. Most importantly, the maximum electric field-induced polarization was significantly increased, whereas the critical field was decreased with increasing PMM content, suggesting a remarkable softening effect of the anti-ferroelectric order in PZ due to some degree of dipole frustration. PZ-based energy storage ceramics have gained increasing attention since Chen et al.[202] reported a U_d of 7.1 J/cm^3 for PZ thin films, but mostly on ceramic films.

It is widely accepted that the phase stability of the perovskite ABO_3 structure is closely related to the tolerance factor t, which is expressed as:

$$t = \frac{r_A + r_O}{\sqrt{2}\left(r_A + r_O\right)}$$

where r_A, r_B, and r_O denote the ion radii of A, B, and O, respectively. In general, an AFE phase can be stabilized for $t \approx 1$, and reducing t can enhance to some extent the stability of the AFE phase. For example, Hao et al.[203] substituted Pb^{2+}(1.20 Å) with smaller Sr^{2+}(1.12 Å) to reduce t, and obtained a slim PE loop together with increased $E_{A\text{-}F}$ and a decreased ΔE. As a consequence, $(Pb_{0.95}Sr_{0.05})ZrO_3$ (PSZ$_5$) thin films exhibited a U_d of 14.5 J/cm^3. Substitution with a smaller aliovalent of La^{3+} as donor dopant had a similar effect.[204] In addition, controlling the orientation of PZ thin film also achieved the purpose of stabilizing the antiferroelectric phase. PZ thin film of (100)-orientation needs a higher electric field than the (111)-orientation film to realize the antiferroelectric–ferroelectric phase transition.[205]

Tailoring the local electric field and stress to enhance U_d is a hard but interesting approach. As is well known, introducing new nanoparticles into a material often brings about improved properties. For example, Sa et al.[206] used α-Fe_2O_3 nanoparticles to modify PZ thin films, and obtained a U_d of 17.4 J/cm^3 and a P_{max} of as high as 78 μC/cm^2, which was attributed to the local field effect. By comparison, Chen et al.[207] fabricated a self-assembled PZ:NiO nano-columnar composite using PLD, and proposed that tensile stress mainly came from the interface of the two phases. At $E_b = 1,000$ kV/cm, the 5 vol% NiO thin films possess a high P_{max} of ~91 μC/cm^2, and a U_d of 24.6 J/cm^3. Meanwhile, Ge et al.[208, 209] carried out a series of work on optimizing the polarization behavior, such as enhancing P_{max}, and reducing $E_{A\text{-}F}-E_{F\text{-}A}$, using the stress engineering method. To the best of our knowledge, it is difficult to directly measure and characterize local electric field and stress, and thus the corresponding methods need to be developed. Furthermore, the gradient sequence of the thin film is significant for enhancing U_d. Ye et al.[210] reported a U_d of 16.3 J/cm^3 for down-graded PZ-based thin film, which was higher than that of 9.7 J/cm^3 for up-graded thin films.

It should be mentioned that $PbHfO_3$ (PHO) antiferroelectric ceramics have some similarities with PZO, and the related studies have been concentrated on phase structures.[211–213] Nevertheless, the energy storage properties of PHO-based ceramics have rarely been reported. In 2019, Ye et al.[214] reported a novel solid solution of $(1-x)$ $PbHfO_3-xPb(Mg_{1/2}W_{1/2})O_3$ [$(1-x)$PHf$-x$PMW] with a high U_d of 3.7 J/cm^3 and a η of

TABLE 2.11

Energy storage performance of some PZ-based AFE materials

Composition	Type	E_b (kV/cm)	U_d (J/cm³)	η (%)	Ref.
$Pb_{0.98}La_{0.02}(Hf_{0.45}Sn_{0.55})_{0.995}O_3$	Bulk	380	7.63	94	[215]
$0.92PbZrO_3-0.08Pb(Mg_{1/2}Mo_{1/2})O_3$	Bulk		0.58		[201]
$0.9PbHfO_3-0.1Pb(Mg_{1/2}W_{1/2})O_3$	Bulk	155	3.7	72.5	[214]
$PbHfO_3$	Bulk	235	2.7		[216]
$0.95PbHfO_3-0.05Pb(Zn_{1/2}W_{1/2})O_3$	Bulk	200	5.03	62	[217]
$PbZrO_3$	Film		7.1		[202]
$PbZrO_3 + 5\%La$	Film	6,000	14.9		[204]
$PbZrO_3 + Sr$	Film		14.5		[203]
$PbZrO_3 + Eu$	Film		16.3		[210]
(110)- or (111)-oriented $PbZrO_3$	Film	700	5.3		[209]
$PbZrO_3 + Fe_2O_3 NPs$	Film	600	17.4		[206]
NiO nano-columns embedded $PbZrO_3$	Film		24.6		[207]

72.5% at $x = 0.10$. In 2020, Chao et al.[215] prepared $Pb_{0.98}La_{0.02}(Hf_xSn_{1-x})_{0.995}O_3$ anti-ferroelectric ceramics, and achieved a good energy storage performance: U_d and η were of 7.63 J/cm³ and 94% for $x = 0.45$ composition, respectively. Chauhan et al.[216] prepared high-quality $PbHfO_3$ ceramics via the solid-state synthesis route, and a typical AFE loop revealed an optimal recoverable energy storage density of 2.7 J/cm³, which is 286% higher than the reported data with an efficiency (η) of 65% at 235 kV/cm at RT. Wan et al.[217] obtained a high U_d of 5.03 J/cm³ in $(1-x)PbHfO_3-xPb(Zn_{1/2}W_{1/2})O_3$ under a relatively low electric field of 200 kV/cm by inducing ferro-electrically active ions into the $PbHfO_3$ structural matrix to soften the AFE order and induce a large maximum polarization. Recently, Huang et al.[218] reported pure PHO ceramic films by a sol-gel method followed by annealing at 650 °C, and achieved a U_d of 24.9 J/cm³. The energy storage performance of some PZ-based AFE materials is listed in Table 2.11.

5.2 (Pb, La)(Zr, Sn, Ti)O₃-Based Antiferroelectric Ceramics

As discussed in Section 3.2.1, $Pb(Zr_{1-x}Ti_x)O_3$ (PZT, $x \approx 0.05$) solid solutions located at the Zr-rich region of the ferroelectric–antiferroelectric (FE–AFE) phase boundary, exhibit rich phase structures and good electrical properties.[89] It is worth noting that PZT compositions at the FE–AFE phase boundary differ from those of the MPB in terms of electric properties. For instance, FE and AFE cannot transform into each other under a stress force or an electric field, while temperature, stress, or other factors can induce the FE–AFE phase transition. Meanwhile, the region of the antiferroelectric phase for PZT is relatively narrow, and a slight composition fluctuation (e.g. around PZT 95/5) would easily cause deviation from the AFE to FE phase.

La^{3+} and Sn^{4+} are very popular A and B site dopants to the PZT 95/5 ceramics, having similar effects.[219–223] La^{3+}, as a donor dopant, substitutes Pb^{2+} to disturb long-range ordering of domains by vacancy defect, which would strengthen the relaxor characteristic and expand the stability region of the antiferroelectric phase. Sn^{4+} as an equal valence dopant of Ti^{4+} can not only expand the stability region of the antiferroelectric phase but also adjust the Zr/Ti ratio to accommodate Ti up to 10 mol%. Currently, the relevant works on stabilizing the Pb-based antiferroelectric phase mainly focus on adjusting suitable Zr/Sn/Ti ratios.[224–226] Liu et al.[227] found that E_{A-F} linearly increased and the squareness of the PE loop slightly improved when the Ti content was reduced from 0.11 to 0.07 mol at a fixed Zr content of 0.58 mol. The U_d was enhanced, thereby, from 0.28 to 2.35 J/cm^3 for $Pb_{0.97}La_{0.02}(Zr_{0.58}Sn_{0.35}Ti_{0.07})O_3$ antiferroelectric ceramic bulks. In addition, it should be particularly noted that it is difficult to obtain a high U_d when E_{A-F} exceeds E_b.

To improve the breakdown behavior of Pb-based antiferroelectric ceramic bulks, Zhang et al.[228–230] carried out a series of works using some special sintering techniques. For instance, the $(Pb_{0.87}Ba_{0.1}La_{0.02})(Zr_{0.68}Sn_{0.24}Ti_{0.08})O_3$ (PBLZST) ceramics exhibited a high U_d of 3.2 J/cm^3 at $E_b = 180$ kV/cm due to smaller grain size and well insulation.[229] Considering that special sintering techniques usually require expensive equipment, these methods are not suitable for large-scale production. Thus, some low-cost and convenient solutions were also proposed. Bian et al.[231] used amorphous SiO_2 to coat $Pb_{0.97}La_{0.02}(Zr_{0.33}Sn_{0.55}Ti_{0.12})O_3$ (PLZST 2/33/55/12), and obtained a high U_d of 2.68 J/cm^3 due to an E_b enhancement from 12.2 to 23.8 kV/mm. It should be noticed that the introduction of non-antiferroelectric phase typically reduces the content of the original antiferroelectric phase, even leading to the disappearance of E_{A-F}. Wang et al.[232] used a rolling process to enhance the mechanical strength of $(Pb_{0.98}La_{0.02})(Zr_{0.55}Sn_{0.45})_{0.995}O_3$ (PLZS) antiferroelectric ceramics. At 400 kV/cm, the PLZS ceramics exhibited a high U_d of 10.4 J/cm^3 and a η of 87%. Similarly, Zhang et al.[233] and Liu et al.[234] utilized the tape-casting method to fabricate antiferroelectric thick films. For example, the tape-casted $(Pb_{0.98}-xLa_{0.02}Srx)(Zr_{0.9}Sn_{0.1})_{0.995}O_3$ (PLSZS) thick films displayed a U_d of 11.18 J/cm^3 and a η of 82.2% for $x = 0.04$.

Due to the phase structure complexity near the FE–AFE boundary, PZT-based ceramics usually display different polarization behavior especially in the form of ceramic films. Gao et al.[235] deposited differently oriented $(Pb_{0.98}La_{0.02})(Zr_{0.95}Ti_{0.05})O_3$ (PLZT) thin films using PLD, and obtained a U_d of ~40 J/cm^3 but a η of only ~50% for the (111) PLZT film, which may be related to the ferroelectric-like behavior (i.e., with a high P_r) under high electric fields. Despite this result, the combination of relaxor ferroelectric and antiferroelectric is considered an effective approach to enhance U_d. Note that the antiferroelectric behavior can be observed at different formulas such as PLZT at Zr/Ti $\approx$ 52:48 and PLZST at Zr/(Sn+Ti) $\approx$ 65:55.[85, 236] For the development of micro-electronic devices, flexible substrates such as Ti, Si, Ni foils, etc., are required. Ma et al.[237, 238] carried out a series of meaningful works in the low-cost integration development of PLZT-based thin films. For instance, $Pb_{0.92}La_{0.08}Zr_{0.95}Ti_{0.05}O_3$ (PLZT 8/95/5) antiferroelectric ceramic films deposited on metal foils by CSD possess a high U_d of 53 J/cm^3, and a 5,000 h effective work time at

TABLE 2.12

Energy storage properties of some PLZST-based antiferroelectric materials

Composition	Type	E_b (kV/cm)	U_d (J/cm³)	η (%)	Ref.
$(Pb_{0.858}Ba_{0.1}La_{0.02}Y_{0.008})(Zr_{0.65}Sn_{0.3}Ti_{0.05})O_3–(Pb_{0.97}La_{0.02})(Zr_{0.9}Sn_{0.05}Ti_{0.05})O_3$ (SPS)	Bulk	162	6.5		[228]
$(Pb_{0.87}Ba_{0.1}La_{0.02})(Zr_{0.68}Sn_{0.24}Ti_{0.08})O_3$ (HP)	Bulk	180	3.2		[229]
$Pb_{0.97}La_{0.02}(Zr_{0.50}Sn_{0.46}Ti_{0.04})O_3$	Bulk		3.2	86.5	[222]
$Pb_{0.97}La_{0.02}(Zr_{0.33}Sn_{0.55}Ti_{0.12})O_3$ @ SiO_2	Bulk	238	2.68		[231]
$(Pb_{0.98}La_{0.02})(Zr_{0.55}Sn_{0.45})_{0.995}O_3$	Bulk	400	10.4	87	[232]
$(Pb_{0.98-x}La_{0.02}Sr_x)(Zr_{0.9}Sn_{0.1})O_3$	Bulk		11.18	82.2	[234]
$(Pb_{0.97}La_{0.02})(Zr_{0.93}Sn_{0.03}Ti_{0.04})O_3$	Bulk		2.0–3.4		[223]
$Pb_{0.97}La_{0.02}(Zr_xSn_{0.05}Ti_{0.95-x})O_3$	Bulk	376	12.1	90	[233]
$Pb_{0.92}La_{0.08}Zr_{0.95}Ti_{0.05}O_3$	Film	300	37		[237]
$(Pb_{0.97}La_{0.02})(Zr_{1-x-y}Sn_xTi_y)O_3$	Film		13.7		[220]
PLZT, 8/52/48	Film	1,600	22	77	[239]
$(Pb_{0.97}La_{0.02})(Zr_{0.66}Sn_{0.23}Ti_{0.11})O_3$	Film	4,000	46.3	84	[236]
$(Pb_{0.98}La_{0.02})(Zr_{0.95}Ti_{0.05})O_3$	Film		30		[235]
$Pb_{0.97}La_{0.02}(Zr_{0.95}Ti_{0.05})O_3 + SrTiO_3$	Film		101	62	[221]

RT.[237] The energy storage properties of some PLZST-based antiferroelectric ceramics and films are summarized and listed in Table 2.12.

5.3 NaNbO₃-Based Antiferroelectric Ceramics

NaNbO₃ (NN) is a well-documented nonpolar antiferroelectric material that possesses complicated crystal structures due to the rotation of oxygen octahedrons and the off-centered displacement of Nb⁵⁺.[240, 241] In addition to the temperature-induced transitions, antiferroelectric–ferroelectric phase transition can also be induced by mechanical stress, grain size, and electric field.[242–244] Note that an electric field-induced antiferroelectric–ferroelectric phase transition is usually irreversible at RT because of the small free energy difference between the antiferroelectric and ferroelectric phases.

Generally, reducing the tolerance factor t is a widely accepted method to stabilize the antiferroelectric phase of NN ceramics. For instance, Shimizu et al.[245] used CaZrO₃ (CZ), with similar electronegativity but a smaller tolerance factor t, to modify NN ceramics. As the content of CZ increased, an obvious double PE loop was displayed accompanied by an $E_{A–F}$ gradually increasing while P_{max} decreased. Other compounds such as SrZrO₃ (SZ),[246] CaHfO₃ (CH),[247] and BiScO₃ (BS),[248] were also utilized to stabilize the antiferroelectric phase, but only the energy storage performances of a few systems were reported. It should be pointed out that the double P–E loop of NN-based ceramics typically shows a significant hysteresis at

RT, resulting in low U_d. Zhou et al.[249] reported Bi_2O_3-modified NN energy storage ceramic bulks, and obtained a high U_d of 4.03 J/cm³ for $Na_{0.7}Bi_{0.1}NbO_3$ ceramics at E_b = 250 kV/cm, which was attributed to the disrupted random electric fields and reduced domain sizes. The energy storage properties of the $Na_{0.7}Bi_{0.1}NbO_3$ system showed good temperature stability (20–100°C), fatigue endurance (10⁵ cycles), and charge-discharge properties. Subsequently, more Bi-based compounds including $Bi(Mg_{1/3}Nb_{2/3})O_3$ (BMN),[250] $Bi(Ni_{1/2}Sn_{1/2})O_3$ (BNS),[251] and so on, were used to improve the relaxor characteristics and acquire high U_d. It is particularly noticeable that the P–E loop of $BiMeO_3$-modified system became slimmer than that of pure NN ceramic, but the characteristic double hysteresis loop failed to be observed. In 2019, Qi et al.[3] introduced BNT to NN ceramics, and obtained an orthorhombic AFE phase and relaxor characteristics (nanoscale domains), which synergistically contributed to the record-high energy-storage density U_d of ≈12.2 J cm⁻³ and acceptable energy efficiency η ≈ 69% at 68 kV mm⁻¹. Wei et al.[252] fabricated the NN-SBT R-AFE ceramics with a high U_d of 4.5 J/cm³ and a η of 90.3% at 288 kV/cm. Tan et al.[253] utilized SPS to prepare NN-24SBT and obtained fine-grained and highly densified ceramics with an ultrahigh U_d of 12.2 J/cm³ and a η of 88% at 660 kV/cm. Recently, Jiang et al.[254] reported the NN–BF system, with the U_d and η improved to 18.5 J/cm³ and 83.3%, respectively, as shown in Figure 2.9, setting the record of U_d for ceramic bulks of AFE materials.

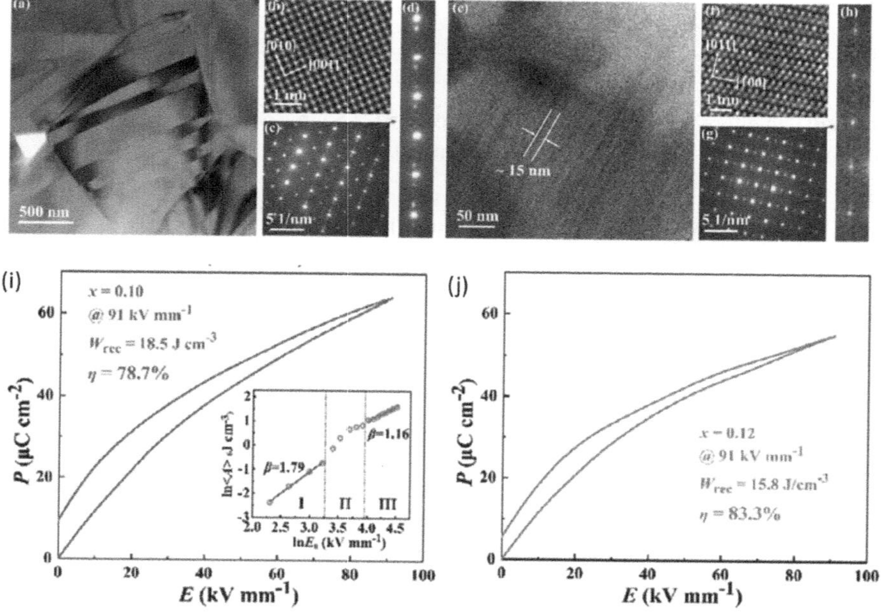

FIGURE 2.9   (a) Domain morphology of NN–xBF ceramic with x = 0.04, and corresponding (b) HR-TEM image, (c–d) SAED patterns. (e) Domain morphology of x = 0.10, and the corresponding (f) HR-TEM image, (g–h) SAED patterns; (i) P–E loop of x = 0.1; (j) P–E loop of x = 0.12.[254]

TABLE 2.13

The energy storage properties of NN-based dielectric materials

Composition	Type	E_b (kV/cm)	U_d (J/cm³)	η (%)	Ref.
$Na_{0.7}Bi_{0.1}NbO_3$	Bulk	250	4.03	85.4	[249]
$NaNbO_3-0.1Bi(Mg_{2/3}Nb_{1/3})NbO_3$	Bulk	300	2.8	82	[250]
$NaNbO_3-0.15Bi(Mg_{2/3}Nb_{1/3})NbO_3$	Bulk	300	2.4	90	[250]
$(1-x)NaNbO_3-xBi(Ni_{1/2}Sn_{1/2})O_3$	Bulk		5	68	[251]
$0.76NaNbO_3-0.24(Bi_{0.5}Na_{0.5})TiO_3$	Bulk	680	12.2	69	[3]
$NaNbO_3-Sr_{0.7}Bi_{0.2}TiO_3$	Bulk	288	4.5	90.3	[252]
$0.76NaNbO_3-0.24Sr_{0.7}Bi_{0.2}TiO_3(SPS)$	Bulk	660	12.2	88	[253]
$0.90NaNbO_3-0.10BiFeO_3$	Bulk	995	18.5	83.3	[254]
$NaNbO_3-0.04CaZrO_3 + Mn$	Film		19.64		[257]
$0.92NaNbO_3-0.08SrZrO_3$	Film		2.9	67	[256]

A double P–E hysteresis loop could be displayed in high-quality NN single crystal and ceramic films as opposed to ceramic bulks, and their energy storage properties, piezoelectricity, and dielectric tunability were studied. In 2018, Fujii et al.[255] deposited $0.92NaNbO_3-0.08SrZrO_3$ (0.92NN–0.08SZ) antiferroelectric thin films on $SrRuO_3$-buffered ST substrates with different orientations by using PLD. The 0.92NN–0.08SZ thin films grown on (110)-oriented ST substrate exhibited antiferroelectric behavior, while those on (001)-oriented substrate remained ferroelectric. Recently, Beppu et al.[256] obtained a U_d of only 2.9 J/cm³ for the 0.92NN–0.08SZ thin films at $E_b \approx 400$ kV/cm. In addition, Luo et al.[257] fabricated Mn-doped $0.96NaNbO_3-0.04CaZrO_3$ (0.96NN–0.04CZ) thin films using a sol-gel method, and reduced the leakage current of the sample with 1 mol% Mn by an order of 10^3-10^4 compared to the pure compositions. The U_d and η values of 1 mol% Mn thin films are 19.64 J/cm³ and 64.5%, respectively. Overall, the simple composition and non-toxic metal elements make the NN-based materials an interesting topic for energy storage research. The energy storage properties of some NN-based dielectric materials are summarized in Table 2.13.

5.4 AgNbO₃-Based Antiferroelectric Ceramics

$AgNbO_3$ (AN) exhibits a series of complicated phase structures and a relatively low bandgap (~2.8 eV), and it is difficult to display a double hysteresis loop at RT. Early studies of AN ceramics were concentrated on their properties for microwave communication and photocatalysis.[258, 259] In addition, the fabrication of AN-based ceramics usually requires an oxygen-rich environment due to the reduction of Ag_2O at high temperatures. Similar to NN, AN-based ceramic bulks are the main research direction for energy storage capacitors.

In 2007, Fu et al.[260] fabricated AN ceramics that displayed a double hysteresis loop with a P_{max} of 52 μC/cm^2 at 220 kV/cm, exceeding other dielectric materials at the same electric field. These results have stimulated the studies of AN ceramics for energy storage applications. Tian et al.[261] synthesized pure AN antiferroelectric ceramic bulks, found two polar structures by TEM and variable temperature P–I–E loops measurement, and attained a high U_d of 2.1 J/cm^3. In general, AN ceramics experience a series phase transition:[261]

$$M_1 \xrightarrow{67°C} M_2 \xrightarrow{267°C} M_3 \xrightarrow{353°C}$$

$$O_1 \xrightarrow{361°C} O_2 \xrightarrow{387°C} T \xrightarrow{579°C} C$$

where M_1, M_2, and M_3 denote the orthorhombic phases in a rhombic orientation, O_1 and O_2 are the orthorhombic phases in a parallel orientation, while T and C denote the tetragonal and cubic phases, respectively. It is known that the stability of the antiferroelectric phase of AN ceramics is closely related to phase transition among M_1, M_2, and M_3. Therefore, shifting the phase transitions among M_1, M_2, and M_3 to lower temperatures would enhance the U_d of AN-based ceramics.[262–265]

Chemical doping is a simple and effective method to enhance the U_d of AN-based ceramic bulks. For example, Ta^{5+} and W^{6+} were used to substitute Nb^{5+} on the B site, and Bi^{3+}, Ca^{2+}, and La^{3+} to replace Ag$^+$ on the A site.[264, 266–269] In 2017, Zhao et al.[262] fabricated Ag(Nb$_{1-x}$Ta$_x$)O$_3$ antiferroelectric ceramics, and found that P_r was decreased from 3.6 to 1.2 μC/cm^2 whereas E_b was enhanced from 175 to 242 kV/cm with increasing Ta content. Ag(Nb$_{0.85}$Ta$_{0.15}$)O$_3$ exhibited a high U_d of 4.2 J/cm^3, together with a η of 69%, which was attributed to the low polarizability of Ta^{5+}. Besides, the electronegativity difference was also thought to play a crucial role in influencing the functional behavior of antiferroelectric materials.[270]

Note that the fabrication of fine-grain ceramics is a good choice to enhance E_b due to an inverse relationship between E_b and grain size. Recently, Wang et al.[271] synthesized AN ceramics using a hydrothermal process, and obtained the maximum E_b = 250 kV/cm among pure AN ceramics so far. However, the P_{max} was decreased compared with other AN ceramics prepared using the solid-state method, and thus the U_d and η were only 1.8 J/cm^3 and 40%, respectively. It is noteworthy that little attention has been paid to other different methods apart from ion doping to enhance the E_b of AN-based ceramics. The energy storage properties of some AN antiferroelectric ceramic bulks and films are listed in Table 2.14.

In summary, for antiferroelectric ceramics, the tolerance factor t, electronegative difference, and polarizability can all influence the stability of the antiferroelectric phase. Appropriate physical (special sintering techniques, rolling process, etc.) and chemical (coating, hydrothermal process, etc.) methods can enhance the E_b by improving the sintering behavior or mechanical strength of the ceramics. Further studies need to be carried out to explore the mechanisms of the antiferroelectric to ferroelectric switching so as to display the characteristic double PE loop of the pure antiferroelectric phase.

TABLE 2.14

Energy storage properties of some AN antiferroelectric materials

Composition	Type	E_b (kV/cm)	U_d (J/cm³)	η (%)	Ref.
$AgNbO_3$	Bulk		2.1		[261]
Ta-modified $AgNbO_3$	Bulk	240	4.2		[262]
$Ag_{1-3x}Bi_xNbO_3$	Bulk		2.6	86	[267]
W-modified $AgNbO_3$	Bulk		3.3		[266]
$Sm_{0.03}Ag_{0.91}NbO_3$	Bulk	290	5.2	68.5	[265]
Bi^{3+} on the A-site and Zn^{2+} on the B-site in $AgNbO_3$	Bulk		4.6		[264]
$Ag_{0.9}Sr_{0.05}NbO_3$	Bulk	190	2.9	56	[270]
2 mol% La-doped $AgNbO_3$	Bulk		4.4	70	[269]
$Ag_{0.92}Ca_{0.04}NbO_3$	Bulk	220	3.55		[268]
$(1-x)AgNbO_3-xAgTaO_3$	Bulk		6.3	90	[263]
$Ag_{0.97}Nd_{0.01}Ta_{0.20}Nb_{0.80}O_3$	Bulk	350	6.5		[272]

REFERENCES

1. Palneedi, H.; Peddigari, M.; Hwang, G. T.; Jeong, D. Y.; Ryu, J. J. A. F. M. High-performance dielectric ceramic films for energy storage capacitors: Progress and outlook, *Adv. Funct. Mater.* **2018**, 28, 1803665. DOI: 10.1002/adfm.201803665.

2. Tan, H.; Yan, Z.; Chen, S.-G.; Samart, C.; Takesue, N.; Salamon, D.; Liu, Y.; Zhang, H. J. S. M. SPS prepared NN-24SBT lead-free relaxor-antiferroelectric ceramics with ultrahigh energy-storage density and efficiency, *Scr. Mater.* **2022**, 210, 114428. DOI: 10.1016/j.scriptamat.2021.114428.

3. Qi, H.; Zuo, R. Z.; Xie, A. W.; Tian, A.; Fu, J.; Zhang, Y.; Zhang, S. J. Ultrahigh energy-storage density in NaNbO3-based lead-free relaxor antiferroelectric ceramics with nanoscale domains, *Adv. Funct. Mater.* **2019**, 29. DOI: 10.1002/adfm.201903877.

4. Hu, W.; Liu, Y.; Withers, R. L.; Frankcombe, T. J.; Norén, L.; Snashall, A.; Kitchin, M.; Smith, P.; Gong, B.; Chen, H.; Schiemer, J.; Brink, F.; Wong-Leung, J. Electron-pinned defect-dipoles for high-performance colossal permittivity materials, *Nat. Mater.* **2013**, 12, 821–826. DOI: 10.1038/nmat3691.

5. Wang, Y.; Shen, Z.-Y.; Li, Y.-M.; Wang, Z.-M.; Luo, W.-Q.; Hong, Y. J. C. I. Optimization of energy storage density and efficiency in BaxSr1-xTiO3 (x≤ 0.4) paraelectric ceramics, *Ceram. Int.* **2015**, 41, 8252–8256. DOI: 10.1016/j.ceramint.2015.02.156.

6. Pu, Y.; Zhang, Q.; Li, R.; Chen, M.; Du, X.; Zhou, S. J. A. P. L. Dielectric properties and electrocaloric effect of high-entropy (Na0. 2Bi0. 2Ba0. 2Sr0. 2Ca0. 2) TiO3 ceramic, *Appl. Phys. Lett.* **2019**, 115, 223901. DOI: 10.1063/1.5126652.

7. Zhang, J.; Wang, J.; Gao, D.; Liu, H.; Xie, J.; Hu, W. J. o. t. E. C. S. Enhanced energy storage performances of CaTiO3-based ceramic through A-site Sm3+ doping and A-site vacancy, *J. Eur. Ceram. Soc.* **2021**, 41, 352–359. DOI: 10.1016/j.jeurceramsoc.2020.09.017.

8. Yang, H.; Sun, Y.; Gao, H.; Zhou, X.; Tan, H.; Shu, C.; Salamon, D.; Guan, S.; Chen, S.; Zhang, H. J. I. T. o. U., Ferroelectrics,; Control, F. Lead-free BF-BT ceramics with ultra-high Curie temperature for piezoelectric accelerometer, *IEEE Transactions on Ultrasonics, Ferroelectrics, and Frequency Control* **2022**. DOI: 10.1109/TUFFC.2022.3143575.

9. Fan, P.; Zhang, S.-T.; Xu, J.; Zang, J.; Samart, C.; Zhang, T.; Tan, H.; Salamon, D.; Zhang, H.; Liu, G. J. J. o. M. C. C. Relaxor/antiferroelectric composites: A solution to achieve high energy storage performance in lead-free dielectric ceramics, *J. Mater. Chem. C* **2020**, 8, 5681–5691. DOI: 10.1039/d0tc00589d.

10. Li, M.; Fan, P.; Ma, W.; Liu, K.; Zang, J.; Samart, C.; Zhang, T.; Tan, H.; Salamon, D.; Zhang, H. J. J. o. M. C. C. Constructing layered structures to enhance the breakdown strength and energy density of Na 0.5 Bi 0.5 TiO3-based lead-free dielectric ceramics, *J. Mater. Chem. C* **2019**, 7, 15292–15300. DOI: 10.1039/c9tc05637h.

11. Shi, R.; Pu, Y.; Ji, J.; Li, J.; Guo, X.; Wang, W.; Yang, M. Correlation between flash sintering and dielectric breakdown behavior in donor-doped barium titanate ceramics, *Ceram. Int.* **2020**, 46, 12846–12851. DOI: 10.1016/j.ceramint.2020.02.055.

12. Chen, W. T.; Gurdal, A. E.; Tuncdemir, S.; Guo, J.; Guo, H.; Randall, C. A. Considering the possibility of bonding utilizing cold sintering for ceramic adhesives, *J. Am. Ceram. Soc.* **2017**, 100, 5421–5432. DOI: 10.1111/jace.15098.

13. Ma, J. P.; Chen, X. M.; Ouyang, W. Q.; Wang, J.; Li, H.; Fang, J. L. Microstructure, dielectric, and energy storage properties of BaTiO3 ceramics prepared via cold sintering, *Ceram. Int.* **2018**, 44, 4436–4441. DOI: 10.1016/j.ceramint.2017.12.044.

14. Li, J.; Shen, Z.; Chen, X.; Yang, S.; Zhou, W.; Wang, M.; Wang, L.; Kou, Q.; Liu, Y.; Li, Q.; Xu, Z.; Chang, Y.; Zhang, S.; Li, F. Grain-orientation-engineered multilayer ceramic capacitors for energy storage applications, *Nat. Mater.* **2020**, 19, 999–1005. DOI: 10.1038/s41563-020-0704-x.

15. Zhou, H. Y.; Liu, X. Q.; Zhu, X. L.; Chen, X. M. CaTiO3 linear dielectric ceramics with greatly enhanced dielectric strength and energy storage density, *J. Am. Ceram. Soc.* **2018**, 101, 1999–2008. DOI: 10.1111/jace.15371.

16. Huebner, W.; Zhang, S. C.; Gilmore, B.; Krogh, M. L.; Schultz, B. C.; Pate, R. C.; Rinehart, L. F.; Lundstrom, J. M. High breakdown strength, multilayer ceramics for compact pulsed power applications, In Digest of Technical Papers-IEEE International Pulsed Power Conference; IEEE, **1999** pp. 1242–1245.

17. Gupta, S. M.; Tripathi, M. A review of TiO2 nanoparticles, *Chin. Sci. Bull.* **2011**, 56, 1639–1657. DOI: 10.1007/s11434-011-4476-1.

18. Parker, R. A.; Wasilik, J. H. Dielectric constant and dielectric loss of TiO2 (Rutile) at low frequencies, *Phys. Rev.* **1960**, 120, 1631–1637. DOI: 10.1103/PhysRev.120.1631.

19. Guo, D.; Ito, A.; Goto, T.; Tu, R.; Wang, C.; Shen, Q.; Zhang, L. Preparation of rutile TiO2 thin films by laser chemical vapor deposition method, *J. Adv. Ceram.* **2013**, 2, 162–166. DOI: 10.1007/s40145-013-0056-y.

20. Hu, W.; Liu, Y.; Withers, R. L.; Frankcombe, T. J.; Norén, L.; Snashall, A.; Kitchin, M.; Smith, P.; Gong, B.; Chen, H.; Schiemer, J.; Brink, F.; Wong-Leung, J. Electron-pinned defect-dipoles for high-performance colossal permittivity materials, *Nat. Mater.* **2013**, 12, 821–826. DOI: 10.1038/nmat3691.

21. Yu, Y.; Zhao, Y.; Qiao, Y. L.; Feng, Y.; Li, W. L.; Fei, W. D. Defect engineering of rutile TiO2 ceramics: Route to high voltage stability of colossal permittivity, *J. Mater. Sci. Technol.* **2021**, 84, 10–15. DOI: 10.1016/j.jmst.2020.12.046.

22. Tuichai, W.; Danwittayakul, S.; Chanlek, N.; Thongbai, P. Nonlinear current-voltage and giant dielectric properties of Al3+ and Ta5+ co-doped TiO2 ceramics, *Mater. Res. Bull.* **2019**, 116, 137–142. DOI: 10.1016/j.materresbull.2019.04.026.

23. Fan, J.; Chen, Y.; Long, Z.; He, G.; Hu, Z. Grain-boundary engineering inducing thermal stability, low dielectric loss and high energy storage in Ta+Ho co-doped TiO2 ceramics, *Ceram. Int.* **2022**, 48, 21584–21592. DOI: 10.1016/j.ceramint.2022.04.124.

24. Li, J.; Li, F.; Zhuang, Y.; Jin, L.; Wang, L.; Wei, X.; Xu, Z.; Zhang, S. Microstructure and dielectric properties of (Nb + In) co-doped rutile TiO2 ceramics, *J. Appl. Phys.* **2014**, 116, 2–11. DOI: 10.1063/1.4893316.

25. Li, J.; Li, F.; Li, C.; Yang, G.; Xu, Z.; Zhang, S. Evidences of grain boundary capacitance effect on the colossal dielectric permittivity in (Nb + In) co-doped TiO2 ceramics, *Sci. Rep.* **2015**, 5, 8295. DOI: 10.1038/srep08295.

26. Chao, S.; Petrovsky, V.; Dogan, F. Effects of sintering temperature on the microstructure and dielectric properties of titanium dioxide ceramics, *J. Mater. Sci.* **2010**, 45, 6685–6693. DOI: 10.1007/s10853-010-4761-4.

27. Ye, Y.; Zhang, S. C.; Dogan, F.; Schaimiloglu, E.; Gaudet, J.; Castro, P.; Roybal, M.; Joler, M.; Christodoulou, C. Influence of nanocrystalline grain size on the breakdown strength of ceramic dielectrics, In Digest of Technical Papers-IEEE International Pulsed Power Conference; IEEE, **2003** pp. 719–722.

28. Liu, J.; Zhang, J.; Wei, M.; Yao, Z.; Chen, H.; Yang, C. Dielectric properties of manganese-doped TiO2 with different alkali-free glass contents for energy storage application, *J. Mater. Sci.: Mater. Electron.* **2016**, 27, 7680–7684. DOI: 10.1007/s10854-016-4753-5.

29. Wei, M.; Zhang, J.; Huang, J.; Chen, H.; Yang, C. Microstructure and electrical properties of TiO2–CaO–MgO–Al2O3–SiO2 glass-ceramic with sol–gel method, *J. Mater. Sci.: Mater. Electron.* **2016**, 27, 11623–11627. DOI: 10.1007/s10854-016-5294-7.

30. Chao, S.; Dogan, F. Processing and dielectric properties of TiO2 Thick films for high-energy density capacitor applications, *Int. J. Appl. Ceram. Technol.* **2011**, 8, 1363–1373. DOI: 10.1111/j.1744-7402.2010.02592.x.

31. Yadav, A. K.; Gautam, C. R. A review on crystallisation behaviour of perovskite glass ceramics, *Adv. Appl. Ceram.* **2014**, 113, 193–207. DOI: 10.1179/1743676113Y.0000000134.

32. Pai, Y.-Y.; Tylan-Tyler, A.; Irvin, P.; Levy, J. Physics of SrTiO3 -based heterostructures and nanostructures: A review, *Rep. Prog. Phys.* **2018**, 81, 036503. DOI: 10.1088/1361-6633/aa892d.

33. Hu, Q. G.; Shen, Z. Y.; Li, Y. M.; Wang, Z. M.; Luo, W. Q.; Xie, Z. X. Enhanced energy storage properties of dysprosium doped strontium titanate ceramics, *Ceram. Int.* **2014**, 40, 2529–2534. DOI: 10.1016/j.ceramint.2013.07.126.

34. Fergus, J. W. Oxide materials for high temperature thermoelectric energy conversion, *J. Eur. Ceram. Soc.* **2012**, 32, 525–540. DOI: 10.1016/j.jeurceramsoc.2011.10.007.

35. Kong, X.; Yang, L.; Cheng, Z.; Zhang, S. Bi-modified SrTiO3-based ceramics for high-temperature energy storage applications, *J. Am. Ceram. Soc.* **2020**, 103, 1722–1731. DOI: 10.1111/jace.16844.

36. Ang, C.; Yu, Z.; Cross, L. Oxygen-vacancy-related low-frequency dielectric relaxation and electrical conduction, *Phys. Rev. B: Condens. Matter Mater. Phys.* **2000**, 62, 228–236. DOI: 10.1103/PhysRevB.62.228.

37. Ang, C.; Yu, Z. Dielectric relaxor and ferroelectric relaxor: Bi-doped paraelectric SrTiO3, *J. Appl. Phys.* **2002**, 91, 1487–1494. DOI: 10.1063/1.1428799.

38. Ang, C.; Yu, Z. High capacitance-temperature sensitivity and "giant" dielectric constant in SrTi O3, *Appl. Phys. Lett.* **2007**, 90, 202903. DOI: 10.1063/1.2736298.

39. Shen, Z. Y.; Hu, Q. G.; Li, Y. M.; Wang, Z. M.; Luo, W. Q.; Hong, Y.; Xie, Z. X.; Liao, R. H. Structure and dielectric properties of Re0.02Sr0.97 TiO3 (Re = La, Sm, Gd, Er) ceramics for high-voltage capacitor applications, *J. Am. Ceram. Soc.* **2013**, 96, 2551–2555. DOI: 10.1111/jace.12364.

40. Shen, Z. Y.; Li, Y. M.; Luo, W. Q.; Wang, Z. M.; Gu, X. Y.; Liao, R. H. Structure and dielectric properties of Nd x Sr1-x TiO3 ceramics for energy storage application, *J. Mater. Sci.: Mater. Electron.* **2013**, 24, 704–710. DOI: 10.1007/s10854-012-0798-2.

41. Shen, Z. Y.; Luo, W. Q.; Li, Y. M.; Hu, Q. G.; Wang, Z. M.; Gu, X. Y. Electrical hetero-structure of Nd0.1Sr0.9TiO3 ceramic for energy storage applications, *J. Mater. Sci.: Mater. Electron.* **2013**, 24, 607–612. DOI: 10.1007/s10854-012-0871-x.

42. Wang, Y.; Shen, Z. Y.; Li, Y. M.; Wang, Z. M.; Luo, W. Q.; Hong, Y. Optimization of energy storage density and efficiency in BaxSr1-xTiO3 (x≤0.4) paraelectric ceramics, *Ceram. Int.* **2015**, 41, 8252–8256. DOI: 10.1016/j.ceramint.2015.02.156.

43. Nishigaki, S.; Murano, K.; Ohkoshi, A. Dielectric properties of ceramics in the system (Sr0.50Pb0.25Ca0.25)TiO3-Bi2O3·3TiO2 and their applications in a high-voltage capacitor, *J. Am. Ceram. Soc.* **1982**, 65, 554–560. DOI: 10.1111/j.1151-2916.1982.tb10781.x.

44. Kong, X.; Yang, L.; Cheng, Z.; Liang, G.; Zhang, S. (Ba,Sr)TiO3-Bi(Mg,Hf)O3 Lead-free ceramic capacitors with high energy density and energy efficiency, *ACS Appl. Energy Mater.* **2020**, 3, 12254–12262. DOI: 10.1021/acsaem.0c02320.

45. Fletcher, N. H.; Hilton, A. D.; Ricketts, B. W. Optimization of energy storage density in ceramic capacitors, *J. Phys. D: Appl. Phys.* **1996**, 29, 253–258. DOI: 10.1088/0022-3727/29/1/037.

46. Song, Z.; Liu, H.; Zhang, S.; Wang, Z.; Shi, Y.; Hao, H.; Cao, M.; Yao, Z.; Yu, Z. Effect of grain size on the energy storage properties of (Ba0.4Sr0.6)TiO3 paraelectric ceramics, *J. Eur. Ceram. Soc.* **2014**, 34, 1209–1217. DOI: 10.1016/j.jeurceramsoc.2013.11.039.

47. Wu, Y. J.; Huang, Y. H.; Wang, N.; Li, J.; Fu, M. S.; Chen, X. M. Effects of phase constitution and microstructure on energy storage properties of barium strontium titanate ceramics, *J. Eur. Ceram. Soc.* **2017**, 37, 2099–2104. DOI: 10.1016/j.jeurceramsoc.2016.12.052.

48. Zhang, Q.; Wang, L.; Luo, J.; Tang, Q.; Du, J. Improved energy storage density in barium strontium titanate by addition of BaO-SiO2-B2O3 glass, *J. Am. Ceram. Soc.* **2009**, 92, 1871–1873. DOI: 10.1111/j.1551-2916.2009.03109.x.

49. Kim, S. H.; Koh, J. H. ZnBO-doped (Ba, Sr)TiO3 ceramics for the low-temperature sintering process, *J. Eur. Ceram. Soc.* **2008**, 28, 2969–2973. DOI: 10.1016/j.jeurceramsoc.2008.04.034.

50. Shen, Z. Y.; Wang, Y.; Tang, Y.; Yu, Y.; Luo, W. Q.; Wang, X.; Li, Y.; Wang, Z.; Song, F. Glass modified barium strontium titanate ceramics for energy storage capacitor at elevated temperatures, *J. Materiomics* **2019**, 5, 641–648. DOI: 10.1016/j.jmat.2019.06.003.

51. Yang, X.; Li, W.; Qiao, Y.; Zhang, Y.; He, J.; Fei, W. High energy-storage density of lead-free (Sr1-1.5xBix)Ti0.99Mn0.01O3 thin films induced by Bi3+-VSr dipolar defects, *Phys. Chem. Chem. Phys.* **2019**, 21, 16359–16366. DOI: 10.1039/c9cp01368g.

52. Zhang, Y.; Li, W.; Wang, Z.; Qiao, Y.; Yu, Y.; Zhao, Y.; Song, R.; Xia, H.; Fei, W. Ultrahigh energy storage and electrocaloric performance achieved in SrTiO3 amorphous thin films: Via polar cluster engineering, *J. Mater. Chem. A* **2019**, 7, 17797–17805. DOI: 10.1039/c9ta05446d.

53. Pan, H.; Zeng, Y.; Shen, Y.; Lin, Y. H.; Ma, J.; Li, L.; Nan, C. W. BiFeO3-SrTiO3 thin film as a new lead-free relaxor-ferroelectric capacitor with ultrahigh energy storage performance, *J. Mater. Chem. A* **2017**, 5, 5920–5926. DOI: 10.1039/c7ta00665a.

54. Hou, C.; Huang, W.; Zhao, W.; Zhang, D.; Yin, Y.; Li, X. Ultrahigh Energy Density in SrTiO3 Film Capacitors, *ACS Appl. Mater. Interfaces* **2017**, 9, 20484–20490. DOI: 10.1021/acsami.7b02225.

55. Gao, W.; Yao, M.; Yao, X. Improvement of energy density in SrTiO3 film capacitor via self-repairing behavior, *Ceram. Int.* **2017**, 43, 13069–13074. DOI: 10.1016/j.ceramint.2017.06.162.

56. Gao, W.; Yao, M.; Yao, X. Achieving ultrahigh breakdown strength and energy storage performance through periodic interface modification in SrTiO3 thin film, *ACS Appl. Mater. Interfaces* **2018**, 10, 28745–28753. DOI: 10.1021/acsami.8b07151.

57. Chen, X.; Peng, B.; Ding, M. J.; Zhang, X.; Xie, B.; Mo, T.; Zhang, Q.; Yu, P.; Wang, Z. L. Giant energy storage density in lead-free dielectric thin films deposited on Si wafers with an artificial dead-layer, *Nano Energy* **2020**, 78, 105390. DOI: 10.1016/j.nanoen.2020.105390.

58. Li, F.; Zhai, J.; Shen, B.; Zeng, H. Recent progress of ecofriendly perovskite-type dielectric ceramics for energy storage applications, *J. Adv. Dielectr.* **2018**, 8, 1830005. DOI: 10.1142/S2010135X18300050.

59. Pu, Y.; Wang, W.; Guo, X.; Shi, R.; Yang, M.; Li, J. Enhancing the energy storage properties of Ca0.5Sr0.5TiO3-based lead-free linear dielectric ceramics with excellent stability through regulating grain boundary defects, *J. Mater. Chem. C* **2019**, 7, 14384–14393. DOI: 10.1039/c9tc04738g.

60. Zhou, H. Y.; Zhu, X. N.; Ren, G. R.; Chen, X. M. Enhanced energy storage density and its variation tendency in CaZrxTi1–xO3ceramics, *J. Alloys Compd.* **2016**, 688, 687–691. DOI: 10.1016/j.jallcom.2016.07.078.

61. Luo, B.; Wang, X.; Tian, E.; Song, H.; Wang, H.; Li, L. Enhanced energy-storage density and high efficiency of lead-free CaTiO3-BiScO3 linear dielectric ceramics, *ACS Appl. Mater. Interfaces* **2017**, 9, 19963–19972. DOI: 10.1021/acsami.7b04175.

62. Wang, W.; Pu, Y.; Guo, X.; Shi, R.; Shi, Y.; Yang, M.; Li, J.; Peng, X.; Li, Y. Enhanced energy storage density and high efficiency of lead-free Ca1-xSrxTi1-yZryO3 linear dielectric ceramics, *J. Eur. Ceram. Soc.* **2019**, 39, 5236–5242. DOI: 10.1016/j.jeurceramsoc.2019.08.020.

63. Ouyang, T.; Pu, Y.; Ji, J.; Zhou, S.; Li, R. Ultrahigh energy storage capacity with superfast discharge rate achieved in Mg-modified Ca0.5Sr0.5TiO3-based lead-free linear ceramics for dielectric capacitor applications, *Ceram. Int.* **2021**, 47, 20447–20455. DOI: 10.1016/j.ceramint.2021.04.054.

64. Zhang, J.; Wang, J.; Gao, D.; Liu, H.; Xie, J.; Hu, W. Enhanced energy storage performances of CaTiO3-based ceramic through A-site Sm3+ doping and A-site vacancy, *J. Eur. Ceram. Soc.* **2021**, 41, 352–359. DOI: 10.1016/j.jeurceramsoc.2020.09.017.

65. Chu, B.; Zhou, X.; Ren, K.; Neese, B.; Lin, M.; Wang, Q.; Bauer, F.; Zhang, Q. M. A dielectric polymer with high electric energy density and fast discharge speed, *Science* **2006**, 313, 334–336. DOI: 10.1126/science.1127798.

66. Randall, C. A.; Ogihara, H.; Kim, J. R.; Yang, G. Y.; Stringer, C. S.; Trolier-McKinstry, S.; Lanagan, M. High temperature and high energy density dielectric materials, In PPC2009 - 17th IEEE International Pulsed Power Conference **2009**, 346–351. DOI: 10.1109/PPC.2009.5386292.

67. Shay, D. P.; Podraza, N. J.; Donnelly, N. J.; Randall, C. A. High energy density, high temperature capacitors utilizing Mn-doped 0.8CaTiO3-0.2CaHfO3 ceramics, *J. Am. Ceram. Soc.* **2012**, 95, 1348–1355. DOI: 10.1111/j.1551-2916.2011.04962.x.

68. Hao, X.; Zhai, J.; Kong, L. B.; Xu, Z. A comprehensive review on the progress of lead zirconate-based antiferroelectric materials, *Prog. Mater. Sci.* **2014**, 63, 1–57. DOI: 10.1016/j.pmatsci.2014.01.002.

69. Xiu, S.; Xiao, S.; Xue, S.; Shen, B.; Zhai, J. Crystallization kinetics behaviour and dielectric properties of strontium barium niobate-based glass–ceramics, *J. Mater. Sci.: Mater. Electron.* **2016**, 27, 5324–5330. DOI: 10.1007/s10854-016-4431-7.

70. Wang, S.; Tian, J.; Jiang, T.; Zhai, J.; Shen, B. Effect of phase structures on dielectric properties and energy storage performances in Na2O-BaO-PbO-Nb2O5-SiO2-Al2O3 glass-ceramics, *Ceram. Int.* **2018**, 44, 23109–23115. DOI: 10.1016/j.ceramint.2018.09.118.

71. Li, C.; Zhang, Q.; Tang, Q.; Zhou, H.; Tan, F.; Du, J. Dielectric and energy storage properties of BaO-SrO-Na2O-Nb2O5-SiO2 glass–ceramics with different crystallization times, *J. Electron. Mater.* **2016**, 45, 3025–3029. DOI: 10.1007/s11664-015-4298-z.

72. Zhou, Y.; Li, Y.; Qiao, Y.; Feng, R. Investigation of structure, dielectric and energy-storage properties of lead-free niobate glass and glass-ceramics, *J. Alloys Compd.* **2018**, 747, 55–59. DOI: 10.1016/j.jallcom.2018.03.016.

73. Chen, K.; Jiang, T.; Shen, B.; Zhai, J. Effects of crystalline temperature on microstructures and dielectric properties in BaO-Na2O-Bi2O3-Nb2O5-Al2O3-SiO2 glass-ceramics, *Mater. Sci. Eng. B: Solid-State Mater. Adv. Technol.* **2021**, 263, 114885. DOI: 10.1016/j.mseb.2020.114885.

74. Song, J.; Chen, G. Dielectric behavior and energy storage properties in BaO-SrO-Nb2O5-B2O3 system glass-ceramics with Gd2O3 addition, *J. Mater. Sci.: Mater. Electron.* **2014**, 25, 349–354. DOI: 10.1007/s10854-013-1593-4.

75. Zhou, Y.; Zhang, Q.; Luo, J.; Tang, Q.; Du, J. Structural and dielectric characterization of Gd2O3-added BaO-Na2O-Nb2O5-SiO2 glass-ceramic composites, *Scr. Mater.* **2011**, 65, 296–299. DOI: 10.1016/j.scriptamat.2011.04.027.

76. Liu, T. Y.; Chen, G. H.; Song, J.; Yuan, C. L. Crystallization kinetics and dielectric characterization of CeO2-added BaO-SrO-Nb2O5-B2O3-SiO2 glass-ceramics, *Ceram. Int.* **2013**, 39, 5553–5559. DOI: 10.1016/j.ceramint.2012.12.069.

77. Tian, J.; Wang, S.; Jiang, T.; Chen, K.; Zhai, J.; Shen, B. Ultrahigh energy storage density and excellent charge–discharge properties of Bi2O3-Nb2O5-SiO2-Al2O3 glass-ceramic with CeO2 doping, *J. Electron. Mater.* **2019**, 48, 6183–6188. DOI: 10.1007/s11664-019-07432-y.

78. Zhao, Z.; Liang, X.; Zhang, T.; Hu, K.; Li, S.; Zhang, Y. Effects of cerium doping on dielectric properties and defect mechanism of barium strontium titanate glass-ceramics, *J. Eur. Ceram. Soc.* **2020**, 40, 712–719. DOI: 10.1016/j.jeurceramsoc.2019.10.023.

79. Jiang, T.; Chen, K.; Shen, B.; Zhai, J. Enhanced energy-storage density in sodium-barium-niobate based glass-ceramics realized by doping CaF2 nucleating agent, *J. Mater. Sci.: Mater. Electron.* **2019**, 30, 15277–15284. DOI: 10.1007/s10854-019-01900-1.

80. Zheng, J.; Chen, G. H.; Yuan, C. L.; Zhou, C. R.; Chen, X.; Feng, Q.; Li, M. Dielectric characterization and energy-storage performance of lead-free niobate glass-ceramics added with La2O3, *Ceram. Int.* **2016**, 42, 1827–1832. DOI: 10.1016/j.ceramint.2015.09.146.

81. Chakrabarti, A.; Molla, A. R. Zirconia assisted crystallization of ferroelectric BaBi2Nb2O9 based glass-ceramics: Kinetics, optical and dielectrical properties, *J. Alloys Compd.* **2020**, 844, 156181. DOI: 10.1016/j.jallcom.2020.156181.

82. Michael, E. K.; Trolier-Mckinstry, S. Amorphous-nanocrystalline lead titanate thin films for dielectric energy storage, *J. Ceram. Soc. Jpn.* **2014**, 122, 250–255. DOI: 10.2109/jcersj2.122.250.

83. Michael-Sapia, E. K.; Li, H. U.; Jackson, T. N.; Trolier-Mckinstry, S. Nanocomposite bismuth zinc niobate tantalate for flexible energy storage applications, *J. Appl. Phys.* **2015**, 118, 234102. DOI: 10.1063/1.4937560.

84. Yao, M.; Li, Q.; Li, F.; Peng, Y.; Su, Z.; Yao, X. Leakage current and breakdown behavior of bismuth-doped amorphous strontium titanate thin film, *Mater. Chem. Phys.* **2018**, 206, 48–55. DOI: 10.1016/j.matchemphys.2017.12.005.

85. Xie, S.; Liu, C.; Bai, H.; Chen, K.; Shen, B.; Zhai, J. Simultaneously ultra-low dielectric loss and rapid discharge time in Ta2O5 doped niobate-based glass–ceramics, *J. Mater. Sci.* **2021**, 56, 16278–16289. DOI: 10.1007/s10853-021-06353-8.

86. EI-Desoky, M. M.; Wally, N. K.; Ali, A. M.; Harby, A. E.; Hannora, A. E. Relaxor ferroelectric-like behavior in 10PbTiO3–10Fe2O3–30V2O5–50B2O3 glass for energy storage applications, *J. Mater. Sci.: Mater. Electron.* **2021**, 32, 22408–22416. DOI: 10.1007/s10854-021-06727-3.

87. Liu, C.; Xie, S.; Bai, H.; Yan, F.; Fu, T.; Shen, B.; Zhai, J. Excellent energy storage performance of niobate-based glass-ceramics via introduction of nucleating agent, *J. Materiomics* **2022**, 1–9. DOI: 10.1016/j.jmat.2022.03.001.

88. Chen, Y.; Wang, S.; Zhou, H.; Xu, Q.; Wang, Q.; Zhu, J. A systematic analysis of the radial resonance frequency spectra of the PZT-based (Zr/Ti = 52/48) piezoceramic thin disks, *J. Adv. Ceram.* **2020**, 9, 380–392. DOI: 10.1007/s40145-020-0378-5.

89. Shrout, T. R.; Zhang, S. J. Lead-free piezoelectric ceramics: Alternatives for PZT?, *J. Electroceram.* **2007**, 19, 111–124. DOI: 10.1007/s10832-007-9047-0.

90. Gao, J.; Liu, Y.; Wang, Y.; Wang, D.; Zhong, L.; Ren, X. High temperature-stability of (Pb0.9La0.1)(Zr0.65Ti0.35)O3 ceramic for energy-storage applications at finite electric field strength, *Scr. Mater.* **2017**, 137, 114–118. DOI: 10.1016/j.scriptamat.2017.05.011.

91. Zhang, T. F.; Tang, X. G.; Liu, Q. X.; Jiang, Y. P.; Huang, X. X.; Zhou, Q. F. Energy-storage properties and high-temperature dielectric relaxation behaviors of relaxor ferroelectric Pb(Mg1/3Nb2/3)O3-PbTiO3 ceramics, *J. Phys. D: Appl. Phys.* **2016**, 49, 095302. DOI: 10.1088/0022-3727/49/9/095302.

92. Dai, X.; Viehland, D. Effects of lanthanum modification on the antiferroelectric-ferroelectric stability of high zirconium-content lead zirconate titanate, *J. Appl. Phys.* **1994**, 76, 3701–3709. DOI: 10.1063/1.357439.

93. Gupta, S. M.; Li, J. F.; Viehland, D. Coexistence of relaxor and normal ferroelectric phases in morphotropic phase boundary compositions of lanthanum-modified lead zirconate titanate, *J. Am. Ceram. Soc.* **1998**, 81, 557–564. DOI: 10.1111/j.1151-2916.1998.tb02374.x.

94. Kumar, A.; Kim, S. H.; Peddigari, M.; Jeong, D. H.; Hwang, G. T.; Ryu, J. High energy storage properties and electrical field stability of energy efficiency of (Pb0.89La0.11)(Zr0.70Ti0.30)0.9725O3 relaxor ferroelectric ceramics, *Electron. Mater. Lett.* **2019**, 15, 323–330. DOI: 10.1007/s13391-019-00124-z.

95. Hu, Z.; Ma, B.; Liu, S.; Narayanan, M.; Balachandran, U. Relaxor behavior and energy storage performance of ferroelectric PLZT thin films with different Zr/Ti ratios, *Ceram. Int.* **2014**, 40, 557–562. DOI: 10.1016/j.ceramint.2013.05.139.

96. Liu, Y.; Hao, X.; An, S. Significant enhancement of energy-storage performance of (Pb 0.91La0.09)(Zr0.65Ti0.35)O3 relaxor ferroelectric thin films by Mn doping, *J. Appl. Phys.* **2013**, 114, 174102. DOI: 10.1063/1.4829029.

97. Peng, B.; Xie, Z.; Yue, Z.; Li, L. Improvement of the recoverable energy storage density and efficiency by utilizing the linear dielectric response in ferroelectric capacitors, *Appl. Phys. Lett.* **2014**, 105, 052904. DOI: 10.1063/1.4892454.

98. Zhang, T.; Li, W.; Hou, Y.; Yu, Y.; Song, R.; Cao, W.; Fei, W. High-energy storage density and excellent temperature stability in antiferroelectric/ferroelectric bilayer thin films, *J. Am. Ceram. Soc.* **2017**, 100, 3080–3087. DOI: 10.1111/jace.14876.

99. Zhang, T.; Li, W.; Zhao, Y.; Yu, Y.; Fei, W. High energy storage performance of opposite double-heterojunction ferroelectricity–insulators, *Adv. Funct. Mater.* **2018**, 28, 1706211. DOI: 10.1002/adfm.201706211.

100. Zhang, L.; Hao, X.; Yang, J.; An, S.; Song, B. Large enhancement of energy-storage properties of compositional graded (Pb1-xLax)(Zr0.65Ti0.35)O3 relaxor ferroelectric thick films, *Appl. Phys. Lett.* **2013**, 103, 113902. DOI: 10.1063/1.4821209.

101. Nguyen, M. D.; Houwman, E. P.; Rijnders, G. Energy storage performance and electric breakdown field of thin relaxor ferroelectric PLZT films using microstructure and growth orientation control, *J. Phys. Chem. C* **2018**, 122, 15171–15179. DOI: 10.1021/acs.jpcc.8b04251.

102. Nguyen, M. D.; Nguyen, C. T. Q.; Vu, H. N.; Rijnders, G. Controlling microstructure and film growth of relaxor-ferroelectric thin films for high break-down strength and energy-storage performance, *J. Eur. Ceram Soc.* **2018**, 38, 95–103. DOI: 10.1016/j.jeurceramsoc.2017.08.027.

103. Ma, B.; Hu, Z.; Koritala, R. E.; Lee, T. H.; Dorris, S. E.; Balachandran, U. PLZT film capacitors for power electronics and energy storage applications, *J. Mater. Sci.: Mater. Electron.* **2015**, 26, 9279–9287. DOI: 10.1007/s10854-015-3025-0.

104. Peng, B.; Tang, S.; Lu, L.; Zhang, Q.; Huang, H.; Bai, G.; Miao, L.; Zou, B.; Liu, L.; Sun, W.; Wang, Z. L. Low-temperature-poling awakened high dielectric breakdown strength and outstanding improvement of discharge energy density of (Pb,La)(Zr,Sn,Ti)O3 relaxor thin film, *Nano Energy* **2020**, 77, 105132. DOI: 10.1016/j.nanoen.2020.105132.

105. Isupov, V.; Smolenskii, G.; Agranovskaya, A.; Krainik, N. J. S. P.-S. S. New ferroelectrics of complex composition, *Soviet Physics-Solid State* **1961**, 2, 2651–2654.

106. Rao, B. N.; Datta, R.; Chandrashekaran, S. S.; Mishra, D. K.; Sathe, V.; Senyshyn, A.; Ranjan, R. Local structural disorder and its influence on the average global structure and polar properties in Na0.5Bi0.5TiO3, *Phys. Rev. B: Condens. Matter Mater. Phys.* **2013**, 88, 224103. DOI: 10.1103/PhysRevB.88.224103.

107. Reichmann, K.; Feteira, A.; Li, M. Bismuth Sodium Titanate based materials for piezo-electric actuators, *Materials* **2015**, 8, 8467–8495. DOI: 10.3390/ma8125469.

108. Suchanicz, J.; Kluczewska-Chmielarz, K.; Sitko, D.; Jagło, G. Electrical transport in lead-free Na0.5Bi0.5TiO3 ceramics, *J. Adv. Ceram.* **2021**, 10, 152–165. DOI: 10.1007/s40145-020-0430-5.

109. Qiao, X.; Zhang, F.; Wu, D.; Chen, B.; Zhao, X.; Peng, Z.; Ren, X.; Liang, P.; Chao, X.; Yang, Z. Superior comprehensive energy storage properties in Bi0.5Na0.5TiO3-based relaxor ferroelectric ceramics, *Chem. Eng. J.* **2020**, 388, 124158. DOI: 10.1016/j.cej.2020.124158.

110. Yang, F.; Pan, Z.; Ling, Z.; Hu, D.; Ding, J.; Li, P.; Liu, J.; Zhai, J. Realizing high comprehensive energy storage performances of BNT-based ceramics for application in pulse power capacitors, *J. Eur. Ceram. Soc.* **2021**, 41, 2548–2558. DOI: 10.1016/j.jeurceramsoc.2020.11.049.

111. Zhang, X.; Hu, D.; Pan, Z.; Lv, X.; He, Z.; Yang, F.; Li, P.; Liu, J.; Zhai, J. Enhancement of recoverable energy density and efficiency of lead-free relaxor-ferroelectric BNT-based ceramics, *Chem. Eng. J.* **2021**, 406, 126818. DOI: 10.1016/j.cej.2020.126818.

112. Zhu, C.; Cai, Z.; Luo, B.; Guo, L.; Li, L.; Wang, X. High temperature lead-free BNT-based ceramics with stable energy storage and dielectric properties, *J. Mater. Chem. A* **2020**, 8, 683–692. DOI: 10.1039/c9ta10347c.

113. Ma, C.; Tan, X. In situ transmission electron microscopy study on the phase transition-sin lead-free (1-x)(Bi1/2Na1/2)TiO3-xBaTiO3 ceramics, *J. Am. Ceram. Soc.* **2011**, 94, 4040–4044. DOI: 10.1111/j.1551-2916.2011.04670.x.

114. Jo, W.; Schaab, S.; Sapper, E.; Schmitt, L. A.; Kleebe, H. J.; Bell, A. J.; Rödel, J. On the phase identity and its thermal evolution of lead free (Bi 1/2Na1/2)TiO3-6 mol BaTiO3, *J. Appl. Phys.* **2011**, 110, 074106. DOI: 10.1063/1.3645054.

115. Ye, H.; Yang, F.; Pan, Z.; Hu, D.; Lv, X.; Chen, H.; Wang, F.; Wang, J.; Li, P.; Chen, J.; Liu, J.; Zhai, J. Significantly improvement of comprehensive energy storage performances with lead-free relaxor ferroelectric ceramics for high-temperature capacitors applications, *Acta Mater.* **2021**, 203, 116484. DOI: 10.1016/j.actamat.2020.116484.

116. Gao, F.; Dong, X.; Mao, C.; Liu, W.; Zhang, H.; Yang, L.; Cao, F.; Wang, G. Energy-storage properties of 0.89Bi0.5Na0.5TiO3-0.06BaTiO3-0.05K0.5Na0.5NbO3 lead-free anti-ferroelectric ceramics, *J. Am. Ceram. Soc.* **2011**, 94, 4382–4386. DOI: 10.1111/j.1551-2916.2011.04731.x.

117. Cao, W.; Li, W.; Feng, Y.; Bai, T.; Qiao, Y.; Hou, Y.; Zhang, T.; Yu, Y.; Fei, W. Defect dipole induced large recoverable strain and high energy-storage density in lead-free Na 0.5 Bi 0.5 TiO3 -based systems, *Appl. Phys. Lett.* **2016**, 108, 202902. DOI: 10.1063/1.4950974.

118. Ren, X. Large electric-field-induced strain in ferroelectric crystals by point-defect-mediated reversible domain switching, *Nat. Mater.* **2004**, 3, 91–94. DOI: 10.1038/nmat1051.

119. Li, F.; Zhai, J.; Shen, B.; Liu, X.; Yang, K.; Zhang, Y.; Li, P.; Liu, B.; Zeng, H. Influence of structural evolution on energy storage properties in Bi0.5Na0.5TiO3-SrTiO3-NaNbO3 le ad-free ferroelectric ceramics, *J. Appl. Phys.* **2017**, 121, 054103. DOI: 10.1063/1.4975409.

120. Li, D.; Shen, Z. Y.; Li, Z.; Luo, W.; Wang, X.; Wang, Z.; Song, F.; Li, Y. P-E hysteresis loop going slim in Ba0.3Sr0.7TiO3-modified Bi0.5Na0.5TiO3 ceramics for energy storage applications, *J. Adv. Ceram.* **2020**, 9, 183–192. DOI: 10.1007/s40145-020-0358-9.

121. Li, J.; Li, F.; Xu, Z.; Zhang, S. Multilayer lead-free ceramic capacitors with ultrahigh energy density and efficiency, *Adv. Mater.* **2018**, 30, 1802155. DOI: 10.1002/adma.201802155.

122. Yang, H.; Liu, P.; Yan, F.; Lin, Y.; Wang, T. A novel lead-free ceramic with layered structure for high energy storage applications, *J. Alloys Compd.* **2019**, 773, 244–249. DOI: 10.1016/j.jallcom.2018.09.252.

123. Jia, W.; Hou, Y.; Zheng, M.; Xu, Y.; Yu, X.; Zhu, M.; Yang, K.; Cheng, H.; Sun, S.; Xing, J. Superior temperature-stable dielectrics for MLCCs based on Bi0.5Na0.5TiO3-NaNbO3 system modified by CaZrO3, *J. Am. Ceram. Soc.* **2018**, 101, 3468–3479. DOI: 10.1111/jace.15519.

124. Wang, H.; Zhao, P.; Chen, L.; Li, L.; Wang, X. Energy storage properties of 0.87BaTiO3-0.13Bi(Zn2/3(Nb0.85Ta0.15)1/3)O3 multilayer ceramic capacitors with thin dielectric layers, *J. Adv. Ceram.* **2020**, 9, 292–302. DOI: 10.1007/s40145-020-0367-8.

125. Feng, C.; Yang, C. H.; Li, S. X.; Han, Y. J.; Hu, X. Q.; Jiao, F. Y.; Qian, J.; Du, X. B. Reduced leakage current and large polarization of Na0.5Bi0.5Ti0.98Mn0.02O3 thin film annealed at low temperature, *Ceram. Int.* **2015**, 41, 14179–14183. DOI: 10.1016/j.ceramint.2015.07.041.

126. Wang, J.; Sun, N.; Li, Y.; Zhang, Q.; Hao, X.; Chou, X. Effects of Mn doping on dielectric properties and energy-storage performance of Na0.5Bi0.5TiO3 thick films, *Ceram. Int.* **2017**, 43, 7804–7809. DOI: 10.1016/j.ceramint.2017.03.094.

127. Feng, C.; Yang, C. H.; Sui, H. T.; Geng, F. J.; Han, Y. J. Effect of Fe doping on the crystallization and electrical properties of Na0.5Bi0.5TiO3 thin film, *Ceram. Int.* **2015**, 41, 4214–4217. DOI: 10.1016/j.ceramint.2014.10.066.

128. Zhang, Y. L.; Li, W. L.; Cao, W. P.; Zhang, T. D.; Bai, T. R. G. L.; Yu, Y.; Hou, Y. F.; Feng, Y.; Fei, W. D. Enhanced energy-storage performance of 0.94NBT-0.06BT thin films induced by a Pb0.8La0.1Ca0.1Ti0.975O3 seed layer, *Ceram. Int.* **2016**, 42, 14788–14792. DOI: 10.1016/j.ceramint.2016.06.110.

129. Peng, B.; Zhang, Q.; Li, X.; Sun, T.; Fan, H.; Ke, S.; Ye, M.; Wang, Y.; Lu, W.; Niu, H.; Scott, J. F.; Zeng, X.; Huang, H. Giant electric energy density in epitaxial lead-free thin films with coexistence of ferroelectrics and antiferroelectrics, *Adv. Electron. Mater.* **2015**, 1, 1500052. DOI: 10.1002/aelm.201500052.

130. Guo, Y.; Li, M.; Zhao, W.; Akai, D.; Sawada, K.; Ishida, M.; Gu, M. Ferroelectric and pyroelectric properties of (Na0.5Bi0.5)TiO3-BaTiO3 based trilayered thin films, *Thin Solid Films* **2009**, 517, 2974–2978. DOI: 10.1016/j.tsf.2008.11.100.

131. Chen, P.; Wu, S.; Li, P.; Zhai, J.; Shen, B. Great enhancement of energy storage density and power density in BNBT/: X BFO multilayer thin film hetero-structures, *Inorg. Chem. Front.* **2018**, 5, 2300–2305. DOI: 10.1039/c8qi00487k.

132. Qian, J.; Han, Y.; Yang, C.; Lv, P.; Zhang, X.; Feng, C.; Lin, X.; Huang, S.; Cheng, X.; Cheng, Z. Energy storage performance of flexible NKBT/NKBT-ST multilayer film capacitor by interface engineering, *Nano Energy* **2020**, 74, 104862. DOI: 10.1016/j.nanoen.2020.104862.

133. Yang, L.; Kong, X.; Cheng, Z.; Zhang, S. Ultra-high energy storage performance with mitigated polarization saturation in lead-free relaxors, *J. Mater. Chem. A* **2019**, 7, 8573–8580. DOI: 10.1039/c9ta01165j.

134. Pan, M. J.; Randall, C. A brief introduction to ceramic capacitors, *IEEE Electr. Insul. Mag.* **2010**, 26, 44–50. DOI: 10.1109/MEI.2010.5482787.

135. Zheng, L.; Yuan, L.; Liang, G.; Gu, A. An in situ (K 0.5 Na 0.5)NbO3 -doped barium titanate foam framework and its cyanate ester resin composites with temperature-stable dielectric properties and low dielectric loss, *Mater. Chem. Front.* **2019**, 3, 726–736. DOI: 10.1039/c9qm00006b.

136. Matias, A.; Nikola, N.; Virginia, R.; Satyanarayan, P.; Rahul, V.; Jurij, K.; Rossetti Jr., G. A.; Jürgen, R. BaTiO3-based piezoelectrics: Fundamentals, current status, and perspectives, *Appl. Phys. Rev.* **2017**, 4, 1–53.

137. Gao, W.; Zhu, Y.; Wang, Y.; Yuan, G.; Liu, J. M. A review of flexible perovskite oxide ferroelectric films and their application, *J. Materiomics* **2020**, 6, 1–16. DOI: 10.1016/j.jmat.2019.11.001.

138. Bai, F.; Viehland, D.; Jia, Y.; Schlom, D. G.; Wuttig, M.; Roytburd, A.; Ramesh, R.; Zheng, H.; Wang, J.; Lofland, S. E.; Ma, Z.; Mohaddes-Ardabili, L.; Zhao, T.; Salamanca-Riba, L.; Shinde, S. R.; Ogale, S. B. Multiferroic BaTiO3-CoFe2O4 nanostructures, *Science* **2004**, 303, 661–663.

139. Hu, D.; Pan, Z.; Tan, X.; Yang, F.; Ding, J.; Zhang, X.; Li, P.; Liu, J.; Zhai, J.; Pan, H. Optimization the energy density and efficiency of BaTiO3-based ceramics for capacitor applications, *Chem. Eng. J.* **2021**, 409, 127375. DOI: 10.1016/j.cej.2020.127375.

140. Ogihara, H.; Randall, C. A.; Trolier-Mckinstry, S. High-energy density capacitors utilizing 0.7 baTiO3-0.3 BiScO3 ceramics, *J. Am. Ceram. Soc.* **2009**, 92, 1719–1724. DOI: 10.1111/j.1551-2916.2009.03104.x.

141. Yuan, Q.; Li, G.; Yao, F. Z.; Cheng, S. D.; Wang, Y.; Ma, R.; Mi, S. B.; Gu, M.; Wang, K.; Li, J. F.; Wang, H. Simultaneously achieved temperature-insensitive high energy density and efficiency in domain-engineered BaTiO3-Bi(Mg0.5Zr0.5)O3 lead-free relaxor ferroelectrics, *Nano Energy* **2018**, 52, 203–210. DOI: 10.1016/j.nanoen.2018.07.055.

142. Choi, D. H.; Baker, A.; Lanagan, M.; Trolier-Mckinstry, S.; Randall, C. Structural and dielectric properties in (1-x) BaTiO3-x Bi (Mg 1/2 Ti 1/2) O3 ceramics (0.1 ≤ x ≤ 0.5) and potential for high-voltage multilayer capacitors, *J. Am. Ceram. Soc.* **2013**, 96, 2197–2202. DOI: 10.1111/jace.12312.

143. Li, M. D.; Tang, X. G.; Zeng, S. M.; Jiang, Y. P.; Liu, Q. X.; Zhang, T. F.; Li, W. H. Oxygen-vacancy-related dielectric relaxation behaviours and impedance spectroscopy of Bi(Mg1/2Ti1/2)O3 modified BaTiO3 ferroelectric ceramics, *J. Materiomics* **2018**, 4, 194–201. DOI: 10.1016/j.jmat.2018.03.001.

144. Hanani, Z.; Mezzane, D.; Amjoud, M. b.; Razumnaya, A. G.; Fourcade, S.; Gagou, Y.; Hoummada, K.; El Marssi, M.; Gouné, M. Phase transitions, energy storage performances and electrocaloric effect of the lead-free Ba 0.85 Ca 0.15 Zr 0.10 Ti 0.90 O3 ceramic relaxor, *J. Mater. Sci.: Mater. Electron.* **2019**, 30, 6430–6438. DOI: 10.1007/s10854-019-00946-5.

145. Wang, X. W.; Zhang, B. H.; Shi, Y. C.; Li, Y. Y.; Manikandan, M.; Shang, S. Y.; Shang, J.; Hu, Y. C.; Yin, S. Q. Enhanced energy storage properties in Ba0.85Ca0.15Zr0.1Ti0.9O3 ceramics with glass additives, *J. Appl. Phys.* **2020**, 127, 074103. DOI: 10.1063/1.5138948.

146. Hanani, Z.; Merselmiz, S.; Danine, A.; Stein, N.; Mezzane, D.; Amjoud, M. b.; Lahcini, M.; Gagou, Y.; Spreitzer, M.; Vengust, D.; Kutnjak, Z.; El Marssi, M.; Luk'yanchuk, I. A.; Gouné, M. Enhanced dielectric and electrocaloric properties in lead-free rod-like BCZT ceramics, *J. Adv. Ceram.* **2020**, 9, 210–219. DOI: 10.1007/s40145-020-0361-1.

147. Patel, S.; Sharma, D.; Singh, A.; Vaish, R. Enhanced thermal energy conversion and dynamic hysteresis behavior of Sr-added Ba0.85Ca0.15Ti0.9Zr0.1O3 ferroelectric ceramics, *J. Materiomics* **2016**, 2, 75–86. DOI: 10.1016/j.jmat.2016.01.002.

148. Wu, L.; Wang, X.; Li, L. Core-shell BaTiO3@BiScO3particles for local graded dielectric ceramics with enhanced temperature stability and energy storage capability, *J. Alloys Compd.* **2016**, 688, 113–121. DOI: 10.1016/j.jallcom.2016.07.057.

149. Su, X.; Riggs, B. C.; Tomozawa, M.; Nelson, J. K.; Chrisey, D. B. Preparation of BaTiO3/low melting glass core-shell nanoparticles for energy storage capacitor applications, *J. Mater. Chem. A* **2014**, 2, 18087–18096. DOI: 10.1039/c4ta04282d.

150. Zhang, Y.; Cao, M.; Yao, Z.; Wang, Z.; Song, Z.; Ullah, A.; Hao, H.; Liu, H. Effects of silica coating on the microstructures and energy storage properties of BaTiO3 ceramics, *Mater. Res. Bull.* **2015**, 67, 70–76. DOI: 10.1016/j.materresbull.2015.01.056.

151. Wu, L.; Wang, X.; Gong, H.; Hao, Y.; Shen, Z.; Li, L. Core-satellite BaTiO3@SrTiO3 assemblies for a local compositionally graded relaxor ferroelectric capacitor with enhanced energy storage density and high energy efficiency, *J. Mater. Chem. C* **2015**, 3, 750–758. DOI: 10.1039/c4tc02291b.

152. Yu, Z.; Ang, C.; Guo, R.; Bhalla, A. S. Ferroelectric-relaxor behavior of Ba(Ti 0.7Zr 0.3) O3 ceramics, *J. Appl. Phys.* **2002**, 92, 2655–2657. DOI: 10.1063/1.1495069.

153. Hennings, D.; Schnell, A.; Simon, G. Diffuse ferroelectric phase transitions in Ba(Ti1-yZry)O3 ceramics, *J. Am. Ceram. Soc.* **1982**, 65, 539–544. DOI: 10.1111/j.1151-2916.1982.tb10778.x.

154. Instan, A. A.; Pavunny, S. P.; Bhattarai, M. K.; Katiyar, R. S. Ultrahigh capacitive energy storage in highly oriented Ba(Zr x Ti 1-x)O3 thin films prepared by pulsed laser deposition, *Appl. Phys. Lett.* **2017**, 111, 142903. DOI: 10.1063/1.4986238.

155. Cheng, H.; Ouyang, J.; Zhang, Y.-X.; Ascienzo, D.; Li, Y.; Zhao, Y.-Y.; Ren, Y. Demonstration of ultra-high recyclable energy densities in domain-engineered ferroelectric films, *Nat. Commun.* **2017**, 8, 1999. DOI: 10.1038/s41467-017-02040-y.

156. Reddy, S. R.; Prasad, V. V. B.; Bysakh, S.; Shanker, V.; Hebalkar, N.; Roy, S. K. Superior energy storage performance and fatigue resistance in ferroelectric BCZT thin films grown in an oxygen-rich atmosphere, *J. Mater. Chem. C* **2019**, 7, 7073–7082. DOI: 10.1039/c9tc00569b.

157. Ortega, N.; Kumar, A.; Scott, J. F.; Chrisey, D. B.; Tomazawa, M.; Kumari, S.; Diestra, D. G. B.; Katiyar, R. S. Relaxor-ferroelectric superlattices: High energy density capacitors, *J. Phys.: Condens. Matter* **2012**, 24, 445901. DOI: 10.1088/0953-8984/24/44/445901.

158. Sun, Z.; Ma, C.; Liu, M.; Cui, J.; Lu, L.; Lu, J.; Lou, X.; Jin, L.; Wang, H.; Jia, C.-L. Ultrahigh energy storage performance of lead-free oxide multilayer film capacitors via interface engineering, *Adv. Mater.* **2017**, 29, 1604427. DOI: 10.1002/adma.201604427.

159. Zhang, W.; Gao, Y.; Kang, L.; Yuan, M.; Yang, Q.; Cheng, H.; Pan, W.; Ouyang, J. Space-charge dominated epitaxial BaTiO3 heterostructures, *Acta Mater.* **2015**, 85, 207–215. DOI: 10.1016/j.actamat.2014.10.063.

160. Ru, J.; Min, D.; Lanagan, M.; Li, S.; Chen, G. Enhanced energy storage properties of thermostable sandwich-structured BaTiO3/polyimide nanocomposites with better controlled interfaces, *Mater. Des.* **2021**, 197, 109270. DOI: 10.1016/j.matdes.2020.109270.

161. Yang, C.; Qian, J.; Lv, P.; Wu, H.; Lin, X.; Wang, K.; Ouyang, J.; Huang, S.; Cheng, X.; Cheng, Z. Flexible lead-free BFO-based dielectric capacitor with large energy density, superior thermal stability, and reliable bending endurance, *J. Materiomics* **2020**, 6, 200–208. DOI: 10.1016/j.jmat.2020.01.010.

162. Gao, X.; Li, Y.; Chen, J.; Yuan, C.; Zeng, M.; Zhang, A.; Gao, X.; Lu, X.; Li, Q.; Liu, J. M. High energy storage performances of Bi 1–x Sm x Fe 0.95 Sc 0.05 O3 lead-free ceramics synthesized by rapid hot press sintering, *J. Eur. Ceram. Soc.* **2019**, 39, 2331–2338. DOI: 10.1016/j.jeurceramsoc.2019.02.009.

163. Li, Q.; Ji, S.; Wang, D.; Zhu, J.; Li, L.; Wang, W.; Zeng, M.; Hou, Z.; Gao, X.; Lu, X.; Li, Q.; Liu, J. M. Simultaneously enhanced energy storage density and efficiency in novel BiFeO3-based lead-free ceramic capacitors, *J. Eur. Ceram. Soc.* **2021**, 41, 387–393. DOI: 10.1016/j.jeurceramsoc.2020.08.032.

164. Lee, M. H.; Kim, D. J.; Park, J. S.; Kim, S. W.; Song, T. K.; Kim, M. H.; Kim, W. J.; Do, D.; Jeong, I. K. High-performance lead-free piezoceramics with high curie temperatures, *Adv. Mater.* **2015**, 27, 6976–6982. DOI: 10.1002/adma.201502424.

165. Wu, J.; Fan, Z.; Xiao, D.; Zhu, J.; Wang, J. Multiferroic bismuth ferrite-based materials for multifunctional applications: Ceramic bulks, thin films and nanostructures, *Prog. Mater. Sci.* **2016**, 84, 335–402. DOI: 10.1016/j.pmatsci.2016.09.001.

166. Hang, Q.; Zhou, W.; Zhu, X.; Zhu, J.; Liu, Z.; Al-Kassab, T. Structural, spectroscopic, and dielectric characterizations of Mn-doped 0.67BiFeO3-0.33BaTiO3 multiferroic ceramics, *J. Adv. Ceram.* **2013**, 2, 252–259. DOI: 10.1007/s40145-013-0068-7.

167. Liu, N.; Liang, R.; Zhou, Z.; Dong, X. Designing lead-free bismuth ferrite-based ceramics learning from relaxor ferroelectric behavior for simultaneous high energy density and efficiency under low electric field, *J. Mater. Chem. C* **2018**, 6, 10211–10217. DOI: 10.1039/C8TC03855D.

168. Qi, H.; Xie, A.; Tian, A.; Zuo, R. Superior energy-storage capacitors with simultaneously giant energy density and efficiency using nanodomain engineered BiFeO3 -BaTiO3 -NaNbO3 lead-free bulk ferroelectrics, *Adv. Energy Mater.* **2020**, 10, 1903338. DOI: 10.1002/aenm.201903338.

169. Wang, G.; Li, J.; Zhang, X.; Fan, Z.; Yang, F.; Feteira, A.; Zhou, D.; Sinclair, D. C.; Ma, T.; Tan, X.; Wang, D.; Reaney, I. M. Ultrahigh energy storage density lead-free multilayers by controlled electrical homogeneity, *Energy Environ. Sci.* **2019**, 12, 582–588. DOI: 10.1039/c8ee03287d.

170. Correia, T. M.; McMillen, M.; Rokosz, M. K.; Weaver, P. M.; Gregg, J. M.; Viola, G.; Cain, M. G. A lead-free and high-energy density ceramic for energy storage applications, *J. Am. Ceram. Soc.* **2013**, 96, 2699–2702. DOI: 10.1111/jace.12508.

171. Pan, H.; Li, F.; Liu, Y.; Zhang, Q.; Wang, M.; Lan, S.; Zheng, Y.; Ma, J.; Gu, L.; Shen, Y.; Yu, P.; Zhang, S.; Chen, L. Q.; Lin, Y. H.; Nan, C. W. Ultrahigh–energy density lead-free dielectric films via polymorphic nanodomain design, *Science* **2019**, 365, 578–582. DOI: 10.1126/science.aaw8109.

172. Clauset, A.; Shalizi, C. R.; Newman, M. E. J. Power-law distributions in empirical data, *SIAM Rev.* **2009**, 51, 661–703. DOI: 10.1137/070710111.

173. McMillen, M.; Douglas, A. M.; Correia, T. M.; Weaver, P. M.; Cain, M. G.; Gregg, J. M. Increasing recoverable energy storage in electroceramic capacitors using "dead-layer" engineering, *Appl. Phys. Lett.* **2012**, 101, 242909. DOI: 10.1063/1.4772016.

174. Hou, Y.; Han, R.; Li, W.; Luo, L.; Fei, W. Significantly enhanced energy storage performance in BiFeO3/BaTiO3/BiFeO3 sandwich-structured films through crystallinity regulation, *Phys. Chem. Chem. Phys.* **2018**, 20, 21917–21924. DOI: 10.1039/c8cp04072a.

175. Zhu, H.; Liu, M.; Zhang, Y.; Yu, Z.; Ouyang, J.; Pan, W. Increasing energy storage capabilities of space-charge dominated ferroelectric thin films using interlayer coupling, *Acta Mater.* **2017**, 122, 252–258. DOI: 10.1016/j.actamat.2016.09.051.

176. Li, J. F.; Wang, K.; Zhu, F. Y.; Cheng, L. Q.; Yao, F. Z. (K, Na) NbO3-based lead-free piezoceramics: Fundamental aspects, processing technologies, and remaining challenges, *J. Am. Ceram. Soc.* **2013**, 96, 3677–3696. DOI: 10.1111/jace.12715.

177. EGERTON, L.; DILLON, D. M. Piezoelectric and dielectric properties of ceramics in the system potassium—Sodium niobate, *J. Am. Ceram. Soc.* **1959**, 42, 438–442. DOI: 10.1111/j.1151-2916.1959.tb12971.x.

178. Yang, Z.; Du, H.; Qu, S.; Hou, Y.; Ma, H.; Wang, J.; Wang, J.; Wei, X.; Xu, Z. Significantly enhanced recoverable energy storage density in potassium-sodium niobate-based lead free ceramics, *J. Mater. Chem. A* **2016**, 4, 13778–13785. DOI: 10.1039/c6ta04107h.

179. Shao, T.; Du, H.; Ma, H.; Qu, S.; Wang, J.; Wang, J.; Wei, X.; Xu, Z. Potassium-sodium niobate based lead-free ceramics: Novel electrical energy storage materials, *J. Mater. Chem. A* **2017**, 5, 554–563. DOI: 10.1039/c6ta07803f.

180. Qu, B.; Du, H.; Yang, Z.; Liu, Q. Large recoverable energy storage density and low sintering temperature in potassium-sodium niobate-based ceramics for multilayer pulsed power capacitors, *J. Am. Ceram. Soc.* **2017**, 100, 1517–1526. DOI: 10.1111/jace.14728.

181. Qu, B.; Du, H.; Yang, Z.; Liu, Q.; Liu, T. Enhanced dielectric breakdown strength and energy storage density in lead-free relaxor ferroelectric ceramics prepared using transition liquid phase sintering, *RSC Adv.* **2016**, 6, 34381–34389. DOI: 10.1039/c6ra01919f.

182. Yang, Y.; Ji, Y.; Fang, M.; Zhou, Z.; Zhang, L.; Ren, X. Morphotropic relaxor boundary in a relaxor system showing enhancement of electrostrain and dielectric permittivity, *Phys. Rev. Lett.* **2019**, 123, 137601. DOI: 10.1103/PhysRevLett.123.137601.

183. Won, S. S.; Kawahara, M.; Kuhn, L.; Venugopal, V.; Kwak, J.; Kim, I. W.; Kingon, A. I.; Kim, S.-H. BiFeO3 -doped (K 0.5,Na 0.5)(Mn 0.005,Nb 0.995)O3 ferroelectric thin film capacitors for high energy density storage applications, *Appl. Phys. Lett.* **2017**, 110, 152901. DOI: 10.1063/1.4980113.

184. Huang, Y.; Shu, L.; Zhang, S.; Zhou, Z.; Cheng, Y. Y. S.; Peng, B.; Liu, L.; Zhang, Y.; Wang, X.; Li, J. F. Simultaneously achieved high-energy storage density and efficiency in (K,Na)NbO3-based lead-free ferroelectric films, *J. Am. Ceram. Soc.* **2021**, 104, 4119–4130. DOI: 10.1111/jace.17808.

185. Yeh, J. W.; Chen, S. K.; Lin, S. J.; Gan, J. Y.; Chin, T. S.; Shun, T. T.; Tsau, C. H.; Chang, S. Y. Nanostructured high-entropy alloys with multiple principal elements: Novel alloy design concepts and outcomes, *Adv. Eng. Mater.* **2004**, 6, 299–303. DOI: 10.1002/adem.200300567.

186. Cantor, B.; Chang, I. T. H.; Knight, P.; Vincent, A. J. B. Microstructural development in equiatomic multicomponent alloys, *Mat. Sci. Eng., A.* **2004**, 375, 213–218. DOI: 10.1016/j.msea.2003.10.257.

187. Yeh, J. W.; Chen, S. K.; Gan, J. Y.; Lin, S. J.; Chin, T. S.; Shun, T. T.; Tsau, C. H.; Chang, S. Y. Formation of simple crystal structures in Cu-Co-Ni-Cr-Al-Fe-Ti-V alloys with multiprincipal metallic elements, *Metall. Mater. Trans. A* **2004**, 35a, 2533–2536. DOI: 10.1007/s11661-006-0234-4.

188. Miracle, D. B.; Senkov, O. N. A critical review of high entropy alloys and related concepts, *Acta Mater.* **2017**, 122, 448–511. DOI: 10.1016/j.actamat.2016.08.081.

189. Pu, Y. P.; Zhang, Q. W.; Li, R.; Chen, M.; Du, X. Y.; Zhou, S. Y. Dielectric properties and electrocaloric effect of high-entropy (Na0.2Bi0.2Ba0.2Sr0.2Ca0.2)TiO3 ceramic, *Appl. Phys. Lett.* **2019**, 115. DOI: 10.1063/1.5126652.

190. Liu, J.; Ren, K.; Ma, C. Y.; Du, H. L.; Wang, Y. G. Dielectric and energy storage properties of flash-sintered high-entropy (Bi0.2Na0.2K0.2Ba0.2Ca0.2)TiO3 ceramic, *Ceram. Int.* **2020**, 46, 20576–20581. DOI: 10.1016/j.ceramint.2020.05.090.

191. Liu, W.; Li, F.; Chen, G. H.; Li, G. H.; Shi, H. W.; Li, L.; Guo, Y. M.; Zhai, J. W.; Wang, C. C. Comparative study of phase structure, dielectric properties and electrocaloric effect in novel high-entropy ceramics, *J. Mater. Sci.* **2021**, 56, 18417–18429. DOI: 10.1007/s10853-021-06530-9.

192. Xiong, W.; Zhang, H. F.; Cao, S. Y.; Gao, F.; Svec, P.; Dusza, J.; Reece, M. J.; Yan, H. X. Low-loss high entropy relaxor-like ferroelectrics with A-site disorder, *J. Eur. Ceram. Soc.* **2021**, 41, 2979–2985. DOI: 10.1016/j.jeurceramsoc.2020.11.030.

193. Yang, W. T.; Zheng, G. P. High energy storage density and efficiency in nanostructured (Bi0.2Na0.2K0.2La0.2Sr0.2)TiO3 high-entropy ceramics, *J. Am. Ceram. Soc.* **2022**, 105, 1083–1094. DOI: 10.1111/jace.18129.

194. Zhou, S. Y.; Pu, Y. P.; Zhang, X. Q.; Shi, Y.; Gao, Z. Y.; Feng, Y.; Shen, G. D.; Wang, X. Y.; Wang, D. W. High energy density, temperature stable lead-free ceramics by introducing high entropy perovskite oxide, *Chem. Eng. J.* **2022**, 427. DOI: 10.1016/j.cej.2021.131684.

195. Chen, L.; Deng, S. Q.; Liu, H.; Wu, J.; Qi, H.; Chen, J. Giant energy-storage density with ultrahigh efficiency in lead-free relaxors via high-entropy design, *Nat. Commun.* **2022**, 13. DOI: 10.1038/s41467-022-30821-7.

196. Yang, B. B.; Zhang, Y.; Pan, H.; Si, W. L.; Zhang, Q. H.; Shen, Z. H.; Yu, Y.; Lan, S.; Meng, F. Q.; Liu, Y. Q.; Huang, H. B.; He, J. Q.; Gu, L.; Zhang, S. J.; Chen, L. Q.; Zhu, J.; Nan, C. W.; Lin, Y. H. High-entropy enhanced capacitive energy storage, *Nat. Mater.* **2022**, 21, 1074–+. DOI: 10.1038/s41563-022-01274-6.

197. Kittel, C. Theory of antiferroelectric crystals, *Phys. Rev.* **1951**, 82, 729–732. DOI: 10.1103/PhysRev.82.729.

198. Chen, X.; Zhang, H.; Cao, F.; Wang, G.; Dong, X.; Gu, Y.; He, H.; Liu, Y. Charge-discharge properties of lead zirconate stannate titanate ceramics, *J. Appl. Phys.* **2009**, 106, 034105. DOI: 10.1063/1.3187778.

199. Tagantsev, A. K.; Vaideeswaran, K.; Vakhrushev, S. B.; Filimonov, A. V.; Burkovsky, R. G.; Shaganov, A.; Andronikova, D.; Rudskoy, A. I.; Baron, A. Q. R.; Uchiyama, H.; Chernyshov, D.; Bosak, A.; Ujma, Z.; Roleder, K.; Majchrowski, A.; Ko, J.-H.; Setter, N. The origin of antiferroelectricity in PbZrO3, *Nat. Commun.* **2013**, 4, 2229. DOI: 10.1038/ncomms3229.

200. Ren, Z. H.; Zhang, N.; Su, L. W.; Wu, H.; Ye, Z. G. Softening of antiferroelectricity in PbZrO3-Pb(Mn1/2W1/2)O3 complex perovskite solid solution, *J. Appl. Phys.* **2014**, 116. DOI: 10.1063/1.4885935.

201. Gao, P.; Liu, C.; Liu, Z. H.; Wan, H. Y.; Yuan, Y.; Li, H. J.; Pu, Y. P.; Ye, Z. G. Softening of antiferroelectric order in a novel PbZrO3-based solid solution for energy storage, *J. Eur. Ceram. Soc.* **2022**, 42, 1370–1379. DOI: 10.1016/j.jeurceramsoc.2021.12.027.

202. Chen, N.; Bai, G. R.; Auciello, O.; Koritala, R. E.; Lanagan, M. T. Properties and orientation of antiferroelectric lead zirconate thin films grown by MOCVD, *MRS Online Proc. Libr.* **1999**, 541, 345–350. DOI: 10.1557/proc-541-345.

203. Hao, X.; Zhai, J.; Yao, X. Improved energy storage performance and fatigue endurance of Sr-doped PbZrO3 antiferroelectric thin films, *J. Am. Ceram. Soc.* **2009**, 92, 1133–1135. DOI: 10.1111/j.1551-2916.2009.03015.x.

204. Parui, J.; Krupanidhi, S. B. Enhancement of charge and energy storage in sol-gel derived pure and La-modified PbZrO3 thin films, *Appl. Phys. Lett.* **2008**, 92, 192901. DOI: 10.1063/1.2928230.

205. Tani, T.; Li, J. F.; Viehland, D.; Payne, D. A. Antiferroelectric-ferroelectric switching and induced strains for sol-gel derived lead zirconate thin layers, *J. Appl. Phys.* **1994**, 75, 3017–3023. DOI: 10.1063/1.356146.

206. Sa, T.; Cao, Z.; Wang, Y.; Zhu, H. Enhancement of charge and energy storage in PbZrO3 thin films by local field engineering, *Appl. Phys. Lett.* **2014**, 105, 043902. DOI: 10.1063/1.4891768.

207. Chen, M. J.; Ning, X. K.; Wang, S. F.; Fu, G. S. Significant enhancement of energy storage density and polarization in self-assembled PbZrO3:NiO nano-columnar composite films, *Nanoscale* **2019**, 11, 1914–1920. DOI: 10.1039/c8nr08887j.

208. Ge, J.; Remiens, D.; Costecalde, J.; Chen, Y.; Dong, X.; Wang, G. Effect of residual stress on energy storage property in PbZrO3 antiferroelectric thin films with different orientations, *Appl. Phys. Lett.* **2013**, 103, 162903. DOI: 10.1063/1.4825336.

209. Ge, J.; Remiens, D.; Dong, X.; Chen, Y.; Costecalde, J.; Gao, F.; Cao, F.; Wang, G. Enhancement of energy storage in epitaxial PbZrO3 antiferroelectric films using strain engineering, *Appl. Phys. Lett.* **2014**, 105, 112908. DOI: 10.1063/1.4896156.

210. Ye, M.; Sun, Q.; Chen, X.; Jiang, Z.; Wang, F. Electrical and energy storage performance of Eu-doped PbZrO3 thin films with different gradient sequences, *J. Am. Ceram. Soc.* **2012**, 95, 1486–1488. DOI: 10.1111/j.1551-2916.2012.05130.x.

211. Corker, D. L.; Glazer, A. M.; Kaminsky, W.; Whatmore, R. W.; Dec, J.; Roleder, K. Investigation into the crystal structure of the Perovskite Lead Hafnate, PbHfO3, *Acta Crystallogr., Sect. B: Struct. Sci.* **1998**, 54, 18–28. DOI: 10.1107/S0108768197009208.

212. Madigou, V.; Baudour, J. L.; Bouree, F.; Favotto, C. L.; Roubin, M.; Nihoul, G. Crystallographic structure of lead hafnate (PbHfO3) from neutron powder diffraction and electron microscopy, *Philos. Mag. A* **1999**, 79, 847–858. DOI: 10.1080/01418619908210335.

213. Burkovsky, R. G.; Bronwald, I.; Andronikova, D.; Lityagin, G.; Piecha, J.; Souliou, S. M.; Majchrowski, A.; Filimonov, A.; Rudskoy, A.; Roleder, K.; Bosak, A.; Tagantsev, A. Triggered incommensurate transition in PbHfO3, *Phys. Rev. B.* **2019**, 100. DOI: 10.1103/PhysRevB.100.014107.

214. Gao, P.; Liu, Z. H.; Zhang, N.; Wu, H.; Bokov, A. A.; Ren, W.; Ye, Z. G. New Antiferroelectric Perovskite System with ultrahigh energy-storage performance at low electric field, *Chem. Mater.* **2019**, 31, 979–990. DOI: 10.1021/acs.chemmater.8b04470.

215. Chao, W.; Yang, T.; Li, Y. Achieving high energy efficiency and energy density in PbHfO3-based antiferroelectric ceramics, *J. Mater. Chem. C* **2020**, 8, 17016–17024. DOI: 10.1039/d0tc04617e.

216. Chauhan, V.; Wang, B. X.; Ye, Z. G. Structure, Antiferroelectricity and energy-storage performance of lead hafnate in a wide temperature range, *Materials* **2023**, 16. DOI: 10.3390/ma16114144.

217. Wan, H. Y.; Liu, Z. H.; Zhuo, F. P.; Xi, J. W.; Gao, P.; Zheng, K.; Jiang, L. Y.; Xu, J.; Li, J. R.; Zhang, J.; Zhuang, J.; Niu, G.; Zhang, N.; Ren, W.; Ye, Z. G. Synergistic design of a new PbHfO3-based antiferroelectric solid solution with high energy storage and large strain performances under low electric fields, *J. Mater. Chem. A* **2023**, 11, 25484–25496. DOI: 10.1039/d3ta05425j.

218. Huang, X.-X.; Zhang, T.-F.; Wang, W.; Ge, P.-Z.; Tang, X.-G. Tailoring energy-storage performance in antiferroelectric PbHfO3 thin films, *Mater. Des.* **2021**, 204, 109666. DOI: 10.1016/j.matdes.2021.109666.

219. Xu, B.; Moses, P.; Pai, N. G.; Cross, L. E. Charge release of lanthanum-doped lead zirconate titanate stannate antiferroelectric thin films, *Appl. Phys. Lett.* **1998**, 72, 593–595. DOI: 10.1063/1.120817.

220. Sharifzadeh Mirshekarloo, M.; Yao, K.; Sritharan, T. Large strain and high energy storage density in orthorhombic perovskite (Pb0.97La0.02)(Zr1−x−ySnxTiy)O3 antiferroelectric thin films, *Appl. Phys. Lett.* **2010**, 97, 142902. DOI: 10.1063/1.3497193.

221. Zhang, A. H.; Wang, W.; Li, Q. J.; Zhu, J. Y.; Wang, D. D.; Lu, X. B.; Zeng, M.; Yao, L. M.; Pan, Z. B. Internal-strain release and remarkably enhanced energy storage performance in PLZT–SrTiO3 multilayered films, *Appl. Phys. Lett.* **2020**, 117, 252901. DOI: 10.1063/5.0030279.

222. Dan, Y.; Xu, H.; Zou, K.; Zhang, Q.; Lu, Y.; Chang, G.; Huang, H.; He, Y. Energy storage characteristics of (Pb,La)(Zr,Sn,Ti)O3 antiferroelectric ceramics with high Sn content, *Appl. Phys. Lett.* **2018**, 113, 063902. DOI: 10.1063/1.5044712.

223. Liu, P.; Fan, B.; Yang, G.; Li, W.; Zhang, H.; Jiang, S. High energy density at high temperature in PLZST antiferroelectric ceramics, *J. Mater. Chem. C* **2019**, 7, 4587–4594. DOI: 10.1039/c9tc00944b.

224. Xu, B.; Ye, Y.; Cross, L. E. Dielectric properties and field-induced phase switching of lead zirconate titanate stannate antiferroelectric thick films on silicon substrates, *J. Appl. Phys.* **2000**, 87, 2507–2515. DOI: 10.1063/1.372211.

225. Markowski, K.; Park, S. E.; Yoshikawa, S.; Cross, L. E. Effect of compositional variations in the lead lanthanum zirconate stannate titanate system on electrical properties, *J. Am. Ceram. Soc.* **1996**, 79, 3297–3304. DOI: 10.1111/j.1151-2916.1996.tb08108.x.

226. Zheng, Q.; Yang, T.; Wei, K.; Wang, J.; Yao, X. Effect of Sn:Ti variations on electric filed induced AFE–FE phase transition in PLZST antiferroelectric ceramics, *Ceram. Int.* **2012**, 38, S9–S12. DOI: 10.1016/j.ceramint.2011.04.037.

227. Liu, Z.; Bai, Y.; Chen, X.; Dong, X.; Nie, H.; Cao, F.; Wang, G. Linear composition-dependent phase transition behavior and energy storage performance of tetragonal PLZST antiferroelectric ceramics, *J. Alloys Compd.* **2017**, 691, 721–725. DOI: 10.1016/j.jallcom.2016.08.328.

228. Zhang, L.; Jiang, S.; Fan, B.; Zhang, G. Enhanced energy storage performance in (Pb0.858Ba0.1La0.02Y0.008)(Zr0.65Sn0.3Ti0.05)O3-(Pb0.97La0.02)(Zr0.9Sn0.05 Ti0.05)O3 anti-ferroelectric composite ceramics by Spark Plasma Sintering, *J. Alloys Compd.* **2015**, 622, 162–165. DOI: 10.1016/j.jallcom.2014.09.171.

229. Zhang, G.; Zhu, D.; Zhang, X.; Zhang, L.; Yi, J.; Xie, B.; Zeng, Y.; Li, Q.; Wang, Q.; Jiang, S. High-energy storage performance of (Pb0.87Ba0.11a0.02)(Zr0.68Sn0.24Ti0.08)O3 antiferroelectric ceramics fabricated by the hot-press sintering method, *J. Am. Ceram. Soc.* **2015**, 98, 1175–1181. DOI: 10.1111/jace.13412.

230. Zhang, G.; Liu, S.; Yu, Y.; Zeng, Y.; Zhang, Y.; Hu, X.; Jiang, S. Microstructure and electrical properties of (Pb 0.87Ba 0.1La 0.02)(Zr 0.68Sn 0.24Ti 0.08)O3 anti-ferroelectric ceramics fabricated by the hot-press sintering method, *J. Eur. Ceram. Soc.* **2013**, 33, 113–121. DOI: 10.1016/j.jeurceramsoc.2012.08.011.

231. Bian, F.; Yan, S.; Xu, C.; Liu, Z.; Chen, X.; Mao, C.; Cao, F.; Bian, J.; Wang, G.; Dong, X. Enhanced breakdown strength and energy density of antiferroelectric Pb,La(Zr,Sn,Ti) O3 ceramic by forming core-shell structure, *J. Eur. Ceram. Soc.* **2018**, 38, 3170–3176. DOI: 10.1016/j.jeurceramsoc.2018.03.028.

232. Wang, H.; Liu, Y.; Yang, T.; Zhang, S. Ultrahigh energy-storage density in antiferro-electric ceramics with field-induced multiphase transitions, *Adv. Funct. Mater.* **2019**, 29, 1807321. DOI: 10.1002/adfm.201807321.

233. Zhang, Y.; Liu, P.; Kandula, K. R.; Li, W.; Meng, S.; Qin, Y.; Zhang, H.; Zhang, G. Achieving excellent energy storage density of Pb0.97La0.02(ZrxSn0.05Ti0.95-x) O3 ceramics by the B-site modification, *J. Eur. Ceram. Soc.* **2021**, 41, 360–367. DOI: 10.1016/j.jeurceramsoc.2020.08.039.

234. Liu, X.; Li, Y.; Hao, X. Ultra-high energy-storage density and fast discharge speed of (Pb0.98-xLa0.02Srx)(Zr0.9Sn0.1)0.995O3 antiferroelectric ceramics prepared via the tape-casting method, *J. Mater. Chem. A* **2019**, 7, 11858–11866. DOI: 10.1039/ c9ta02149c.

235. Gao, M.; Tang, X.; Leung, C. M.; Dai, S.; Li, J.; Viehland, D. D. Phase transition and energy storage behavior of antiferroelectric PLZT thin films epitaxially deposited on SRO buffered STO single crystal substrates, *J. Am. Ceram. Soc.* **2019**, 102, 5180–5191. DOI: 10.1111/jace.16380.

236. Lin, Z.; Chen, Y.; Liu, Z.; Wang, G.; Rémiens, D.; Dong, X. Large energy storage density, low energy loss and highly stable (Pb0.97La0.02)(Zr0.66Sn0.23Ti0.11)O3 antifer-roelectric thin-film capacitors, *J. Eur. Ceram. Soc.* **2018**, 38, 3177–3181. DOI: 10.1016/j. jeurceramsoc.2018.03.004.

237. Ma, B.; Kwon, D. K.; Narayanan, M.; Balachandran, U. Dielectric properties and energy storage capability of antiferroelectric Pbo.92La0.08Zro.95Ti0.05O3 film-on-foil capacitors, *J. Mater. Res.* **2009**, 24, 2993–2996. DOI: 10.1557/jmr.2009.0349.

238. Ma, B.; Kwon, D. K.; Narayanan, M.; (Balu) Balachandran, U. Fabrication of antifer-roelectric PLZT films on metal foils, *Mater. Res. Bull.* **2009**, 44, 11–14. DOI: 10.1016/j. materresbull.2008.09.006.

239. Tong, S.; Ma, B.; Narayanan, M.; Liu, S.; Koritala, R.; Balachandran, U.; Shi, D. Lead lanthanum zirconate titanate ceramic thin films for energy storage, *ACS Appl. Mater. Inter.* **2013**, 5, 1474–1480. DOI: 10.1021/am302985u.

240. Zhang, M. H.; Fulanović, L.; Egert, S.; Ding, H.; Groszewicz, P. B.; Kleebe, H. J.; Molina-Luna, L.; Koruza, J. Electric-field-induced antiferroelectric to ferroelectric phase transition in polycrystalline NaNbO3, *Acta Mater.* **2020**, 200, 127–135. DOI: 10.1016/j.actamat.2020.09.002.

241. Chen, J.; Feng, D. TEM study of phases and domains in NaNbO3 at room temperature, *Phys. Status Solidi A* **1988**, 109, 171–185. DOI: 10.1002/pssa.2211090117.

242. Saito, T.; Adachi, H.; Wada, T.; Adachi, H. Pulsed-laser deposition of ferroelectric NaNbO3 thin films, *Jap. J. Appl. Phys., Part 1* **2005**, 44, 6969–6972. DOI: 10.1143/ JJAP.44.6969.

243. Koruza, J.; Groszewicz, P.; Breitzke, H.; Buntkowsky, G.; Rojac, T.; Malič, B. Grain-size-induced ferroelectricity in NaNbO3, *Acta Mater.* **2017**, 126, 77–85. DOI: 10.1016/j.actamat.2016.12.049.

244. Antipin, M. Y.; Lindeman, S. V.; Struchkov, Y. T. Crystal structure of the electric-field-induced ferroelectric phase of NaNbO3, *Ferroelectrics* **1993**, 141, 307–311. DOI: 10.1080/00150199308223458.

245. Shimizu, H.; Guo, H.; Reyes-Lillo, S. E.; Mizuno, Y.; Rabe, K. M.; Randall, C. A. Lead-free antiferroelectric: XCaZrO3-(1 - X)NaNbO3 system (0 ≤ x ≤ 0.10), *Dalton Trans.* **2015**, 44, 10763–10772. DOI: 10.1039/c4dt03919j.

246. Guo, H.; Shimizu, H.; Mizuno, Y.; Randall, C. A. Strategy for stabilization of the antiferroelectric phase (Pbma) over the metastable ferroelectric phase (P2 1 ma) to establish double loop hysteresis in lead-free (1– x)NaNbO3 - x SrZrO3 solid solution, *J. Appl. Phys.* **2015**, 117, 214103. DOI: 10.1063/1.4921876.

247. Gao, L.; Guo, H.; Zhang, S.; Randall, C. A. A perovskite lead-free antiferroelectric xCaHfO3 -(1-x) NaNbO3 with induced double hysteresis loops at room temperature, *J. Appl. Phys.* **2016**, 120, 204102. DOI: 10.1063/1.4968790.

248. Gao, L.; Guo, H.; Zhang, S.; Randall, C. A. Stabilized antiferroelectricity in xBiScO3 - (1-x)NaNbO3 lead-free ceramics with established double hysteresis loops, *Appl. Phys. Lett.* **2018**, 112, 092905. DOI: 10.1063/1.5017697.

249. Zhou, M.; Liang, R.; Zhou, Z.; Dong, X. Superior energy storage properties and excellent stability of novel NaNbO3-based lead-free ceramics with A-site vacancy obtained: Via a Bi2O3 substitution strategy, *J. Mater. Chem. A* **2018**, 6, 17896–17904. DOI: 10.1039/c8ta07303a.

250. Ye, J.; Wang, G.; Zhou, M.; Liu, N.; Chen, X.; Li, S.; Cao, F.; Dong, X. Excellent comprehensive energy storage properties of novel lead-free NaNbO3-based ceramics for dielectric capacitor applications, *C* **2019**, 7, 5639–5645. DOI: 10.1039/c9tc01414d.

251. Dong, X.; Li, X.; Chen, X.; Chen, H.; Sun, C.; Shi, J.; Pang, F.; Zhou, H. High energy storage density and power density achieved simultaneously in NaNbO3-based lead-free ceramics via antiferroelectricity enhancement, *J. Materiomics* **2021**, 7, 629–639. DOI: 10.1016/j.jmat.2020.11.016.

252. Wei, T.; Liu, K.; Fan, P.; Lu, D.; Ye, B.; Zhou, C.; Yang, H.; Tan, H.; Salamon, D.; Nan, B. J. C. I. Novel NaNbO3–Sr0. 7Bi0·2TiO3 lead-free dielectric ceramics with excellent energy storage properties, *Ceramics International* **2021**, 47, 3713–3719.

253. Tan, H.; Yan, Z. L.; Chen, S. G.; Samart, C.; Takesue, N.; Salamon, D.; Liu, Y.; Zhang, H. B. SPS prepared NN-24SBT lead-free relaxor-antiferroelectric ceramics with ultrahigh energy-storage density and efficiency, *Scr. Mater.* **2022**, 210. DOI: 10.1016/j.scriptamat.2021.114428.

254. Jiang, J.; Meng, X. J.; Li, L.; Guo, S.; Huang, M.; Zhang, J.; Wang, J.; Hao, X. H.; Zhu, H. G.; Zhang, S. T. Ultrahigh energy storage density in lead-free relaxor antiferroelectric ceramics via domain engineering, *Energy Storage Mater.* **2021**, 43, 383–390. DOI: 10.1016/j.ensm.2021.09.018.

255. Fujii, I.; Shimasaki, T.; Nobe, T.; Adachi, H.; Wada, T. Effects of SrTiO3 substrate orientations on crystal and domain structures and electric properties of NaNbO3 –SrZrO3 films, *Jap. J. Appl. Phys.* **2018**, 57, 11UF13. DOI: 10.7567/JJAP.57.11UF13.

256. Beppu, K.; Shimasaki, T.; Fujii, I.; Imai, T.; Adachi, H.; Wada, T. Energy storage properties of antiferroelectric 0.92NaNbO3-0.08SrZrO3 film on (001)SrTiO3 substrate, *Phys. Lett. A* **2020**, 384, 126690. DOI: 10.1016/j.physleta.2020.126690.

257. Luo, B.; Dong, H.; Wang, D.; Jin, K. Large recoverable energy density with excellent thermal stability in Mn-modified NaNbO3-CaZrO3 lead-free thin films, *J. Am. Ceram. Soc.* **2018**, 101, 3460–3467. DOI: 10.1111/jace.15528.

258. Kania, A.; Kwapuliński, J. Ag1-xNaxNbO3 (ANN) solid solutions: From disordered antiferroelectric AgNbO3 to normal antiferroelectric NaNbO3, *J. Phys.: Condens. Matter* **1999**, 11, 8933–8946. DOI: 10.1088/0953-8984/11/45/316.

259. Wang, D.; Kako, T.; Ye, J. New series of solid-solution semiconductors (AgNbO3) 1-x(SrTiO3)x with modulated band structure and enhanced visible-light photocatalytic activity, *J. Phys. Chem. C* **2009**, 113, 3785–3792. DOI: 10.1021/jp807393a.

260. Fu, D.; Endo, M.; Taniguchi, H.; Taniyama, T.; Itoh, M. AgNbO3: A lead-free material with large polarization and electromechanical response, *Appl. Phys. Lett.* **2007**, 90, 252907. DOI: 10.1063/1.2751136.

261. Tian, Y.; Jin, L.; Zhang, H.; Xu, Z.; Wei, X.; Politova, E. D.; Stefanovich, S. Y.; Tarakina, N. V.; Abrahams, I.; Yan, H. High energy density in silver niobate ceramics, *J. Mater. Chem. A* **2016**, 4, 17279–17287. DOI: 10.1039/c6ta06353e.

262. Zhao, L.; Liu, Q.; Gao, J.; Zhang, S.; Li, J. F. Lead-free antiferroelectric silver niobate tantalate with high energy storage performance, *Adv. Mater.* **2017**, 29, 1701824. DOI: 10.1002/adma.201701824.

263. Luo, N.; Han, K.; Cabral, M. J.; Liao, X.; Zhang, S.; Liao, C.; Zhang, G.; Chen, X.; Feng, Q.; Li, J.-F.; Wei, Y. Constructing phase boundary in AgNbO3 antiferroelectrics: Pathway simultaneously achieving high energy density and efficiency, *Nat. Commun.* **2020**, 11, 4824. DOI: 10.1038/s41467-020-18665-5.

264. Yan, Z.; Zhang, D.; Zhou, X.; Qi, H.; Luo, H.; Zhou, K.; Abrahams, I.; Yan, H. Silver niobate based lead-free ceramics with high energy storage density, *J. Mater. Chem. A* **2019**, 7, 10702–10711. DOI: 10.1039/c9ta00995g.

265. Luo, N.; Han, K.; Zhuo, F.; Xu, C.; Zhang, G.; Liu, L.; Chen, X.; Hu, C.; Zhou, H.; Wei, Y. Aliovalent A-site engineered AgNbO3 lead-free antiferroelectric ceramics toward superior energy storage density, *J. Mater. Chem. A* **2019**, 7, 14118–14128. DOI: 10.1039/c9ta02053e.

266. Zhao, L.; Gao, J.; Liu, Q.; Zhang, S.; Li, J. F. Silver niobate lead-free antiferroelectric ceramics: Enhancing energy storage density by B-site doping, *ACS Appl. Mater. Interfaces* **2018**, 10, 819–826. DOI: 10.1021/acsami.7b17382.

267. Tian, Y.; Jin, L.; Zhang, H.; Xu, Z.; Wei, X.; Viola, G.; Abrahams, I.; Yan, H. Phase transitions in bismuth-modified silver niobate ceramics for high power energy storage, *J. Mater. Chem. A* **2017**, 5, 17525–17531. DOI: 10.1039/c7ta03821f.

268. Luo, N.; Han, K.; Zhuo, F.; Liu, L.; Chen, X.; Peng, B.; Wang, X.; Feng, Q.; Wei, Y. Design for high energy storage density and temperature-insensitive lead-free antiferroelectric ceramics, *J. Mater. Chem. C* **2019**, 7, 4999–5008. DOI: 10.1039/c8tc06549g.

269. Gao, J.; Zhang, Y.; Zhao, L.; Lee, K. Y.; Liu, Q.; Studer, A.; Hinterstein, M.; Zhang, S.; Li, J. F. Enhanced antiferroelectric phase stability in La-doped AgNbO3 : Perspectives from the microstructure to energy storage properties, *J. Mater. Chem. A* **2019**, 7, 2225–2232. DOI: 10.1039/c8ta09353a.

270. Han, K.; Luo, N.; Mao, S.; Zhuo, F.; Chen, X.; Liu, L.; Hu, C.; Zhou, H.; Wang, X.; Wei, Y. Realizing high low-electric-field energy storage performance in AgNbO3 ceramics by introducing relaxor behaviour, *J. Materiomics* **2019**, 5, 597–605. DOI: 10.1016/j.jmat.2019.07.006.

271. Wang, J.; Wan, X.; Rao, Y.; Zhao, L.; Zhu, K. Hydrothermal synthesized AgNbO3 powders: Leading to greatly improved electric breakdown strength in ceramics, *J. Eur. Ceram. Soc.* **2020**, 40, 5589–5596. DOI: 10.1016/j.jeurceramsoc.2020.06.031.

272. Lu, Z.; Bao, W.; Wang, G.; Sun, S. K.; Li, L.; Li, J.; Yang, H.; Ji, H.; Feteira, A.; Li, D.; Xu, F.; Kleppe, A. K.; Wang, D.; Liu, S. Y.; Reaney, I. M. Mechanism of enhanced energy storage density in AgNbO3-based lead-free antiferroelectrics, *Nano Energy* **2021**, 79, 105423. DOI: 10.1016/j.nanoen.2020.105423.

3 Organic Dielectric Materials for Capacitive Energy Storage

Haibo Zhang, Hua Tan, and Shuaikang Huang

1 INTRODUCTION

Ceramic dielectric materials exhibit ultrahigh permittivity, high modulus, and excellent high-temperature consistency.[1–3] However, their practical application is constrained by lower breakdown strength and flexibility. In contrast, polymer dielectric materials possess remarkable attributes such as low cost, high-rated voltage, high-energy storage efficiency, and easy processing, making them more suitable than inorganic materials for constructing large-scale, highly stable dielectric capacitors for energy storage.[4–13] Furthermore, polymer dielectrics demonstrate a unique self-healing behavior in the case of local dielectric breakdown over a large dielectric film. This ensures that in the event of a breakdown or failure, the film forms an open circuit with other components instead of a short circuit, effectively avoiding overall damage.[14, 15]

For instance, biaxially oriented polypropylene (BOPP) stands out as a commercially available dielectric material widely employed across various fields. Its popularity stems from its high breakdown strength, energy efficiency, excellent mechanical properties, and reliable self-healing capability.[15–17] Despite these advantages, the lower energy density (<4 J/cm^3) of traditional polymer dielectrics results in increased weight and volume of capacitors, thereby impeding further development. Notably, capacitors in automotive inverters constitute over 30% of the total volume and approximately 20% of the total weight.[18]

The prospect of achieving both high-energy density and high-power density in dielectric capacitors holds promise for effectively reducing the volume and weight of capacitors. This, in turn, could broaden the application of dielectric capacitors across various fields.

Over the past decades, researchers have dedicated substantial efforts to the study of polymer modification, aiming to enhance the dielectric properties of polymers and enable the production of low-loss, large-scale dielectric capacitors. Various approaches, including polymer blending, multilayer structures, polymer nanocomposites, and the surface deposition of inorganic materials, have been extensively explored. These investigations have resulted in improvements to the breakdown

DOI: 10.1201/9781003454496-3

strength and permittivity of materials, consequently elevating the energy density of the materials.[19–21]

Throughout this research, it has been observed that the dispersion of high permittivity inorganic fillers, such as $BaTiO_3$ and $SrTiO_3$, within the polymer system to create polymer-inorganic composite materials significantly enhances the material's permittivity.[22–26] However, challenges arise due to the distinct mechanical characteristics of the two phases in the polymer-inorganic composite. Uneven stress at the interface during material preparation and post-processing can lead to cracks, jeopardizing the capacitor's safety. Additionally, achieving a stable dispersion of inorganic particles in the matrix proves difficult, often resulting in particle agglomeration, a phenomenon that substantially decreases breakdown strength and compromises capacitor uniformity. Microphase delamination at the interface further accelerates breakdown, limiting the potential increase in energy density. These challenges have hindered the swift application of polymer-inorganic composite films in large-scale commercial production.[27–29]

Despite these obstacles, all-organic intrinsic and modified polymer dielectrics remain the most suitable materials for dielectric energy storage. Their unmatched commercial and research value ensures their continued dominance in the dielectric capacitor energy storage industry.

This section provides a comprehensive overview of intrinsic dielectric polymers and presents recent advancements in the research of all-organic polymer dielectric materials. In Section 2, we delve into the realm of traditional intrinsic dielectric polymers, exploring their dielectric characteristics and the various modification processes associated with different types of intrinsic dielectric polymers. Section 3 then consolidates the recent strides made in all-organic polymer dielectrics. The modified polymers are categorized based on different modification mechanisms, encompassing molecular chain modification, cross-linked polymers, blended polymers, and composite polymers. The corresponding preparation methods and energy storage properties of each category are thoroughly elucidated. Additionally, this section introduces innovative strategies for optimizing polymer properties through structural design and surface modification.

2 INTRINSIC DIELECTRIC POLYMERS

In the evolving landscape of dielectric capacitors, the significance of dielectric materials with outstanding properties has been steadily increasing. Through extensive research, a series of polymer dielectrics exhibiting high-energy storage properties has been identified. Illustrated in Figure 3.1, dielectric materials can be categorized into linear dielectrics and nonlinear dielectrics based on the variations in polarization with the electric field in the hysteresis loop.[30, 31] Linear dielectrics lack ferroelectric domains and permanent dipoles, and their ε_r remains independent of the electric field magnitude. On the other hand, nonlinear dielectrics can be further classified into normal ferroelectric, relaxor ferroelectric, and antiferroelectric categories.[32, 33] The presence of distinct ferroelectric domains in these polymer dielectrics results in diverse polarization behaviors, with ε_r being influenced by the applied electric field.

2.1 LINEAR DIELECTRIC POLYMERS

Capitalizing on their high breakdown strength (E_b), high-energy efficiency (η), cost-effectiveness, and stability, linear dielectric polymers play a crucial role in power systems, high-temperature energy storage, and various other applications.[30, 31] As depicted in Figure 3.1, the electric displacement of linear dielectric polymers exhibits linearity concerning the electric field. While linear dielectric polymers generally possess smaller relative permittivity (ε_r) compared to ferroelectric polymers with higher polarization, their high-energy efficiency (η) and low tan δ contribute to notable energy storage properties. Common commercially available linear dielectric polymer films include BOPP, poly(arylene ether urea) (PEEU), polyethylene terephthalate (PET), polystyrene (PS), polycarbonate (PC), and polymethyl methacrylate (PMMA).[35–38] Additionally, linear dielectric polymers such as polyetherimide (PEI) and polyimide (PI) find applications in high-temperature energy storage due to their excellent performance under elevated temperatures.

BOPP, derived from the polymerization of propylene monomers, is a thermoplastic dielectric polymer. Prepared through a melt extrusion and biaxial stretching process, BOPP films exhibit characteristics such as high breakdown strength (>700 MV/m), low tan δ (<0.02%), cost-effectiveness, and good self-healing ability under an electric field. Consequently, BOPP is widely used as one of the most common commercial dielectrics.[16, 31, 36, 39] However, its low relative permittivity (~2.2) results in a limited energy density (U_d of 3–6 J/cm³) at operating voltage. Moreover, the operating temperature of BOPP is typically below 105°C. Operating in environments exceeding this temperature threshold leads to a significant drop in resistivity, causing a surge in leakage current and a subsequent decrease in U_d and η.[32] For instance, when BOPP capacitors are employed in hybrid electric vehicles, a cooling

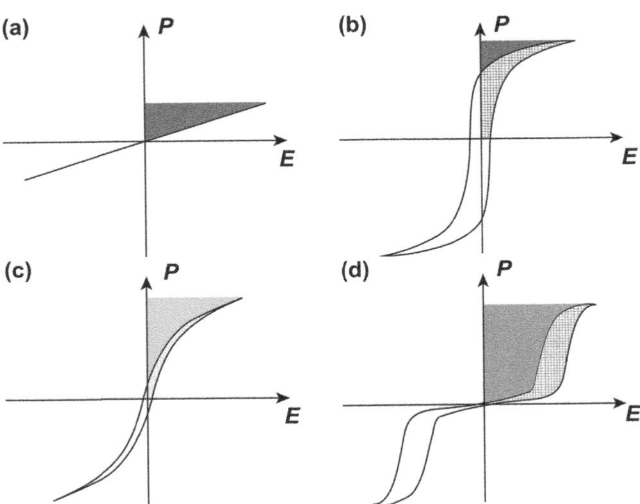

FIGURE 3.1 Hysteresis loops for (a) linear dielectric, (b) normal ferroelectrics, (c) relaxor ferroelectrics, and (d) antiferroelectrics.[34]

system is essential to mitigate conduction loss-induced heating, thereby addressing a major limitation in the energy storage application of BOPP.

Polymethyl methacrylate (PMMA) stands out as a linear dielectric polymer with favorable characteristics. In comparison to BOPP, PMMA exhibits a larger tan δ (<5%) but boasts a higher relative permittivity (ε_r, ~3.5) and a higher maximum operating temperature. These features render PMMA more promising for certain applications.[40, 41] PMMA possesses a high Young's modulus, effectively enhancing its electromechanical breakdown strength (E_b). Due to its compatibility with PVDF-based ferroelectric polymers, PMMA is frequently employed as a high-E_b phase in multiphase blends and as an insulating layer in multilayer structures. The combination of high-insulating PMMA and high-polarity ferroelectric polymers contributes to a significant reduction in the remnant polarization of the ferroelectric phase, leading to the realization of all-organic dielectric polymers with both high U_d and high η.[42]

Beyond homopolymer-type linear dielectrics, several condensation polymers and copolymers exhibit commendable energy storage properties. Polyetherimide (PEI), an amorphous dielectric polymer widely used in high-temperature dielectric capacitors, distinguishes itself with a high glass transition temperature (T_g, 217°C–247°C). This high T_g imparts excellent thermal stability and high-temperature performance to PEI, setting it apart from common dielectric polymers with lower T_g values.[37, 43] PEI, derived from polyimide (PI) modification, incorporates flexible ether bonds for improved processing properties. The material's ability to maintain high-energy storage performance at elevated temperatures is evident, with PEI achieving up to 90% energy efficiency (η) at 150°C and an electric field of 200 MV/m. Even at 200°C, PEI still maintains an impressive η of nearly 80%.[43] This superior performance in high-temperature environments positions PEI as a promising alternative to PI for energy storage applications (Figure 3.2).

Wu et al.[45] employed N-methyl-2-pyrrolidone (NMP) as a solvent and p-toluenesulfonic acid (p-TSOH) as a catalyst in the condensation of 4,4-diphenylmethanediamine (MDA) with thiourea under microwave assistance to produce an aromatic polythiourea (ArPTU), a high-performance linear dielectric polymer. The thiourea unit in ArPTU, characterized by a large dipole moment (4.89 D), results in a substantial relative permittivity (ε_r) of 4.5. ArPTU exhibits good frequency

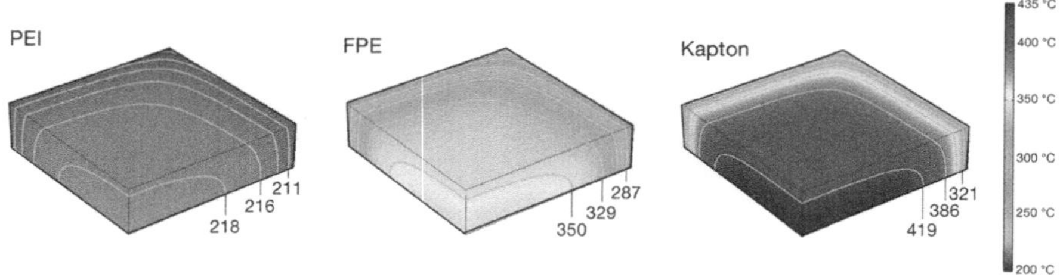

FIGURE 3.2 Internal temperature simulations of three high-temperature dielectric polymers operating continuously at high temperature (200°C) and steady-state electric field (200 MV/m).[33]

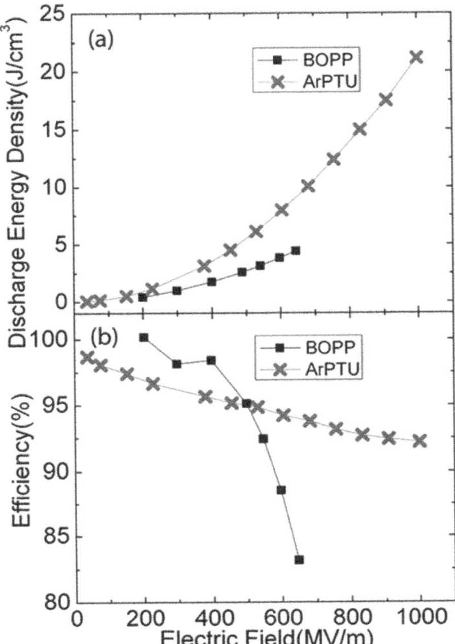

FIGURE 3.3   (a) Discharge energy density and (b) efficiency of ArPTU compared with BOPP.[45]

stability and a low tan δ, registering a value below 0.01 when the frequency is below 100 kHz. Additionally, ArPTU demonstrates excellent temperature stability, with no significant alterations in dielectric properties observed below 150°C. The amorphous, long-range disordered structure of ArPTU films, along with randomly oriented dipoles and Coulombic interactions, enhances carrier scattering, reducing leakage current in the dielectric and increasing breakdown strength (E_b). ArPTU's polar groups function as traps, restricting carrier movement and mitigating the risk of conduction and breakdown. Consequently, ArPTU demonstrates an ultrahigh breakdown strength (1 GV/m). In comparison with the conventional linear dielectric BOPP, ArPTU exhibits lower tan δ, higher E_b, and outstanding energy density (U_d) under a high electric field. Illustrated in Figure 3.3, ArPTU achieves a U_d of 22 J/cm³ at an electric field of 1 GV/m while maintaining an efficiency (η) higher than 90%.

In copolymer research, Wen et al.[46] synthesized poly(acrylonitrile butadiene styrene) (ABS) copolymers. The bulky aromatic groups on the side chains of ABS hinder chain rotation, resulting in high η, while the highly polar cyano groups contribute to an increased ε_r. The solution-cast ABS film achieves a U_d of 7.3 J/cm³ at 525 MV/m. Similarly, poly(styrene acrylonitrile) (PSAN) copolymers exhibit a larger electric displacement due to the presence of cyano groups.[47] The benzene ring group impedes the aggregation of acrylonitrile (AN) units, increasing the disorder of polar groups and reducing remnant polarization. Consequently, PSAN attains a high U_d (6.8 J/cm³) at 400 MV/m while maintaining a high η of 90%.

Dipole glasses and paraelectrics, distinct from linear dielectrics, are unique dielectric materials that lack ferroelectric domains. Dipole glasses feature isolated dipoles, while paraelectrics have interacting dipoles. This distinction allows them to exhibit elongated hysteresis loops while maintaining a substantial polarization.[48] However, dipole glasses and paraelectrics are more commonly observed in inorganic dielectrics and pose challenges in realization within polymer materials. Recent advancements have been made by researchers who achieved dipole glass polymers by incorporating polar groups, such as hydroxyl, carbonyl, and sulfonyl groups into polymer chain segments.[48, 49] For instance, Zhang et al.[50] synthesized sulfonated poly(2,6-dimethyl-1,4-phenylene oxide) (SO2-PPO). The inclusion of the polar methyl-sulfonyl group, characterized by a large dipole moment (4.25 D), contributes to an impressive relative permittivity (ε_r) of 8.8. The rigid structure of PPO enhances breakdown strength (E_b) and increases free volume to facilitate the rotation of polar groups, thereby reducing dielectric loss. The tan δ of SO2-PPO is remarkably low at 0.003, and the energy density (U_d) can reach 24 J/cm^3.

In the case of paraelectric behavior, it can be observed in stretch-treated and high-frequency poly(vinylidene fluoride-co-trifluoroethylene) (PVDF-TrFE) films, such as PVDF-TrFE 50/50 mol%, which is associated with the formation of nanostructures. Further details on this phenomenon will be elaborated in the subsequent section.[51, 52]

2.2 NONLINEAR DIELECTRIC POLYMERS

Nonlinear dielectric polymers encompass normal ferroelectrics, relaxor ferroelectrics, and antiferroelectrics. The presence of spontaneously polarized dipoles in these polymers enhances dielectric polarization, resulting in a larger relative permittivity (ε_r) compared to linear dielectrics. The spontaneously polarized dipoles align in the same direction, forming ferroelectric domains. However, these domains exhibit low energy efficiency (η) due to relaxation loss when subjected to an applied electric field.[53] In contrast, relaxor ferroelectrics and antiferroelectrics possess characteristic ferroelectric domains, featuring lower remanent polarization and superior energy storage performance compared to normal ferroelectric materials.[54, 55]

Polyvinylidene fluoride (PVDF) and its derivatives stand out as typical nonlinear dielectric polymers with distinctive dielectric properties. Since the beginning of the 21st century, research in dielectric energy storage has flourished, and PVDF-based dielectric polymers have garnered widespread attention due to their favorable energy storage properties. The following section describes the research progress of PVDF-based nonlinear dielectric polymers.

2.2.1 PVDF Homopolymers

Polyvinylidene fluoride (PVDF) stands out as one of the most widely used ferroelectric polymers, boasting excellent chemical stability, resistance to weathering, and unique dielectric, piezoelectric, and pyroelectric properties. PVDF, with the chemical formula $-(CH_2-CF_2)_n-$, is a flexible polymer synthesized through the copolymerization of vinylidene fluoride (VDF). VDF comprises 3 wt% hydrogen atoms and 59.4 wt% fluorine atoms.[56] The C–F bond formed due to the electronegativity

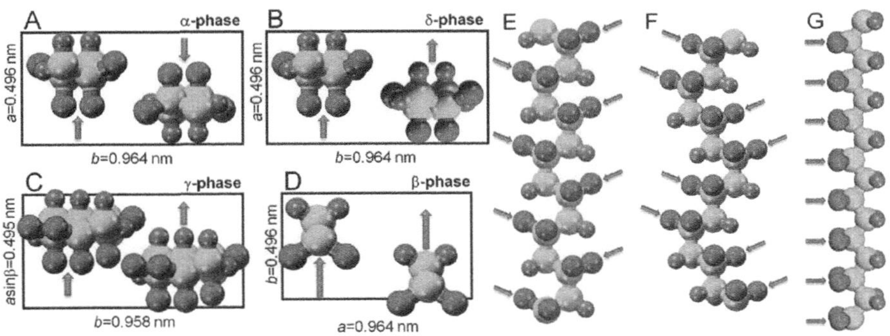

FIGURE 3.4 Unit cells of (A) α, (B) δ, (C) γ, and (D) β forms of PVDF crystals viewed along the c-axes and schematic chain conformations for (E) TGTG′ (α/δ), (F) TTTGTTTG′ (γ), and (G) all-trans (β) rotational sequences. Red, cyan, and blue spheres represent F, C, and H atoms. The projections of dipole directions are indicated by green arrows.[57]

difference between hydrogen and fluorine has a larger dipole moment (1.92 D), resulting in PVDF having a higher relative permittivity (ε_r, 6–12) and energy density (U_d) compared to general polymers. PVDF polymers also exhibit a high DC breakdown strength (E_b) of 700 MV/m. The preparation process allows for the production of PVDF films with varying crystallinity (50%–70%).[57]

Covalent bonds between adjacent CH_2 and CF_2 units in PVDF tend to form stable gauche (G) and trans (T) linkages with bond angles of ±60° or 180°. PVDF exhibits four different crystal forms, as depicted in Figure 3.4. The α-phase and δ-phase share the same 3D dimensions with a chain conformation of trans-gauche-trans-gauche′ (TGTG′). The β-phase adopts an all-trans (TTTT) chain conformation, while the chain conformation of the γ-phase is trans-trans-trans-gauche-trans-trans-gauche′ (TTTGTTTG′). Among these phases, α, γ, and δ exhibit a trans–gauche conformation. The T–G conformation has a dipole moment parallel to the chain axis of 1.02 D and perpendicular to the chain axis of 1.20 D. The different orientations of the polar CF_2 groups in the T–G conformation result in relatively low polarization, with the α-phase exhibiting a macroscopic net dipole moment of zero due to the cancellation of dipole moments. In contrast, the β-phase displays the highest polarization. In the all-trans conformation, the polar groups align in the same direction, resulting in a large dipole moment (2.10 D) perpendicular to the chain axis.[57]

Gong et al [58] reviewed the controlled/living radical polymerization (CRP) and related techniques to tailoring the chemical and physical properties of materials. Li et al.[59] conducted a study on the dielectric properties of three crystalline forms of PVDF—α-phase, β-phase, and γ-phase—under DC and AC electric fields, as illustrated in Figure 3.5. The β-phase, with its large polarization and remnant polarization, exhibits typical ferroelectric D–E loops. In contrast, the α-phase and γ-phase show relatively similar dielectric properties. The dipole moments of the TGTG′ conformation of the α-phase cancel each other out, resulting in less polarity than the TTTG conformation of the γ-phase. Consequently, the α-phase displays smaller

 Dielectric Materials for Capacitive Energy Storage

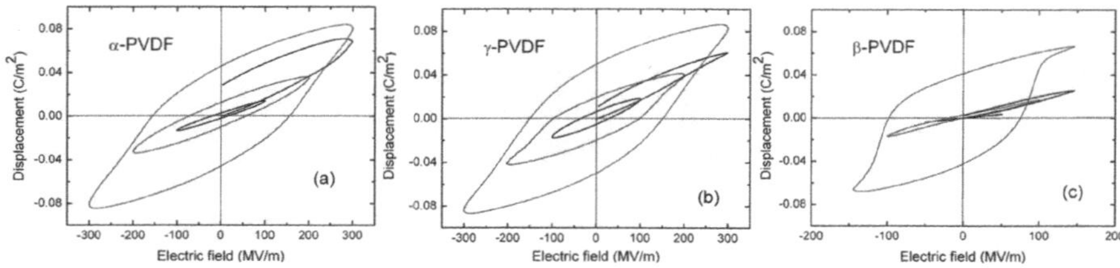

FIGURE 3.5 D–E hysteresis loops for (a) α-PVDF; (b) γ-PVDF; (c) β-PVDF.[59]

maximum and remnant polarization in the D–E loops. Despite different conformations, the total dipole moment remains consistent for various crystalline forms in PVDF.

The D–E loops measured by PVDF under DC and AC electric fields differ significantly due to the presence of irreversible polarization in the dielectric. The β-phase exhibits an elongated D–E loop pattern resembling a linear dielectric at low electric fields. At around 150 MV/m, the polarization increases rapidly and saturates, attributed to the polarization of small-sized dipoles in β-PVDF at low electric fields. These dipoles, having a small content and large free volume, can orient in the presence of an electric field, with the resulting polarization being reversible. Beyond the coercive electric field, the larger-sized dipoles move directionally, producing a larger polarization. However, due to the small free volume of large-sized dipoles, orienting them freely after removing the electric field is challenging. The irreversible polarization rate of β-PVDF is notably high, exceeding 80% of the remnant polarization.

The γ-phase crystals, being larger with fewer amorphous/crystalline interfaces, experience a slower orientation of crystal domains, resulting in a larger irreversible polarization. At an electric field of 300 MV/m, the irreversible polarization of the γ-phase constitutes 60% of the remnant polarization, while the α-phase accounts for only 30%. Li et al.[60] reported on the energy storage properties of the three crystalline phases (α, β, and γ). They observed similar energy storage properties at low electric fields, with the β-phase saturating at 150 MV/m and achieving a maximum U_d of 1.5 J/cm³. The U_d of the α-phase slightly surpasses that of the γ-phase when the electric field is below 250 MV/m. However, as the electric field approaches 300 MV/m, the growth rate of U_d for the α-phase significantly slows down. Beyond 300 MV/m, the D–E loop of the α-phase exhibits a clear trend of transitioning to the γ-phase. This two-phase transition is further confirmed through X-ray diffraction (XRD) characterization. The α-phase displays significant energy loss at 150–350 MV/m, indicating that part of the energy is utilized for the transition from the α-phase TGTG' conformation to the γ-phase TTTG' conformation. Once the transition is complete, the loss curves of the α-phase and γ-phase align well. Among the three crystalline phases of PVDF, only the γ-phase can persist at high electric fields (500 MV/m), demonstrating the highest U_d of 14 J/cm³.

The choice of different preparation processes allows the attainment of PVDF with various crystalline phases. The lowest-energy α-phase can be obtained by direct cooling at a normal speed after melting PVDF. The α-phase can be polarized to achieve the δ-phase at an electric field of 100–200 MV/m. On the other hand, the β-phase and γ-phase can be obtained at high temperatures or high electric fields. High-temperature annealing operations promote the rearrangement of chain segment movement, facilitating the transformation of various crystalline phases.[61] The selection of highly polar solvents also influences the formation of crystalline phases. For instance, α-phase and β-phase are more likely to be generated in cyclohexanone solvent, while α- or γ-phase is favored when using *N*, *N*-Dimethylformamide (DMF) solvent.[62]

The β-phase, characterized by high polarization, is considered an ideal energy storage material. Mechanical stretching of the α-phase or δ-phase, rapid cooling and crystallization of PVDF, and the use of specific nucleating agents or fillers are methods to prepare the β-phase.[63] Ren et al.[64] introduced a novel Press & Folding (P&F) technique to enhance the β-phase. In this approach, PVDF powder is initially hot-pressed and water-cooled to prepare the PVDF film, and a PTFE mold release agent is sprayed on the folded surface to promote the peeling of the single-layer film. The PVDF is then folded, and the P&F cycle is initiated at a specific temperature and pressure. The D–E loops obtained with varying numbers of P&F cycles show a substantial increase in PVDF polarization with an increasing number of cycles, linked to the content of β-phase in PVDF. This is confirmed by Infrared and Raman spectroscopy, where, after six P&F cycles, the β-phase content in PVDF increases from 8% to 98%. The reduction of remnant polarization and the relaxor ferroelectric behavior at a high electric field are associated with the internal stress generated during pressing, which increases the molecular chain spacing of the β-phase, leading to the presence of residual stresses within the material. These residual stresses facilitate domain flipping when the electric field changes, forming reversible highly polarized nanostructures. Finally, after 6 P&F cycles at a temperature of 140°C, the prepared PVDF film achieved an ultrahigh U_d of 39.8 J/cm^3 at 880 MV/m, as illustrated in Figure 3.6(d). Furukawa et al.[65] observed transient antiferroelectric behavior in β-PVDF at low temperatures (−60°C), attributed to the depolarization field causing dipole depolarization reorientation. Besides the control of preparation conditions and physical treatment, it is also possible to modulate the content of each phase in PVDF through multicomponent copolymerization to achieve different dielectric properties.

2.2.2 Binary Copolymers

In pure PVDF, the β-phase exhibits large polarization and high ε_r, but the saturation of polarization at lower electric fields limits the increase in its U_d. Incorporating other monomers into PVDF to form PVDF-based copolymers can prevent premature saturation of electric displacement at low electric fields. Modulating the content of copolymer groups can alter the phase content in PVDF, increase the maximum electric displacement, and reduce the remnant polarization of the dielectric. Additionally, copolymers can achieve relaxor ferroelectrics and antiferroelectrics with different dielectric properties compared to conventional ferroelectrics.

Poly(vinylidene fluoride-hexafluoropropylene) (P(VDF-HFP)) is formed by copolymerizing PVDF with amorphous hexafluoropropylene (HFP). The mechanical properties of the copolymer are significantly influenced by the HFP content. A low HFP content (5–15 mol%) imparts good flexibility to the copolymer, while an

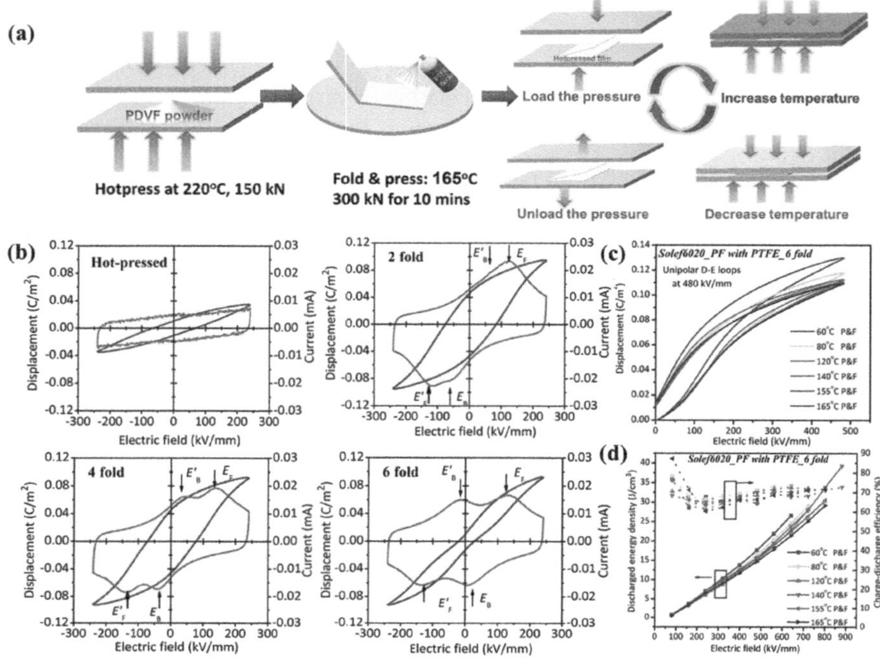

FIGURE 3.6 (a) Schematic demonstration of the P&F process of spraying PTFE release agent on the surface. (b) D–E and current-electric field (*I-E*) loops of hot-pressed PVDF film and P&F PVDF films with different P&F cycles. (c) D–E loops of 6-fold films for P&F operation at different temperatures. (d) Discharged energy density and charge-discharge efficiency of P&F films.[64]

HFP content exceeding 20 mol% results in high elasticity.[66] The ferroelectricity of P(VDF-HFP) is tied to the HFP content, with increased HFP inhibiting the formation of large-sized grains. The presence of CF_3 side groups reduces material crystallinity, decreases remnant polarization, and imparts weak ferroelectric properties.[67, 68] Ferroelectricity in P(VDF-HFP) is also affected by preparation conditions. For example, films prepared by solution casting at room temperature or melt-pressed & quenched exhibit distinct ferroelectric D–E loops, while slow-cooled samples exhibit relaxor ferroelectric properties.[69] Optimizing film processing conditions has led to achieving an U_d of over 25 J/cm³ at room temperature under 700 MV/m in P(VDF-HFP) (95.5/4.5 mol%).[70] Stretching treatments of P(VDF-HFP) films facilitate the transformation of the internal crystalline phase from the nonpolar α-phase to the β-phase. Investigations on different preparation methods and stretching effects on the energy storage properties of P(VDF-HFP) films have shown that solution-cast & stretched samples have the highest U_s and U_d (23 J/cm³ and 13.5 J/cm³, respectively), followed by hot-pressed & stretched samples, while solution-cast samples have the lowest U_d.[71] Despite substantial losses, P(VDF-HFP) exhibits a U_d of 11–13.5 J/cm³ at 600 MV/m (Figure 3.7).

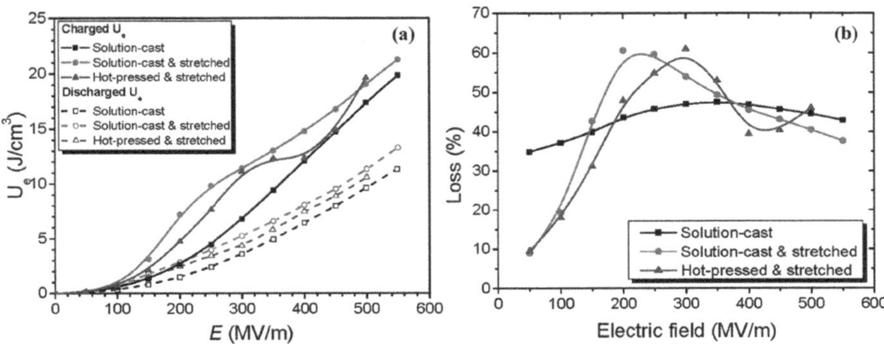

FIGURE 3.7   (a) Charge/discharge energy density and (b) loss of solution-cast, solution-cast & stretched, hot-pressed & stretched samples.[71]

Poly(vinylidene fluoride-chlorotrifluoroethylene) (P(VDF-CTFE)) is produced by incorporating chlorotrifluoroethylene (CTFE) into PVDF. The glass transition temperature (T_g) of P(VDF-CTFE) can vary between $-40°C$ and $45°C$.[72] P(VDF-CTFE) displays an amorphous state at higher CTFE content. A study by Chu et al.[73] on P(VDF-CTFE) with a 91/9 mol% composition achieved a high U_d of 17 J/cm^3 through uniaxial stretching at 575 MV/m. The larger size of CTFE compared to VDF increases chain spacing, stabilizes the TGTG' conformation in the α-phase, and prevents phase transition to the γ-phase at high electric fields. Subsequent research by Zhou et al.[74] further investigated P(VDF-CTFE)'s dielectric properties, revealing a large ε_r of 13 and tan δ of 0.03 at 1 kHz. With optimized film preparation processes and polymer film quality, P(VDF-CTFE) achieved a high-Eb of >700 MV/m at a 500% stretching ratio. P(VDF-CTFE) films exhibit relaxor ferroelectric behavior, achieving a U_d of 25 J/cm^3 at 600 MV/m.

To increase the β-phase in PVDF, trifluoroethylene (TrFE) monomers are added for copolymerization, producing poly(vinylidene fluoride-trifluoroethylene) (P(VDF-TrFE)) copolymers.[75] The TrFE unit destabilizes the TGTG' conformation and enhances the crystal stability of the β phase. P(VDF-TrFE) exhibits stronger ferroelectric properties than PVDF, with an ε_r of about 18 and a remnant polarization of up to 9 μC/cm^2 for P(VDF-TrFE) at 75/25 mol%.[78] The high β-phase content results in copolymers with large remnant polarization and lower energy storage properties (1.13 J/cm^3 at 130 MV/m). Copolymers with varying α, β, and γ phases can be obtained by adjusting TrFE content (20%–50%).[76, 77] Besides direct radical copolymerization, P(VDF-TrFE) can be prepared through catalytic hydrogenation of P(VDF-CTFE) or reduction of P(VDF-CTFE-TrFE), where the interchain linkage generated is a head-to-head linkage. Hydrogenation of P(VDF-CTFE) with 6, 9, 12, and 20 mol% TrFE content resulted in enhanced ferroelectricity with increased maximum electric displacement and remnant polarization.[79] Studies by Su et al.[52] demonstrated diverse dielectric behaviors of P(VDF-TrFE) based on temperature, polarization frequency, and applied electric field, including normal ferroelectric, antiferroelectric-like, and paraelectric properties under varying

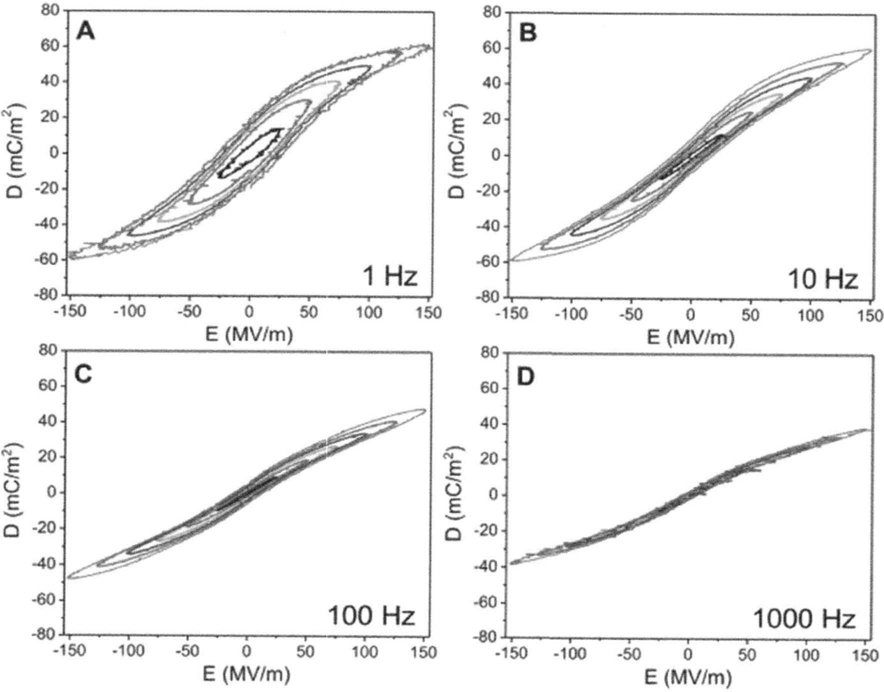

FIGURE 3.8   D-E hysteresis loops of P(VDF-TrFE) at 100°C under different frequencies.[52]

conditions. In Figure 3.8(a), it is evident that P(VDF-TrFE) (50/50 mol%) at 100°C behaves as a typical ferroelectric when the frequency is 1 Hz. As the frequency increases, there is a suppression of ferroelectric domain growth in the copolymer, resulting in reduced polarization loss and the manifestation of distinct D-E loops (Figure 3.8(b–d)). The antiferroelectric-like behavior observed at 10 Hz and 100°C might be attributed to the equilibrium state between depolarization and polarization fields during reverse polarization. At 1 kHz and 100°C, paraelectric properties are evident, possibly explained by the electric field-induced nucleation of nanodomains within the matrix.

High-energy electron irradiation enables operations such as polymer chain breakage, conformational rearrangement, cross-linking, and other modifications.[55, 80–82] Zhang et al.[83] reported that P(VDF-TrFE) exhibited relaxor ferroelectric behavior under high-energy electron irradiation. They irradiated P(VDF-TrFE) (50/50 mol%) in a nitrogen atmosphere with 3-MeV electrons at doses ranging from 4×10^5 to 10^6 Gy (1 Gy = 100 rads). At a dose of 80 Mrad, the ε_r of P(VDF-TrFE) measured was 28, approximately 50% higher than without high-energy irradiation. High-energy electron irradiation induces the attachment of short side groups (–CF3–CH<) to the main chain of P(VDF-TrFE), resulting in a chemical pinning of the chain structure. This pinning, coupled with increased chain spacing, reduces the size of the crystal region in the all-trans conformation and minimizes polarization hysteresis. The crystalline region obtained through irradiation comprises trans-nano-microregions interrupted by trans (T) and gauche (G) chain conformations, leading to the transition of P(VDF-TrFE) from ferroelectrics to relaxor ferroelectrics, as depicted in Figure 3.9(c).

While the binary copolymer of P(VDF-TrFE) shows significantly improved U_d compared to PVDF homopolymer, its strong ferroelectricity results in high energy loss. Introducing relaxor ferroelectric behavior in PVDF-based polymer dielectrics through the pinning effect aims to reduce energy loss effectively. The pinning effect induced by electron beam irradiation produces P(VDF-TrFE) with single hysteresis loop (SHL) behavior, which is more conducive to high-energy storage research than the double hysteresis loop (DHL) observed in relaxor ferroelectrics[84–86]. However, challenges in achieving high-dose and high-energy electron beams in a high-temperature industrial environment, film damage, and a notable increase in carriers causing elevated leakage current and reduced E_b limit the universal application of this method.[87, 88] To address these challenges, researchers are focusing on pinning through ternary copolymerization by introducing bulky monomers to avoid adverse effects from electron irradiation.

2.2.3 Ternary Copolymers

The incorporation of a bulky comonomer in ternary copolymerization achieves physical pinning and increases interchain spacing. This pinning effect leads to the reduction of ferroelectric domain size to the nanometer scale, promoting dipole flipping and diminishing coupling. The pinning effect is crucial for achieving low-hysteresis relaxor ferroelectric behavior, enhancing the efficiency of energy storage in dielectrics. To realize pinning, a copolymer monomer with a larger volume than TrFE is added to P(VDF-TrFE), such as Chlorofluoroethylene (CFE), CTFE, and HFP monomers.[89, 90] It's important to note that direct copolymerization of PVDF with these monomers does not result in chain spacing expansion and pinning. This is due to the bulk effect between VDF and bulky monomers that cannot form a continuous copolymer conformation. For instance, P(VDF-CTFE) does not exhibit chain spacing expansion or pinning, making it challenging to achieve relaxor ferroelectric behavior.[91, 92] The TrFE monomer is crucial for pre-extending the interchain spacing, facilitating the presence of the third comonomer in the crystal structure.

Poly(vinylidene fluoride-trifluoroethylene-chlorofluoroethylene) (P(VDF-TrFE-CFE)) is typically prepared by the suspension polymerization of three monomers.[93, 94] P(VDF-TrFE-CFE) demonstrates distinct relaxor ferroelectric behavior. The addition of CFE introduces physical pinning, forming nanodomains, and reducing frictional loss during dipole movement. Yang et al.[95] investigated P(VDF-TrFE-CFE) with a molar ratio of 59.2/33.6/7.2 and found that the addition of CFE increased the intergranular spacing from $d110/200 = 0.442$ nm in P(VDF-TrFE) to 0.484 nm (15 ≈ 13 > 12). Although the addition of CFE increases chain spacing, its physical pinning effect is weak due to insufficient chain spacing expansion and the large dipole moment (~1.8 D) of CFE. At high electric fields, weakly pinned CFE transforms to the all-trans conformation. P(VDF-TrFE-CFE) exhibits reversible transitions between relaxor ferroelectric and ferroelectric conformations under changing electric fields, resulting in DHL behavior, as depicted in Figure 3.9(a). The relaxor ferroelectric structure of

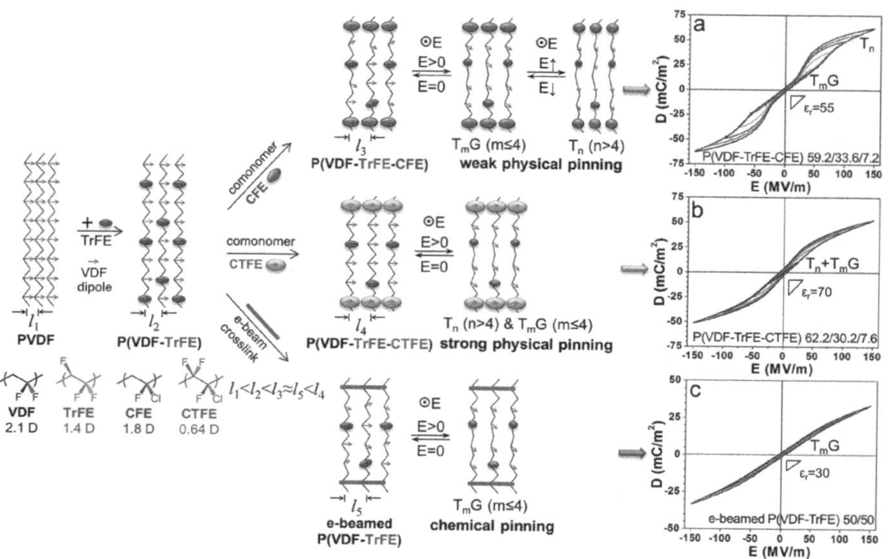

FIGURE 3.9 Schematic of the nanostructures of P(VDF-TrFE)-based ternary copolymers and electron beam irradiation P (VDF-TrFE). (a) P(VDF-TrFE-CFE) (59.2/33.6/7.2 mol%); (b) P(VDF-TrFE-CTFE) (62.2/30.2/7.6 mol%), and (c) electron beam irradiation P(VDF-TrFE) (50/50 mol%) bipolar D–E hysteresis loops at 10 Hz.[90]

P(VDF-TrFE-CFE) primarily comprises TmG ($m \leq 4$) conformations, while the ferroelectric structure consists of long T_n ($n > 4$) conformations.[96] Li et al.[97] prepared P(VDF-TrFE-CFE) at a molar ratio of 56/36.5/7.5 using suspension polymerization. At frequencies below 1 kHz, it exhibits ε_r greater than 50 and tan δ less than 0.05. Chu et al.[73] reported the energy storage performance of P(VDF-TrFE-CFE) with a molar ratio of 58.3/34.2/7.5, achieving a U_d of about 9.5 J/cm³ at 425 MV/m.

To enhance physical pinning, Chlorotrifluoroethylene (CTFE) was introduced to increase chain spacing. Poly(vinylidene fluoride-trifluoroethylene-chlorotrifluoroethylene) (P(VDF-TrFE-CTFE)) can be prepared through direct radical polymerization or by reducing some chlorine atoms in P(VDF-CTFE).[98–100] CTFE, being larger than CFE and having a lower dipole moment (~0.64 D), imparts strong physical pinning. Yang et al.[101] reported a P(VDF-TrFE-CTFE) terpolymer (62.2/30.2/7.6 mol%). This terpolymer, with a larger spacing of d110/200 = 0.490 nm (l4 > l5 ≈ l3), exhibits a narrow SHL relaxor behavior similar to irradiated P(VDF-TrFE), as illustrated in Figure 3.9(b). Most of the relaxor ferroelectric structures in this terpolymer consist of random TmG or TmS ($m \leq 4$) sequences (S stands for skew linkage conformation) and some long T_n ($n > 4$) conformations. Chung et al.[100] studied the properties of P(VDF-TrFE-CTFE) with different molar ratios and found that increasing the amount of VDF, while keeping the VDF/TrFE molar ratio constant, reduces the crystallinity and crystal size of the copolymer.[96] CTFE is unable to co-crystallize in the VDF or VDF-TrFE crystal region, splitting large-sized crystal regions and reducing domain wall area and crystal size. The VDF

TABLE 3.1

Dielectric and energy storage parameters of intrinsic dielectric polymers

Category	Intrinsic Polymers	ε_r @1 kHz	tan δ @1 kHz	E_b (MV/m)	U_d (J/cm³)	η (%)	Ref.
Linear dielectric polymers	PP	~2.2	<0.0002	450–560	3–6	>90	[36]
	PMMA	~3.5	–	500	~4	~85	[41]
	PEI	~3.25	~0.02	550	4	>90	[37]
	PI	~3.2	~0.02	600	~5	~80	[44]
	ArPTU	~4.5	0.06	1,000	22	~92.5	[45]
	PEEU	~4.8	~0.015	~600	~8	~90	[106]
	PET	3.3	<0.005	570	1–1.5	–	[35]
	PC	~2.9	~0.002	–	4.39	94.38	[38]
Nonlinear dielectric polymers	α-PVDF	~13.5	~0.03	331.8	6.8	46.6	[59]
	β-PVDF	~13	~0.03	306	5.6	65.9	[59]
	γ-PVDF	~9.5	~0.03	400.7	9.5	62.5	[59]
	P&F-PVDF	~13	~0.03	880	39.8	~72	[64]
	P(VDF-HFP) (96/4 mol%)	~9.5	–	~550	13.5	58.7	[71]
	P(VDF-CTFE) (91/9 mol%)	13	0.03	600	25	–	[72]
	P(VDF-TrFE) (80/20 mol%)	~16	~0.03	300	–	–	[79]
	Irradiated P(VDF-TrFE) (50/50 mol%)	28	0.06	–	–	–	[83]
	P(VDF-TrFE-CFE) (58.3/34.2/7.5 mol%)	>50	–	425	9.5	–	[73]
	P(VDF-TrFE-CTFE) (65.6/26.7/7.7 mol%)	~40	–	~500	~13	~62	[104]
	P(VDF-TrFE-HFP) (62/38/2.5 mol%)	~30	~0.06	–	–	–	[105]

unit enhances polarization, and the TrFE unit expands chain spacing while inducing the formation of the all-trans conformation. At a certain VDF content, as the TrFE content increases, the material's melting point rises, and ε_r decreases.[102, 103] Zhang et al.[104] investigated different compositions of P(VDF-TrFE-CTFE), obtaining an E_b of more than 500 MV/m in P(VDF-TrFE-CTFE) with a molar ratio of 65.6/26.7/7.7. Its ε_r at the Curie temperature (35°C–40°C) is approximately 60, achieving a U_d of >13 J/cm³. Additionally, terpolymers like P(VDF-TrFE-HFP) and P(VDF-TrFE-DB) also exhibit good performance.[105] However, ternary copolymers face challenges such as complex procedures, high cost, and relatively low E_b, making large-scale applications difficult in a short time.[106–109]

In summary, intrinsic dielectric polymers have demonstrated commendable energy storage performance, as presented in Table 3.1, showcasing the dielectric and energy storage parameters of representative intrinsic dielectric polymers. Linear dielectric polymers, exemplified by BOPP, exhibit exceptional energy storage properties, with high E_b ensuring a certain U_d and low tan δ addressing safety and stability concerns in practical applications. Consequently, linear dielectric polymers remain the predominant choice for commercial applications. Nonlinear dielectric polymers, particularly those based on PVDF, boast high polarization and diverse dielectric behaviors, catering to the demands of high-energy storage applications. Nevertheless, the intrinsic challenges of low ε_r in linear dielectric polymers and the presence of substantial tan δ in nonlinear dielectric polymers pose difficulties for further enhancing energy storage performance through intrinsic methods alone. Modification techniques, such as molecular chain modification, multiphase blending, multilayer structure design, and surface treatment, emerge as potentially the most effective approaches for concurrently achieving high U_d and low tan δ.[110–112] Section 4 primarily delves into these common modification treatments and explores the energy storage properties of the most representative dielectric polymers.

3 RESEARCH PROGRESS OF DIELECTRIC POLYMER MODIFICATION

3.1 MOLECULAR CHAIN MODIFICATION

Molecular chain modification stands out as an effective strategy for enhancing the dielectric performance of polymers, achieved through graft copolymerization or block copolymerization to strategically regulate matrix defects and properties. Numerous studies in molecular chain modification have yielded substantial success, markedly improving material properties.

The incorporation of linear dielectric polymers into polymer chains proves effective in reducing relaxation loss and achieving high-energy density (η). For instance, Gong et al.[113] introduced various poly(methacrylic ester)s (PXMAs) into P(VDF-TrFE-CTFE) through graft polymers. These PXMA side chains efficiently decreased crystal size and crystallinity, resulting in dielectric polymers exhibiting antiferroelectric or linear-like behavior with increasing PXMA content. The addition of PXMA side chains, with their high modulus, further contributed to enhanced mechanical and breakdown resistance of the film. The resulting P(VDF-TrFE-CTFE)-g-PMMA (24 wt%) demonstrated antiferroelectric-like behavior with a DHL after stretching, achieving a U_d of 23.3 J/cm^3 at 675 MV/m. Similarly, grafting PS groups into ferroelectric polymers, such as P(VDF-CTFE) and P(VDF-TrFE-CTFE), led to relaxor ferroelectric or antiferroelectric-like behavior, with substantial improvements in η and U_d.[114–115]

Another approach in molecular chain modification involves introducing polar groups into the side groups or main chains of polymers to enhance the ε_r of the polymer. Polar groups, with large dipole moments, significantly increase polarization under an electric field.[116] Treufeld et al.[117] grafted CN (cyano) groups onto a series of PI graft polymers, resulting in an increased ε_r from 3 to 3.8 at 1 kHz, leading to improved U_d. Similarly, Yuan et al.[118] introduced OH groups on the side chains of PP-based dimers, enhancing ε_r and crystallinity. PP-g-OH (4.2 mol%) achieved an U_d of 7.42 J/cm^3 at 600 MV/m. Additionally, a novel high polarization block copolymer, PFPNP100-b-PFBHD25,

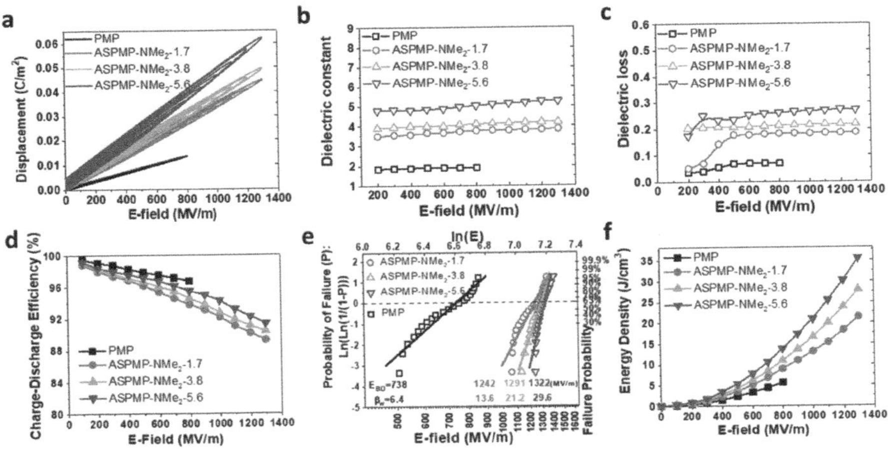

FIGURE 3.10　Dielectric properties of cross-linked polymers with different degrees of functionalization under high electric fields. (a) D–E loops, (b) dielectric constant, (c) dielectric loss, (d) charge-discharge efficiency, (e) Weibull distribution of breakdown strength, and (f) discharge energy density of cross-linked polymers.[122]

combining insulating polynorbornene and conductive polyacetylene segments, exhibited ultrahigh η (99%) and a U_d of 11.35 J/cm³ at 450 MV/m.

Recent advancements in molecular chain modification include the preparation of P(VDF-TrFE-CTFE)-g-PVA graft polymers using reversible addition-fragmentation chain transfer polymerization. The resulting copolymer exhibited a U_d of 13.6 J/cm³ at 500 MV/m, showcasing improved dielectric properties. Feng et al. introduced a polythiourea-b-polydimethylsiloxane (PTU-b-PDMS) block copolymer, where the nonpolar PDMS component reduced the dipole region, leading to superior dielectric properties ($\varepsilon_r = 5$, tan $\delta < 0.01$) and an U_d of 29.8 J/cm³ at 1,166 MV/m. Additionally, a zwitterion-functionalized PMP-based copolymer, ASPMP-NMe2-y, demonstrated enhanced energy storage performance with increasing degrees of functionalization, achieving a high U_d of 35 J/cm³ under 1,300 MV/m.

These advancements underscore the potential of molecular chain modification as a versatile strategy for tailoring the dielectric properties of polymers, paving the way for improved energy storage performance (Figure 3.10).

3.2　Polymer Cross-Linking Modification

Physical or chemical cross-linking plays a crucial role in creating covalent bonds between polymer chains, leading to the formation of stable bulk or network polymers. This cross-linked network can be strategically utilized to tune the dielectric and mechanical properties of the material. Common physical cross-linking methods include photo cross-linking, thermal cross-linking, and electron irradiation cross-linking. For instance, Chen et al.[126] employed a bisazide compound (BA) as a cross-linking agent for ultraviolet irradiation to form cross-linked polymers of

P(VDF-CTFE). Photo cross-linking influenced polymer crystallization by reducing grain size and enhancing interfacial polarization. Smaller grains facilitated dipole deflection and minimized hysteresis. The strong interaction of cross-linking fixed carriers, suppressing the formation of leakage current and partial discharge paths. Cross-linked samples with 2% and 15% cross-linking agents exhibited polarizations of 12.2 and 17.1 $\mu C/cm^2$ at 350 MV/m, compared to 8.5 $\mu C/cm^2$ for untreated P(VDF-CTFE). At 400 MV/m, P(VDF-CTFE) cross-linked with 10% BA achieved a U_d of 22.5 J/cm^3, approaching the 25 J/cm^3 of untreated P(VDF-CTFE) at a higher electric field (600 MV/m).

In another study, Khanchaitit et al.[127] produced cross-linked P(VDF-CTFE) using hot pressing with triallyl isocyanurate (TAIC) as the cross-linking agent and 1,4-bis(t-butyl peroxy) diisopropylbenzene as the initiator. Thermal cross-linking reduced tan δ without compromising polarization. The cross-linked network facilitated the transition between polar and nonpolar conformations, accelerated dipole orientation after removing the electric field, and minimized polarization loss. The cross-linked film exhibited significantly lower ferroelectric and conduction loss compared to the pure film. However, the C3 film, with higher cross-linking, showed elevated tan δ compared to C1 and C2 films. This was attributed to the strong interaction force hindering the free movement of dipoles after removing the electric field, resulting in a larger remnant polarization. Cross-linked P(VDF-CTFE) achieved a U_d of 17 J/cm^3 at 400 MV/m and an energy density (η) of 83%, surpassing other treated P(VDF-CTFE) films under the same electric field (Figure 3.11).

Liu et al.[128] conducted a study where they synthesized a chemically cross-linked polymer, PMGS-ArTU. The incorporation of aromatic thiourea (ArTU), with its large polar dipole of 4.89 D, contributed to a high dielectric constant (ε_r) in the cross-linked polymers. However, the inherent brittleness of ArTU posed challenges to the film-forming properties. To address this, the film's flexibility was significantly enhanced by cross-linking with poly(methylhydrosiloxane) (PMHS). The PMHS, grafted with allyl glycidyl ether, served as a source of high polarization. The combination of ArTU and grafted side chains led to increased material polarization, while the cross-linked structure enhanced the breakdown strength (E_b). The resulting flexible PMGS-ArTU cross-linked polymer demonstrated an impressive energy density (η) of 87.5% and an energy density (U_d) of 3.15 J/cm^3 at 350 MV/m.

3.3 Polymer Blend Modification

Blending is a straightforward and highly effective modification method, allowing for the combination of different dielectric properties from multiple polymers. This approach involves blending various polymers to achieve a composite material that surpasses the intrinsic dielectric polymers in terms of energy storage performance.[129–130] As polarization and breakdown strength (E_b) generally exhibit a negative correlation in polymers, blending serves as a solution to simultaneously enhance both properties.

The blending of linear dielectric polymers with ferroelectric polymers has been demonstrated to reduce the remnant polarization of ferroelectrics, thereby improving the energy storage efficiency (η). For instance, Li et al.[131] utilized the solution casting method to prepare ArPTU/PVDF blends. In this case, a small amount of ArPTU acted as a

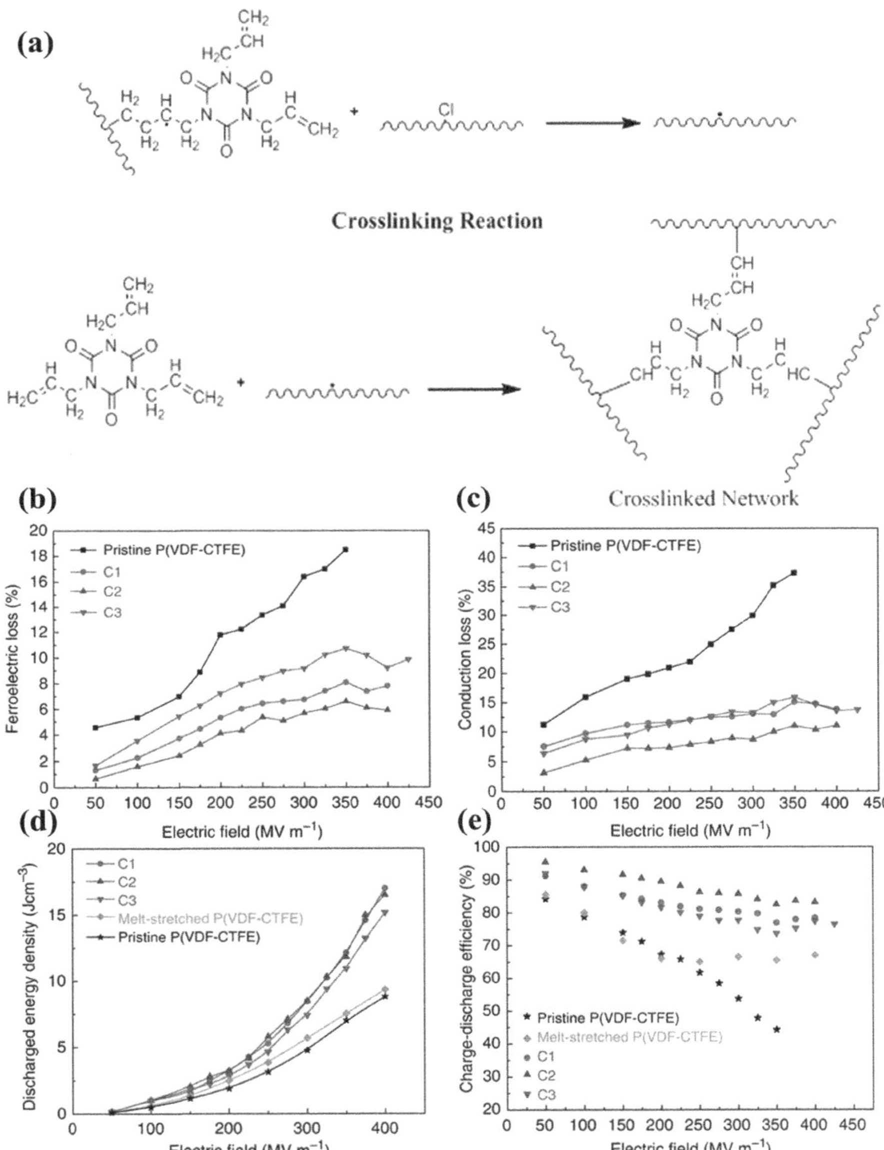

FIGURE 3.11 (a) Cross-linking reaction mechanism (b) ferroelectric loss, (c) conduction loss, (d) discharge energy density, (f) charge-discharge efficiency of P(VDF-CTFE) films under different electric fields.[127]

domain defect, suppressing the saturation polarization of PVDF at low electric fields. This blending strategy induced an increase in the β-phase, leading to enhanced polarization and reduced hysteresis loss. The resulting ArPTU/PVDF blend (10/90 vol%) exhibited an impressive U_d of 10.8 J/cm³, achieving an η of over 83%. Similar benefits

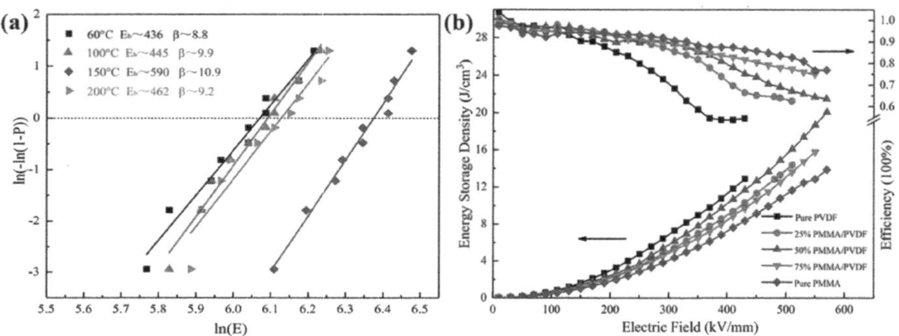

FIGURE 3.12   (a) Breakdown strength of PMMA/PVDF (50/50 vol%) blended films at different heat treatment temperatures. (b) Energy storage properties of PMMA/PVDF blended films with different PMMA contents after heat treatment at 150°C.[135]

were observed when adding PMMA to PVDF, P(VDF-TrFE), and P(VDF-TrFE-CFE). Chi et al. explored the properties of PVDF blended films with different PMMA contents, finding that PMMA had excellent compatibility with PVDF, significantly reducing polarization loss. The crystallinity of PMMA/PVDF blended films increased with higher annealing temperatures, favoring an increase in dielectric constant (ε_r). However, excessive heat treatment temperatures could lead to ε_r mismatch and thermal damage, impacting the film's dielectric properties. The optimal annealing temperature was found to be 150°C, resulting in the highest E_b. The PMMA/PVDF blend (50/50 vol%) achieved a η of 63.5% at 570 MV/m, with a U_d of 20.1 J/cm³ (Figure 3.12).

Moreover, successful outcomes have been achieved through the blending of two-phase blends with similar dielectric properties. The blending of two linear dielectric polymers, in particular, has demonstrated the potential to achieve ultra-high E_b and η, primarily attributed to the dielectric properties and interchain forces of the linear dielectric polymer. For instance, Zhang et al.[137] developed a method to mitigate the impact of defects on film breakdown performance through electrostatic forces in blended polymers. The blending of polyetherimide (PEI) and polyimide (PI) generates a strong interchain electrostatic force, expanding the chain packing morphology and reducing free volume and porosity. This approach proved highly effective, resulting in a remarkable E_b of 1 GV/m, with the blended film achieving an E_b of 550 MV/m at 200°C, surpassing results reported in the literature.

The blending of polyether methyl ether urea (PEMEU) and PEI polymers, studied by Zhang et al. leveraged the breakage of hydrogen bonds to facilitate the free movement of dipoles. The resulting blend (50/50 wt%) exhibited an ε_r of 5.8, marking a 48.6% increase compared to PEMEU and an 80.4% increase compared to PEI. Ferroelectric material blends, especially those involving PVDF-based polymers, have also shown promise in enhancing ferroelectric behavior and achieving high U_d. Rahimabady et al. blended P(VDF-HFP)/PVDF (50/50 wt%), leading to a high U_d of 30.1 J/cm³ at 854 MV/m after quenching, annealing, and hot-pressing operations. The structural changes and high crystallinity of the blended polymers contributed to the larger E_b and ε_r observed in the blended films.[139]

In recent work, a novel ternary blend strategy was employed to blend two ferroelectric polymers, PVDF and P(VDF-HFP), with the addition of PMMA as a linear polymer. This ternary blend demonstrated enhanced E_b and reduced tan δ, with PVDF and PVDF-HFP serving as high-ε_r phases. By adjusting the ratio of the ferroelectric phase, effective interaction forces between PVDF and PVDF-HFP led to denser interchain stacking and promoted the formation of the γ-phase. Furthermore, smaller grain sizes and increased amorphous phase content were achieved by adjusting the quenching temperature. These modifications collectively contributed to an improved E_b of the material. Ultimately, the PMMA/PVDF/PVDF-HFP (40/30/30 wt%) ternary blend film achieved an impressive U_d of 30 J/cm^3 and a η of 80% at 850 MV/m after quenching at 130°C.[142] The feasibility of industrial production was confirmed through a large-area film preparation process, highlighting the potential of multiphase blending as a key technology for addressing challenges in the large-scale preparation of polymer capacitors.

3.4 POLYMER COMPOSITE MODIFICATION

In the preceding section, we introduced various blend strategies aimed at improving the dielectric properties of polymers, where the two phases of the blended polymer exhibited good chain entanglement, and performance enhancement primarily stemmed from interchain forces. In practical terms, blended polymers can be categorized as a type of composite polymer. To distinguish between the two, we define a composite dielectric polymer as a stable combination of two phases with microscale interfaces. The second phase in composite polymers faces challenges in forming chain-level entanglement with the matrix or achieving uniform dispersion in the matrix as an insoluble filler. Performance improvements in composite dielectric polymers arise from the enhanced properties of the second phase and the interfacial interactions formed between the well-dispersed second phase and the matrix.

Unlike polymer-inorganic composites, which may compromise breakdown performance due to compatibility issues and dielectric mismatch, all-organic composites offer good compatibility and low dielectric mismatch, maintaining or even enhancing material properties. The addition of fillers with distinct properties to the matrix allows targeted resolution of matrix shortcomings, leading to improved energy storage performance.[143–146]

In recent decades, researchers have proposed various composite strategies, achieving notable results. Feng et al. incorporated a core-shell structure methyl methacrylate-butadiene-styrene (MBS) filler into P(VDF-HFP) to create a composite film with exceptional performance. Unlike inorganic fillers, MBS rubber particles exhibited compatibility with the matrix, resisting agglomeration and dispersing well in P(VDF-HFP). The cross-linking at the interface of the two phases effectively constrained the carrier movement, enhancing resistivity. This approach led to a U_d of 12.33 J/cm^3 for MBS/P(VDF-HFP) (8/92 vol%) at 700 MV/m, surpassing pure P(VDF-HFP) film by 1.49 times.[147]

To achieve a substantial increase in ε_r, Zhang et al.[148] developed all-organic nanocomposites by introducing conductive polypyrrole (PPy) nanoclips into a P(VDF-CTFE) matrix. The resulting composite (7 wt% PPy) exhibited an ε_r 23 times higher than that of the polymer matrix, with low tan δ (<0.4) at 1 kHz. Incorporating the conductive polymer polyaniline (PANI) into PVDF, as demonstrated by Yuan et al. increased the content of β-PVDF, enhancing polarization. The composite film with 5 vol% PANI doping achieved a high ε_r of 385. However, the low E_b of the composite film resulted from the formation of conductive paths and high conduction loss of the conductive polymer. The maximum U_d of PANI/PVDF (5/95 vol%) at 60 MV/m reached 6.1 J/cm^3.[149]

Many of the blending strategies discussed earlier have also found application in composite dielectric polymers. Combining highly polarizable polymers with high E_b polymers has proven effective. For instance, the addition of P(VDF-TrFE-CFE) filler to polyurea (PUA) simultaneously increased E_b and ε_r in composite films. The microphase separation created an interface that enhanced interfacial polarization. At 513 MV/m, the U_d of the composite film increased by 80% compared to the pure PUA film. Similarly, the addition of PVDF to PI and PP resulted in an increase in ε_r. ArPTU used as filler in combination with PVDF-based ferroelectric polymers, such as P(VDF-CTFE), P(VDF-TrFE-CFE), and P(VDF-TrFE-CTFE), significantly improved the E_b of the material and reduced tan δ. ArPTU chains, with large dipole moment groups, hindered chain movement and limited irreversible polarization. The U_d of ArPTU/P(VDF-TrFE-CFE) increased to 22.06 J/cm^3, while ArPTU/PVDF-TrFE-CTFE achieved a maximum U_d of 19.2 J/cm^3 at 700 MV/m, enhancing η to 85%.[154–155]

The electrode-limited conduction, represented by Schottky emission, becomes the dominant conduction mechanism at high temperatures. This phenomenon leads to increased leakage current, restricting the application of dielectric polymers at temperatures exceeding 150°C. Yuan et al. achieved optimal high-temperature performance by incorporating molecular semiconductors into high-temperature dielectric polymer PEI to confine electrode-injected carriers. The high electron affinity (EAms) of molecular semiconductors facilitated the capture of charge injection and intrinsically excited carriers in polymers with low electron affinity (EAp). Trapped carriers, confined by a large trap level (Φ_e = EAms − EAp), hindered the formation of conduction in the dielectric. The addition of molecular semiconductors increased dielectric resistivity by two orders of magnitude at high temperatures. The PCBM/PEI (0.5 vol% PCBM) composite polymer achieved a high η of 90% at 200°C, with a U_d of 3 J/cm^3. This outstanding performance, even in polymer nanocomposites, demonstrated uniform discharge behavior and high cycling stability at elevated temperatures (200°C) and sustained electric fields (300 MV/m).

The charge confinement effect of molecular semiconductors extends to other high-temperature dielectric polymers, such as polyethersulfone (PES), polyimide (PI), and fluorine polyester (FPE), underscoring the universality of this approach (Figure 3.13).

3.5　Polymer Structure Design

Polymer structure design encompasses bilayer, sandwich, multilayer, and special composite structures, each offering a unique approach to simultaneously enhance

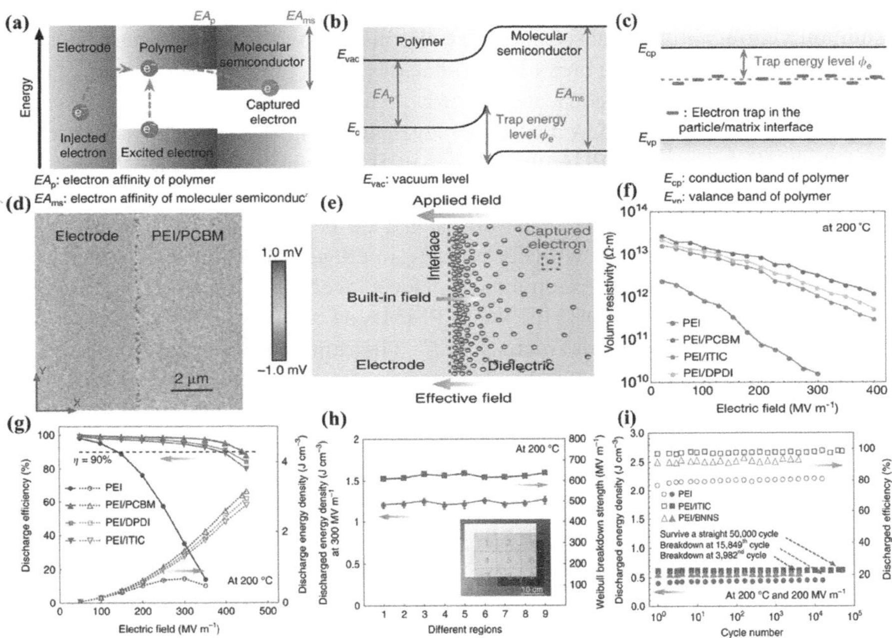

FIGURE 3.13 (a) Schematic diagram of charge transfer in all-organic composites. (b) Schematic diagram of trap energy levels introduced by molecular semiconductors in all-organic composites. (c) Schematic diagram of trap energy levels of insulating particle/polymer composites. (d) Potential distribution at the interface between electrode and composite. (e) Schematic diagram of composite material forming built-in electric field. (f) Volume resistivity of the PEI/PCBM (0.5 vol% PCBM), PEI/ITIC (0.25 vol% ITIC), and PEI/DPDI (0.75 vol% DPDI) composites as a function of the applied electric field at 200°C. (g) Energy storage performance of composites at 200°C. (h) Breakdown strength and discharge energy density uniformity testing of large PEI/ITIC films. (i) Cycling performance of composite films at 200°C and 200 MV/m.[156]

dielectric properties and E_b. These structural designs yield distinct dielectric and energy storage properties. The multilayer structure, formed by alternating different phases, features a large area and a flat interface, amplifying the contribution of interfacial polarization to ε_r. Despite the ε_r mismatch between phases, the electric field redistributes within the polymer film, concentrating on the high-E_b phase. This high-E_b layer becomes the primary contributor to the E_b of the multilayer film, impeding conductive path formation. The high-ε_r layer significantly improves the ε_r of the high-E_b layer, enhancing multilayer polarization.[157–159]

Special structures, such as the linear-transition-nonlinear (LTN) structure and the ferroconcrete-like structure, have been designed to enhance energy storage capacity. Here, we delve into two special structures: the LTN structure and the ferroconcrete-like structure.

The bilayer structure effectively integrates the advantages of two materials while preventing the rapid formation of breakdown paths. Zhou et al.[160] prepared

P(VDF-CTFE) and PUA bilayers, showcasing improved E_b. The electric field redistribution significantly increased the E_b of the bilayer, with the two-phase interface providing protection during breakdown processes. The curvature of conductive paths was increased, limiting electrical tree growth and reducing conduction loss. The bilayer film with 50 vol% PUA content achieved an E_b of 618 MV/m, 24% higher than PUA and 35% higher than PUA alone. Chen et al.[161] reported a P(VDF-TrFE-CFE)/PI bilayer film with a 50 vol% PI content, effectively increasing the E_b of the bilayer films. The U_d of P(VDF-TrFE-CFE)/PI reached 9.6 J/cm³ at 462.4 MV/m.

The sandwich structure, with its high-E_b interlayer and high-ε_r interlayer, has garnered attention for its exceptional performance.[162–163] Chen et al. prepared P(VDF-HFP)/PMMA/P(VDF-HFP) with PMMA as a high-E_b interlayer, effectively blocking the electrical path and increasing E_b. This adjustment improved energy storage performance, yielding a sandwich film with robust mechanical properties and dielectric stability. At 450 MV/m, the U_d reached 20.3 J/cm³, and η was 84%. Wang et al.[164] developed sandwich structures of P(VDF-TrFE-CTFE) and PEI, achieving superior performance. The PEI/P(VDF-TrFE-CTFE)/PEI structure exhibited the best performance, avoiding breakdown at a low electric field and hindering electrode carrier injection. The sandwich film had higher U_d than the two phases, reaching η of 81% and U_d of 8 J/cm³ at 530 MV/m. PVDF\PMMA/DE/PVDF\PMMA sandwich films, prepared by Chen et al.[165–166] demonstrated improved E_b and reduced tan δ. With 30 wt% PMMA content, the sandwich film achieved U_d of 15 J/cm³ and η of 76.5%. Additionally, Wang et al.[167] prepared a PVDF/P(VDF-TrFE-CTFE)/PVDF sandwich film with an interlayer content of 25 vol%, achieving improved performance with higher E_b and U_d than the two constituent phases. The U_d of the sandwich film at 660 MV/m reached 20.86 J/cm³ (Figure 3.14).

Multilayer structured polymer films can be fabricated through various methods, such as solution casting, electrospinning, and multilayer coextrusion. The individual layers in a multilayer structure have smaller thicknesses, enabling more effective charge redistribution and separation of electrical conduction paths.[168–169]

Li et al.[170] utilized stepwise hot pressing to prepare P(TFE-HFP-VDF)/P(VDF-HFP)/P(TFE-HFP-VDF) multilayer structured polymers. P(VDF-HFP) served as a high-ε_r layer, while P(TFE-HFP-VDF), with good thermal stability, acted as an outer layer to increase the electrode/dielectric interface barrier height and impede charge injection. The compatibility between the layers was excellent, and the significant difference in electrical properties hindered charge conduction. The 9-layer film, with P(TFE-HFP-VDF) as the outer layer at 70°C, exhibited an outstanding η of 80% and a high U_d of 15.5 J/cm³, surpassing other dielectric polymers at the same temperature. Jiang et al. employed an improved electrostatic spinning and hot-pressing process to create PMMA/PVDF multilayers with a continuous gradient variation of PMMA content. The nonlinear gradient structure, with more interlayer potential barriers, hindered carrier conduction, suppressed leakage current, and reduced conduction loss. This nonlinear gradient film achieved the highest E_b, along with the highest effective polarization, resulting in an impressive U_d of 38.8 J/cm³ at 802.1 MV/m.[171]

Tseng et al.[172] fabricated polysulfone (PSF)/PVDF multilayer films through coextrusion, varying monolayer thickness to investigate its impact on film performance. By reducing monolayer thickness, conduction loss within PVDF was effectively

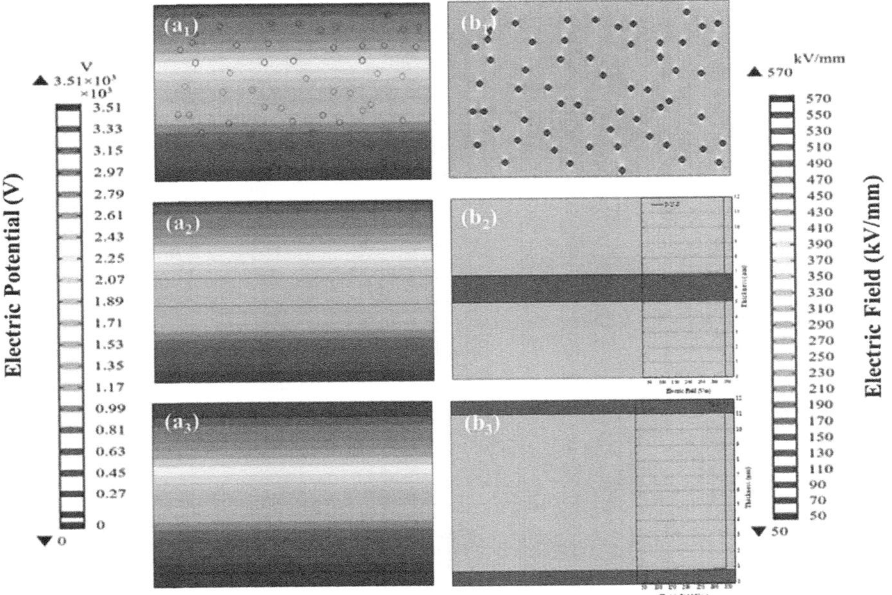

FIGURE 3.14 Distribution of the electric potential (a_1, a_2 and a_3) and electric field (b_1, b_2 and b_3) simulated for the PVTC/PEI blend film, PEI/PVTC/PEI sandwich film, and PVTC/PEI/PVTC sandwich film, respectively.[164]

minimized, enhancing η. However, when the layer thickness fell below 100 nm, effective interfacial polarization became challenging. Similarly, Mackey et al. created PC/PVDF multilayer polymers using micro/nanolayer coextrusion, controlling the total thickness and number of layers. The electric field redistribution and concentration in the PC layer, along with the interlayer interface, impeded charge movement. Optimizing the number of PVDF/PC layers and the thickness of PVDF single layers resulted in the best performance, achieving an U_d of 11 J/cm³ at >600 MV/m. Selecting appropriate monolayer thickness and layer numbers is crucial in regulating multilayer film performance (Figure 3.15).

To address the challenges related to electric field distortion and interlayer bonding arising from significant differences in dielectric properties among adjacent layers in multilayer structures, Sun et al. developed a specialized asymmetric LTN structure using solution casting and hot pressing. This structure includes a transition layer made of a PEI/P(VDF-HFP) blend between two outer layers. The transition layer, being a blend of the outer layer materials, forms a strong interfacial connection with the outer layers. This introduction of a transition layer helps homogenize the internal electric field and reduce electric field distortion. In comparison to blend and bilayer structures, the LTN structure film exhibits significant improvements in U_d and η. Additionally, when compared to conventional symmetric sandwich structures, the LTN structure not only reduces tan δ but also substantially enhances η while maintaining the same level of U_d.

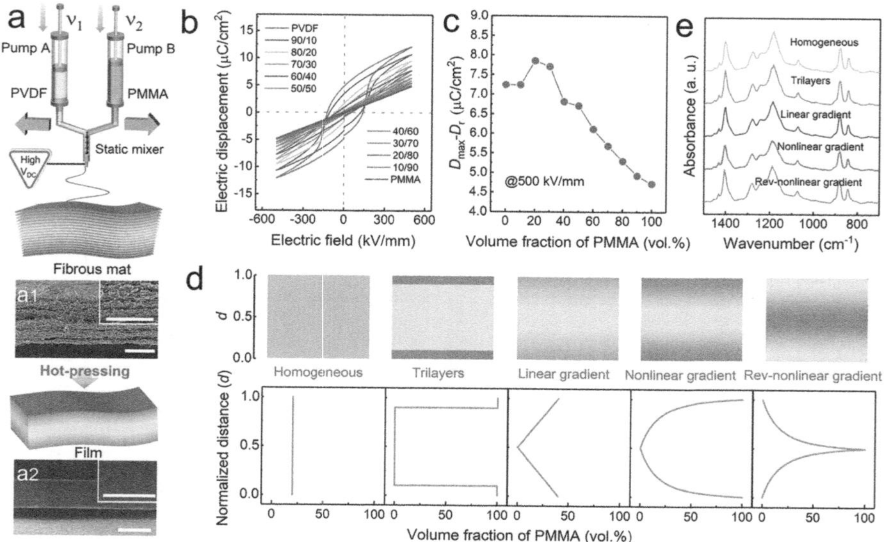

FIGURE 3.15   (a) Schematic diagram of the preparation of PVDF/PMMA multilayer structured film. (b) The distribution of PMMA volume fraction along the out-of-plane direction of the binary polymer films with different structures. (c) Simulation of breakdown path evolution of polymer films with different multilayer structures. (d) Breakdown strength and D_{max}-D_r at 500 MV/m of binary polymer films with different structures. (e) Discharged energy density and efficiency for pure PVDF and binary polymer films with different structures.[171]

In a separate study, Cui et al. created a PMMA/PVDF ferroconcrete-like structure using coaxial spinning and hot pressing. Here, PMMA serves as a shell layer enveloping a PVDF core layer, forming an all-organic polymer film. After hot pressing at a specific temperature, PMMA melts to create a continuous phase, while PVDF maintains a fiber shape due to its higher melting temperature. The intertwining of these two phases results in a ferroconcrete-like structure with a larger interfacial area, enhancing interfacial polarization. Due to the high polarization loss of PVDF, the PMMA/PVDF structure exhibits higher U_d under the electric field. Compared to PMMA/PVDF blend films and sandwich films with different compositions, the new PMMA/PVDF structure demonstrates superior dielectric performance. Specifically, the PMMA/PVDF (45/55 wt%) film achieves the highest η at 73% under 640 MV/m, while PMMA/PVDF (51/49 wt%) attains the highest U_d at 630 MV/m, reaching 20.7 J/cm³ (Figure 3.16).

3.6 POLYMER SURFACE MODIFICATION

The investigation of interfacial effects within dielectric materials has seen significant developments, particularly in the design of interfaces in multiphase composites and multilayer structures to enhance energy storage performance. However, research on the interface between electrodes and dielectrics has been relatively limited. Breakdown and conduction mechanisms have revealed that defects at the dielectric

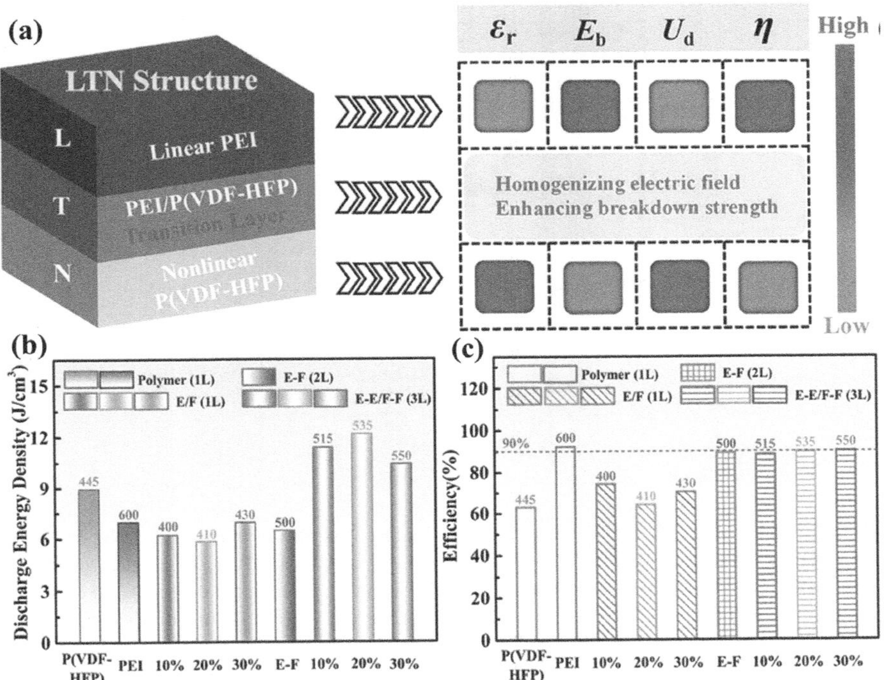

FIGURE 3.16 (a) Schematic illustration of the dielectric energy storage properties of the asymmetric LTN structure. (b) Discharge energy density and (c) efficiency of pure PEI, pure P(VDF-HFP), PEI/P(VDF-HFP) blend, bilayer and LTN structures.[174]

surface can degrade dielectric properties and concentrate electric field distribution, thereby diminishing the breakdown performance of the material. Surface modification to reduce these surface defects represents an advanced approach to improving the dielectric properties of materials.

Luo et al.[175] conducted a study on the impact of surface flatness on the breakdown strength (E_b) of Polytetrafluoroethylene (PTFE) films and proposed a strategic solution. Defects on the surface of polymer films lead to carrier aggregation. To address this issue, the researchers immersed the prepared PTFE in an epoxy resin solution, allowing the dried epoxy resin to fill the valleys on the film's surface, resulting in a flat surface PTFE film. The breakdown strength of pure PTFE and epoxy resin is 415 MV/m and 327 MV/m, respectively. However, the introduction of epoxy resin as an interface modifier in PTFE (0.5 wt% epoxy resin) significantly increases the breakdown strength to 555 MV/m. This modified PTFE film achieves an ultrahigh η of 98% at room temperature with a discharge energy density (U_d) of 3.58 J/cm³. Moreover, the high-temperature performance of the PTFE is also enhanced, with a U_d of 1.93 J/cm³ and η of 99% at 150°C (Figure 3.17).

Pei et al.[176] investigated the impact of different drying temperatures on polyvinylidene fluoride (PVDF) and observed a significant reduction in breakdown strength (E_b) when PVDF was dried at 90°C. Surface morphology analysis using scanning

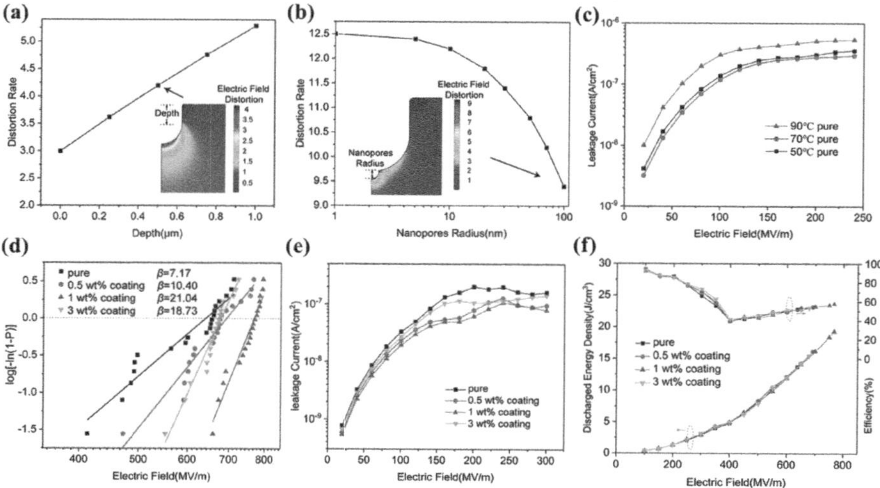

FIGURE 3.17 Electric field strength distortion rate as a function of (a) hole depth and (b) nanopore radius. (c) Variation of leakage current of PVDF films dried at different temperatures. (d) Weibull distribution of breakdown strength, (e) leakage current, and (f) energy storage performance of PVDF films coated with different mass fractions of PMMA.[176]

electron microscopy (SEM) revealed a decrease in defect depth and number with increasing drying temperature. However, the study also observed nanopores at the bottom of surface defects in the polymer film dried at 90°C, which were absent at lower temperatures. Finite element analysis simulated hole-like defects, introducing the concept of the distortion rate as the ratio of the defective electric field to the average electric field. Results indicated that the distortion rate increased with defect depth, deviating from experimental findings. This suggested that the decreased E_b at 90°C might be related to the presence of bottom nanopores. Simulation of nanopores showed that smaller pore size led to a higher distortion rate, causing local electric field concentration and the formation of discharge paths, resulting in increased leakage currents at all drying temperatures.

To further address the effect of nanopores, a nanolayer of different concentrations of polymethyl methacrylate (PMMA) was spin-coated onto the PVDF surface. Three PMMA coating concentrations represented coating of bottom nanopores only (0.5 wt%), complete coating of defects (1 wt%), and overcoating (3 wt%). PMMA, with a high Young's modulus and good compatibility with PVDF, effectively suppressed the generation of leakage currents and improved the breakdown strength of the dielectric polymer film. The optimal energy storage performance was achieved at 1 wt% PMMA coating concentration, resulting in an U_d of 19.08 J/cm³ at 767.05 MV/m for the PMMA/PVDF organic nano-interlayer structure.

The accomplishments and dielectric properties of all-organically modified polymers are summarized in Table 3.2. In conclusion, various modification methods for intrinsic dielectric polymers have proven effective in obtaining advanced dielectric materials and will remain a focus of future research. Polymer chain modification,

TABLE 3.2

Dielectric and energy storage properties of modification polymers

Modification Category	Materials	ε_r @1 kHz	tan δ @1 kHz	E_b (MV/m)	U_d (J/cm^3)	η (%)	Ref.
Chain modification	P(VDF-TrFE-CTFE)-g-PMMA (24 wt% PMMA)	~8	~0.05	675	23.3	71	[113]
	P(VDF-TrFE-CTFE)-g-PEMA (22 wt% PEMA)	~6.5	~0.04	~500	12	~70	[102]
	P(VDF-TrFE-CTFE)-g-PEMA (26 wt% PEMA)	~7	~0.04	675	19.3	83.4	[114]
	[P(VDF-HFP)-g-PDMA (5.9 wt% PDMA)	~15	~0.05	859	33	–	[177]
	P(VDF-CTFE)-g-PS (34 wt% PS)	~5	0.006	600	~9	82.4	[115]
	P(VDF-TrFE-CTFE)-g-PVA (23 mol% PVA)	~12	~0.05	500	13.6	~63	[120]
	PI-g-CN	~3.8	<0.01	@100	~0.22	~70	[117]
	PP-g-OH (4.2 mol% OH)	~4.6	–	600	7.42	–	[118]
	PFPNP$_{100}$-b-PFBHD$_{25}$	8	0.003	450	11.45	99	[119]
	PTU$_{0.95}$-b-PDMS$_{0.05}$	5	<0.01	1,166	29	–	[121]
	P(MMA-MAA) (19 mol% MAA)	~3.8	~0.04	550	13	92	[178]
	ASPMP-NMe$_2$-5.6	5.2	<0.002	~1,300	35	~92	[122]
Cross-linking modification	P (VDF-CTFE)-10 wt% BA	–	–	400	22.5	–	[126]
	P (VDF-TrFE)-15 wt% BA	–	–	350	21	~66	[179]
	XL-P(VDF-TrFE)	15		575	17.5	~70	[124]
	P(VDF-CTFE)-3 wt% TAIC	~9.9	~0.03	400	17	83	[127]
	P(VDF-CTFE)	~17	–	324	17.5	–	[180]
	-(PS-COOH)-15wt% BPO						
	P(VDF-TrFE-VA)-HMDI	–	–	500	~11	~70	[123]
	P(CTFE-VDF)-TAIC	3.2	0.25	550	4.72	90	[181]
	PP-HP	~3.1	~0.002	350	1.75	~98	[125]
	PMGS-35 wt% ArTU	~6	~0.02	350	3.15	87.5	[128]

(Continued)

TABLE 3.2
(Continued)

Modification Category	Materials	ε_r @1 kHz	tan δ @1 kHz	E_b (MV/m)	U_d (J/cm³)	η (%)	Ref.
Blend modification	ArPTU /PVDF (10/90 vol%)	9.2	0.02	700	10.8	83	[131]
	PMMA/PVDF (50/50 vol%)	~5	~0.04	570	20.1	63.5	[135]
	PMMA/P(VDF-HFP) (42.6/57.4 vol%)	~6.5	~0.02	476	11.2	85.8	[133]
	PMMA/P(VDF-TrFE) (30/70 wt%)	~7.4	~0.03	450	10	75	[134]
	PMMA/P(VDF-TrFE-CFE) (40/60 wt%)	~5	~0.03	~700	10.67	63	[132]
	PI/PEI (50/50 wt%)	~3	~0.005	950	~8	~65	[44]
	PEMEU /PEI (50/50 wt%)	5.81	0.0087	–	–	–	[137]
	ArPTU/PI (10/90 wt%)	4.52	0.0034	443.16	4.00	–	[138]
	PEEU/PI	4.8	0.0028	495	5.14	–	[136]
	PVDF /P(VDF-HFP) (50/50 wt%)	9.1	–	854	30.1	–	[139]
	P(VDF-TrFE-CFE)/PVDF (20/80 wt%)	~13.5	~0.032	480	13.63	45	[140]
	P(VDF-TrFE-CFE)/PVDF (40/60 vol%)	~28	~0.05	640	19.6	–	[141]
	PMMA/PVDF/P(VDF-HFP) (40/30/30 wt%)	~6	~0.052	850	30	80	[142]
Composite modification	MBS/ PVDF (12/88 vol%)	~7.8	~0.02	535	9.85	62.4	[143]
	MBS/ P(VDF-HFP) (8/92 vol%)	~6.8	~0.06	700	12.33	62.5	[147]
	PANI/PVDF (5/95 vol%)	385	0.85	60	6.1	–	[149]
	P(VDF-TrFE-CFE)/PUA (20/80 vol%)	4.1	~0.01	450	4.3	~62	[150]
	PVDF/PI (50/50 wt%)	7.745	~0.08	–	–	–	[151]
	PVDF/PP (20/80 vol%)	~3.55	<0.005	–	–	–	[152]
	TPU/ P(DF-HFP) (3/97 vol%)	8.5	0.02	537.8	10.36	–	[146]
	ArPTU/P(VDF-CTFE) (2/98 wt%)	~10	~0.05	450	17.8	~60	[153]
	ArPTU/P(VDF-TrFE-CFE) (10/90 wt%)	30.02	~0.03	407.5	22.06	~72	[154]
	ArPTU/P(VDF-TrFE-CTFE) (15/85 wt%)	11.3	0.01	700	19.2	85	[155]
	P-DB/PMMA (40/60 vol%)	~8	~0.04	400	9.3	85	[145]
	PCBM/PEI (0.5/99.5 vol%) @150°C	~3.4	~0.002	550	~4.8	~88	[156]

Modification Category	Materials	ε_r @1 kHz	tan δ @1 kHz	E_b (MV/m)	U_d (J/cm^3)	η (%)	Ref.
		Structure Design					
Bilayer	PUA/ P(VDF-CTFE) (50/50 vol%)	5.7	~0.02	618	–	77	[160]
	P(VDF-TrFE-CFE)/PI (50/50 vol%)	6.04	~0.02	462.4	9.6	58	[161]
Sandwich	P(VDF-HFP)/PMMA/P(VDF-HFP) (35/30/35 vol%)	~7	~0.06	450	20.3	84	[40]
	PEI/P(VDF-TrFE-CTFE)/PEI (42.5/15/42.5 vol%)	~6.25	~0.01	530	8	81	[164]
	PVDF/PMMA/DE/PVDF/PMMA	~7	~0.08	@350	~15	~76.5	[165]
	PVDF/DE/PVDF	~10.5	~0.05	438	20.92	72	[166]
	PVDF/P(VDF-TrFE-CTFE)/PVDF (37.5/25/37.5 vol%)	12.06	~0.03	660	20.86	62	[167]
Multilayer	P(TFE-HFP-VDF)/P(VDF-HFP)/ P(TFE-HFP-VDF) (9 layers) @70°C	–	~0.01	648	15.5	80.4	[170]
	P(VDF-HFP)/P(VDF-TrFE-CFE) (50/50 vol) (16 layers)	~16	–	637.6	22.8	82.3	[169]
	Nonlinear gradient PMMA/PVDF (20/80 vol%)	~8	~0.03	802.1	38.8	>80	[171]
	PSF/ PVDF (30/70 vol%) (32 layers)	–	–	416	–	–	[172]
	PC/PVDF (50/50 vol%) (32 layers)	–	~0.02	>650	11	~58	[173]
Special Structure	PEI/PEI/P(VDF-HFP) (20/80 vol%)/P(VDF-HFP)	~5.2	~0.02	535	12.15	89.9	[174]
	PMMA/PMMA/P(VDF-HFP) (10/90 wt%)/P(VDF-HFP)	~7.5	~0.05	490	~7.52	~85	[157]
	PMMA/PVDF (51/49 wt%)	~4	~0.05	640	20.7	63	[42]
	PSF/P(VDF-TrFE-CFE) (30/70 vol%)	~30	~0.05	~530	~9	~50	[159]
Surface modification	0.5 wt% epoxy resin /PTFE	~2.3	~0.0005	555	3.58	98	[175]
	1 wt% PMMA/PVDF	~6.8	~0.02	767.05	19.08	~60	[176]
	PVDF/PP	–	–	649.31	–	–	[182]

cross-linking modification, and the design of blends, composites, and multilayer structures offer avenues for enhancing breakdown strength, polarization, and other desirable properties. Additionally, surface modification to reduce defects and minimize surface electric field distortion represents an advanced method for improving breakdown performance across different temperatures.

REFERENCES

1. Kong, L. B.; Zhang, T. S.; Ma, J.; Boey, F. Progress in synthesis of ferroelectric ceramic materials via high-energy mechanochemical technique, *Prog. Mater. Sci.* **2008**, 53, 207–322. DOI: 10.1016/j.pmatsci.2007.05.001.

2. Wang, G.; Lu, Z.; Li, Y.; Li, L.; Ji, H.; Feteira, A.; Zhou, D.; Wang, D.; Zhang, S.; Reaney, I. M. Electroceramics for high-energy density capacitors: Current status and future perspectives, *Chem. Rev.* **2021**, 121, 6124–6172. DOI: 10.1021/acs.chemrev.0c01264.

3. Weyland, F.; Zhang, H.; Novak, N. Enhancement of energy storage performance by criticality in lead-free relaxor ferroelectrics, *Phys. Status Solidi Rapid Res. Lett.* **2018**, 12, 1800165. DOI: 10.1002/pssr.201800165.

4. Chen, X.; Qin, H.; Qian, X.; Zhu, W.; Li, B.; Zhang, B.; Lu, W.; Li, R.; Zhang, S.; Zhu, L.; Domingues Dos Santos, F.; Bernholc, J.; Zhang, Q. M. Relaxor ferroelectric polymer exhibits ultrahigh electromechanical coupling at low electric field, *Science* **2022**, 375, 1418–1422. DOI: 10.1126/science.abn0936.

5. Chen, Y. X.; Lu, H. W.; Shen, Z. W.; Li, Z. L.; Shen, Q. D. Cooling rate controlled microstructure evolution through flash DSC and enhanced energy density in P(VDF-CTFE) for capacitor application, *J. Polym. Sci., Part B: Polym. Phys.* **2017**, 55, 1245–1253. DOI: 10.1002/polb.24382.

6. Feng, M.; Feng, Y.; Zhang, T.; Li, J.; Chen, Q.; Chi, Q.; Lei, Q. Recent advances in multilayer-structure dielectrics for energy storage application, *Adv. Sci. (Weinh)* **2021**, 8, e2102221. DOI: 10.1002/advs.202102221.

7. Huang, X.; Sun, B.; Zhu, Y.; Li, S.; Jiang, P. High-k polymer nanocomposites with 1D filler for dielectric and energy storage applications, *Prog. Mater. Sci.* **2019**, 100, 187–225. DOI: 10.1016/j.pmatsci.2018.10.003.

8. Jiongxin, L.; Kyoung-Sik, M.; Byung-Kook, K.; Wong, C. P. Novel all organic high dielectric constant polyaniline/epoxy composites for embedded capacitor applications, In 2006 11th International Symposium on Advanced Packaging Materials: Processes, Properties and Interface, **2006** pp. 88–92.

9. Li, J.; Liu, X.; Feng, Y.; Yin, J. Recent progress in polymer/two-dimensional nanosheets composites with novel performances, *Prog. Polym. Sci.* **2022**, 126. DOI: 10.1016/j.progpolymsci.2022.101505.

10. Li, Q.; Tan, S.; Gong, H.; Lu, J.; Zhang, W.; Zhang, X.; Zhang, Z. Influence of dipole and intermolecular interaction on the tuning dielectric and energy storage properties of polystyrene-based polymers, *Phys. Chem. Chem. Phys.* **2021**, 23, 3856–3865. DOI: 10.1039/d0cp05233g.

11. Li, S.; Xie, D.; Lei, Q. Understanding insulation failure of nanodielectrics: Tailoring carrier energy, *High Volt.* **2020**, 5, 643–649. DOI: 10.1049/hve.2019.0122.

12. Luo, B.; Wang, X.; Sun, H.; Li, L. Dielectric, ferroelectric, and thermodynamic properties of silicone oil modified PVDF films for energy storage application, *Appl. Phys. Lett.* **2016**, 108. DOI: 10.1063/1.4954174.

13. Shi, P.; Zhu, X.; Lou, X.; Yang, B.; Guo, X.; He, L.; Liu, Q.; Yang, S.; Zhang, X. Bi0.5Na0.5TiO3-based lead-free ceramics with superior energy storage properties at high temperatures, *Compos. B. Eng.* **2021**, 215, 108815. DOI: 10.1016/j.compositesb.2021.108815.

14. Reed, C.; Cichanowskil, S. The fundamentals of aging in HV polymer-film capacitors, *IEEE Trans. Dielectr. Electr. Insul.* **1994**, 1, 904–922. DOI: 10.1109/94.326658.

15. Barshaw, E. J.; White, J.; Chait, M. J.; Cornette, J. B.; Bustamante, J.; Folli, F.; Biltchick, D.; Borelli, G.; Picci, G.; Rabuffi, M. High Energy Density (HED) Biaxially-Oriented Poly-Propylene (BOPP) capacitors for pulse power applications, *IEEE Trans. Magn.* **2007**, 43, 223–225. DOI: 10.1109/TMAG.2006.887682.

16. Ritamaki, M.; Rytoluoto, I.; Lahti, K. Performance metrics for a modern BOPP capacitor film, *IEEE Trans. Dielectr. Electr. Insul.* **2019**, 26, 1229–1237. DOI: 10.1109/tdei.2019.007970.

17. Zhou, Y.; Li, Q.; Dang, B.; Yang, Y.; Shao, T.; Li, H.; Hu, J.; Zeng, R.; He, J.; Wang, Q. A scalable, high-throughput, and environmentally benign approach to polymer dielectrics exhibiting significantly improved capacitive performance at high temperatures, *Adv. Mater.* **2018**, 30, e1805672. DOI: 10.1002/adma.201805672.

18. Chen, Q.; Shen, Y.; Zhang, S.; Zhang, Q. M. Polymer-based dielectrics with high energy storage density, *Annu. Rev. Mater. Res.* **2015**, 45, 433–458. DOI: 10.1146/annurev-matsci-070214-021017.

19. Luo, J.; Mao, J.; Sun, W.; Wang, S.; Zhang, L.; Tian, L.; Chen, Y.; Cheng, Y. Research progress of all organic polymer dielectrics for energy storage from the classification of organic structures, *Macromol. Chem. Phys.* **2021**, 222. DOI: 10.1002/macp.202100049.

20. Yang, Z.; Yue, D.; Yao, Y.; Li, J.; Chi, Q.; Chen, Q.; Min, D.; Feng, Y. Energy storage application of all-organic polymer dielectrics: A review, *Polymers* **2022**, 14. DOI: 10.3390/polym14061160.

21. Marwat, M. A.; Xie, B.; Zhu, Y.; Fan, P.; Liu, K.; Shen, M.; Ashtar, M.; Kongparakul, S.; Samart, C.; Zhang, H. Sandwich structure-assisted significantly improved discharge energy density in linear polymer nanocomposites with high thermal stability, *Colloids Surf., A* **2019**, 581, 123802. DOI: 10.1016/j.colsurfa.2019.123802.

22. Xie, B.; Wang, T.; Cai, J.; Zheng, Q.; Liu, Z.; Guo, K.; Mao, P.; Zhang, H.; Jiang, S. High energy density of ferroelectric polymer nanocomposites utilizing PZT@SiO2 nanocubes with morphotropic phase boundary, *Chem. Eng. J.* **2022**, 434, 134659. DOI: 10.1016/j.cej.2022.134659.

23. Yao, L.; Pan, Z.; Zhai, J.; Zhang, G.; Liu, Z.; Liu, Y. High-energy-density with polymer nanocomposites containing of SrTiO3 nanofibers for capacitor application, *Compos. Part A Appl. Sci. Manuf.* **2018**, 109, 48–54. DOI: 10.1016/j.compositesa.2018.02.040.

24. Zhang, X.; Shen, Y.; Zhang, Q.; Gu, L.; Hu, Y.; Du, J.; Lin, Y.; Nan, C. W. Ultrahigh energy density of polymer nanocomposites containing BaTiO3@TiO2 nanofibers by atomic-scale interface engineering, *Adv. Mater.* **2015**, 27, 819–824. DOI: 10.1002/adma.201404101.

25. Zhou, Y.; Han, S. T.; Roy, V. A. L. 6 - Nanocomposite dielectric materials for organic flexible electronics, In *Nanocrystalline Materials* (Second Edition); Tjong, S. C. Ed.; Elsevier: Oxford, **2014** pp. 195–220.

26. Liu, K.; Marwat, M. A.; Ma, W.; Wei, T.; Li, M.; Fan, P.; Lu, D.; Tian, Y.; Samart, C.; Ye, B.; He, J.; Zhang, H. Enhanced energy storage performance of nanocomposites filled with paraelectric ceramic nanoparticles by weakening the electric field distortion, *Ceram. Int.* **2020**, 46, 21149–21155. DOI: 10.1016/j.ceramint.2020.05.192.

27. Guo, M.; Jiang, J.; Shen, Z.; Lin, Y.; Nan, C. W.; Shen, Y. High-energy-density ferroelectric polymer nanocomposites for capacitive energy storage: Enhanced breakdown strength and improved discharge efficiency, *Mater. Today* **2019**, 29, 49–67. DOI: 10.1016/j.mattod.2019.04.015.

28. Sun, L.; Shi, Z.; Wang, H.; Zhang, K.; Dastan, D.; Sun, K.; Fan, R. Ultrahigh discharge efficiency and improved energy density in rationally designed bilayer polyetherimide–BaTiO3/P(VDF-HFP) composites, *J. Mater. Chem. A* **2020**, 8, 5750–5757. DOI: 10.1039/d0ta00903b.

29. Thakur, Y.; Zhang, T.; Iacob, C.; Yang, T.; Bernholc, J.; Chen, L. Q.; Runt, J.; Zhang, Q. M. Enhancement of the dielectric response in polymer nanocomposites with low dielectric constant fillers, *Nanoscale* **2017**, 9, 10992–10997. DOI: 10.1039/c7nr01932g.

30. Dou, L.; Lin, Y. H.; Nan, C. W. An overview of linear dielectric polymers and their nanocomposites for energy storage, *Molecules* **2021**, 26. DOI: 10.3390/molecules26206148.

31. Ho, J.; Ramprasad, R.; Boggs, S. Effect of alteration of antioxidant by uv treatment on the dielectric strength of bopp capacitor film, *IEEE Trans. Dielectr. Electr. Insul.* **2007**, 14, 1295–1301. DOI: 10.1109/TDEI.2007.4339492.

32. Singh, M.; Apata, I. E.; Samant, S.; Wu, W.; Tawade, B. V.; Pradhan, N.; Raghavan, D.; Karim, A. Nanoscale strategies to enhance the energy storage capacity of polymeric dielectric capacitors: Review of recent advances, *Polym. Rev.* **2021**, 1–50. DOI: 10.1080/15583724.2021.1917609.

33. Li, Q.; Chen, L.; Gadinski, M. R.; Zhang, S.; Zhang, G.; Li, U.; Iagodkine, E.; Haque, A.; Chen, L. Q.; Jackson, N.; Wang, Q. Flexible high-temperature dielectric materials from polymer nanocomposites, *Nature* **2015**, 523, 576–579. DOI: 10.1038/nature14647.

34. Yang, L.; Kong, X.; Li, F.; Hao, H.; Cheng, Z.; Liu, H.; Li, J. F.; Zhang, S. Perovskite lead-free dielectrics for energy storage applications, *Prog. Mater. Sci.* **2019**, 102, 72–108. DOI: 10.1016/j.pmatsci.2018.12.005.

35. Rabuffi, M.; Picci, G. Status quo and future prospects for metallized polypropylene energy storage capacitors, *IEEE Trans. Plasma Sci.* **2002**, 30, 1939–1942. DOI: 10.1109/tps.2002.805318.

36. Xiong, J.; Wang, X.; Zhang, X.; Xie, Y.; Lu, J.; Zhang, Z. How the biaxially stretching mode influence dielectric and energy storage properties of polypropylene films, *J. Appl. Polym. Sci.* **2020**, 138. DOI: 10.1002/app.50029.

37. Wu, X.; Gandla, D.; Lei, L.; Chen, C.; Tan, D. Q. Superior discharged energy density in polyetherimide composites enabled by ultra-low ZnO@BN core-shell fillers, *Mater. Lett.* **2021**, 290. DOI: 10.1016/j.matlet.2021.129434.

38. Liu, G.; Zhang, T.; Feng, Y.; Zhang, Y.; Zhang, C.; Zhang, Y.; Wang, X.; Chi, Q.; Chen, Q.; Lei, Q. Sandwich-structured polymers with electrospun boron nitrides layers as high-temperature energy storage dielectrics, *Chem. Eng. J.* **2020**, 389. DOI: 10.1016/j.cej.2020.124443.

39. Laihonen, S. J.; Gafvert, U.; Schutte, T.; Gedde, U. W. DC breakdown strength of polypropylene films: Area dependence and statistical behavior, *IEEE Trans. Dielectr. Electr. Insul.* **2007**, 14, 275–286. DOI: 10.1109/TDEI.2007.344604.

40. Chen, J.; Wang, Y.; Yuan, Q.; Xu, X.; Niu, Y.; Wang, Q.; Wang, H. Multilayered ferroelectric polymer films incorporating low-dielectric-constant components for concurrent enhancement of energy density and charge–discharge efficiency, *Nano Energy* **2018**, 54, 288–296. DOI: 10.1016/j.nanoen.2018.10.028.

41. Li, L.; Cheng, J.; Cheng, Y.; Han, T.; Liu, Y.; Zhou, Y.; Han, Z.; Zhao, G.; Zhao, Y.; Xiong, C.; Dong, L.; Wang, Q. Significantly enhancing the dielectric constant and breakdown strength of linear dielectric polymers by utilizing ultralow loadings of nanofillers, *J. Mater. Chem. A* **2021**, 9, 23028–23036. DOI: 10.1039/d1ta05408b.

42. Cui, Y.; Feng, Y.; Zhang, T.; Zhang, C.; Chi, Q.; Zhang, Y.; Wang, X.; Chen, Q.; Lei, Q. Excellent energy storage performance of ferroconcrete-like all-organic linear/ferroelectric polymer films utilizing interface engineering, *ACS Appl. Mater. Interfaces* **2020**, 12, 56424–56434. DOI: 10.1021/acsami.0c16197.

43. Li, Q.; Cheng, S. Polymer nanocomposites for high-energy-density capacitor dielectrics: Fundamentals and recent progress, *IEEE Electr. Insul. Mag.* **2020**, 36, 7–28. DOI: 10.1109/MEI.2020.9070113.

44. Zhang, Q.; Chen, X.; Zhang, B.; Zhang, T.; Lu, W.; Chen, Z.; Liu, Z.; Kim, S. H.; Donovan, B.; Warzoha, R. J.; Gomez, E. D.; Bernholc, J.; Zhang, Q. M. High-temperature polymers with record-high breakdown strength enabled by rationally designed chain-packing behavior in blends, *Matter* **2021**, 4, 2448–2459. DOI: 10.1016/j.matt.2021.04.026.

45. Wu, S.; Li, W.; Lin, M.; Burlingame, Q.; Chen, Q.; Payzant, A.; Xiao, K.; Zhang, Q. M. Aromatic polythiourea dielectrics with ultrahigh breakdown field strength, low dielectric loss, and high electric energy density, *Adv. Mater.* **2013**, 25, 1734–1738. DOI: 10.1002/adma.201204072.

46. Wen, F.; Zhang, L.; Wang, P.; Li, L.; Chen, J.; Chen, C.; Wu, W.; Wang, G.; Zhang, S. A high-temperature dielectric polymer poly(acrylonitrile butadiene styrene) with enhanced energy density and efficiency due to a cyano group, *J. Mater. Chem. A* **2020**, 8, 15122–15129. DOI: 10.1039/d0ta03540h.

47. Wen, F.; Xu, Z.; Xia, W.; Ye, H.; Wei, X.; Zhang, Z. High-energy-density Poly(styrene-co-acrylonitrile) thin films, *J. Electron. Mater.* **2013**, 42, 3489–3493. DOI: 10.1007/s11664-013-2764-z.

48. Bonardd, S.; Moreno-Serna, V.; Kortaberria, G.; Diaz Diaz, D.; Leiva, A.; Saldias, C. Dipolar glass polymers containing polarizable groups as dielectric materials for energy storage applications. A minireview, *Polymers (Basel)* **2019**, 11. DOI: 10.3390/polym11020317.

49. Wei, J.; Zhang, Z.; Tseng, J. K.; Treufeld, I.; Liu, X.; Litt, M. H.; Zhu, L. Achieving high dielectric constant and low loss property in a dipolar glass polymer containing strongly dipolar and small-sized sulfone groups, *ACS Appl. Mater. Interfaces* **2015**, 7, 5248–5257. DOI: 10.1021/am508488w.

50. Zhang, Z.; Wang, D. H.; Litt, M. H.; Tan, L. S.; Zhu, L. High-temperature and high-energy-density dipolar glass polymers based on sulfonylated Poly(2,6-dimethyl-1,4-phenylene oxide), *Angew. Chem. Int. Ed Engl.* **2018**, 57, 1528–1531. DOI: 10.1002/anie.201710474.

51. Wei, J.; Zhu, L. Intrinsic polymer dielectrics for high energy density and low loss electric energy storage, *Prog. Polym. Sci.* **2020**, 106, 101254. DOI: 10.1016/j.progpolymsci.2020.101254.

52. Su, R.; Tseng, J. K.; Lu, M. S.; Lin, M.; Fu, Q.; Zhu, L. Ferroelectric behavior in the high temperature paraelectric phase in a poly(vinylidene fluoride-co-trifluoroethylene) random copolymer, *Polymer* **2012**, 53, 728–739. DOI: 10.1016/j.polymer.2012.01.001.

53. Lovinger Andrew, J. Ferroelectric polymers, *Science* **1983**, 220, 1115–1121. DOI: 10.1126/science.220.4602.1115.

54. Bharti, V.; Xu, H. S.; Shanthi, G.; Zhang, Q. M.; Liang, K. Polarization and structural properties of high-energy electron irradiated poly(vinylidene fluoride-trifluoroethylene) copolymer films, *J. Appl. Phys.* **2000**, 87, 452–461. DOI: 10.1063/1.371883.

55. Bharti, V.; Zhang, Q. M. Dielectric study of the relaxor ferroelectric poly(vinylidene fluoride-trifluoroethylene) copolymer system, *Phys. Rev. B* **2001**, 63. DOI: 10.1103/PhysRevB.63.184103.

56. Prateek; Thakur, V. K.; Gupta, R. K. Recent progress on ferroelectric polymer-based nanocomposites for high energy density capacitors: Synthesis, dielectric properties, and future aspects, *Chem. Rev.* **2016**, 116, 4260–4317. DOI: 10.1021/acs.chemrev.5b00495.

57. Zhu, L.; Wang, Q. Novel ferroelectric polymers for high energy density and low loss dielectrics, *Macromolecules* **2012**, 45, 2937–2954. DOI: 10.1021/ma2024057.

58. Gong, H. H.; Zhang, Y.; Cheng, Y. P.; Lei, M. X.; Zhang, Z. C. The application of controlled/living radical polymerization in modification of PVDF-based fluoropolymer, *Chin. J. Polym. Sci.* **2021**, 39, 1110–1126. DOI: 10.1007/s10118-021-2616-x.

59. Li, J.; Meng, Q.; Li, W.; Zhang, Z. Influence of crystalline properties on the dielectric and energy storage properties of poly(vinylidene fluoride), *J. Appl. Polym. Sci.* **2011**, 122, 1659–1668. DOI: 10.1002/app.34020.

60. Li, W.; Meng, Q.; Zheng, Y.; Zhang, Z.; Xia, W.; Xu, Z. Electric energy storage properties of poly(vinylidene fluoride), *Appl. Phys. Lett.* **2010**, 96, 192905. DOI: 10.1063/1.3428656.

61. Guan, F.; Pan, J.; Wang, J.; Wang, Q.; Zhu, L. Crystal orientation effect on electric energy storage in poly (vinylidene fluoride-co-hexafluoropropylene) copolymers, *Macromolecules* **2010**, 43, 384–392.

62. Feng, Q. K.; Zhong, S. L.; Pei, J. Y.; Zhao, Y.; Zhang, D. L.; Liu, D. F.; Zhang, Y. X.; Dang, Z. M. Recent progress and future prospects on all-organic polymer dielectrics for energy storage capacitors, *Chem. Rev.* **2022**, 122, 3820–3878. DOI: 10.1021/acs.chemrev.1c00793.

63. Meng, N.; Ren, X.; Santagiuliana, G.; Ventura, L.; Zhang, H.; Wu, J.; Yan, H.; Reece, M. J.; Bilotti, E. Ultrahigh beta-phase content poly(vinylidene fluoride) with relaxor-like ferroelectricity for high energy density capacitors, *Nat. Commun.* **2019**, 10, 4535. DOI: 10.1038/s41467-019-12391-3.

64. Ren, X.; Meng, N.; Zhang, H.; Wu, J.; Abrahams, I.; Yan, H.; Bilotti, E.; Reece, M. J. Giant energy storage density in PVDF with internal stress engineered polar nanostructures, *Nano Energy* **2020**, 72. DOI: 10.1016/j.nanoen.2020.104662.

65. Furukawa, T.; Date, M.; Fukada, E. Hysteresis phenomena in polyvinylidene fluoride under high electric field, *J. Appl. Phys.* **1980**, 51, 1135–1141. DOI: 10.1063/1.327723.

66. Martins, P.; Lopes, A. C.; Lanceros-Mendez, S. Electroactive phases of poly(vinylidene fluoride): Determination, processing and applications, *Prog. Polym. Sci.* **2014**, 39, 683–706. DOI: 10.1016/j.progpolymsci.2013.07.006.

67. Ameduri, B. From Vinylidene Fluoride (VDF) to the applications of VDF-containing polymers and copolymers: Recent developments and future trends, *Chem. Rev.* **2009**, 109, 6632–6686. DOI: 10.1021/cr800187m.

68. Sousa, R. E.; Nunes-Pereira, J.; Ferreira, J. C. C.; Costa, C. M.; Machado, A. V.; Silva, M. M.; Lanceros-Mendez, S. Microstructural variations of poly(vinylidene fluoride co-hexafluoropropylene) and their influence on the thermal, dielectric and piezoelectric properties, *Polym. Test.* **2014**, 40, 245–255. DOI: 10.1016/j.polymertesting.2014.09.012.

69. Jayasuriya, A. C.; Schirokauer, A.; Scheinbeim, J. I. Crystal-structure dependence of electroactive properties in differently prepared poly(vinylidene fluoride/hexafluoropropylene) copolymer films, *J. Polym. Sci., Part B: Polym. Phys.* **2001**, 39, 2793–2799. DOI: 10.1002/polb.10035.

70. Zhou, X.; Zhao, X.; Suo, Z.; Zou, C.; Runt, J.; Liu, S.; Zhang, S.; Zhang, Q. M. Electrical breakdown and ultrahigh electrical energy density in poly(vinylidene fluoride-hexafluoropropylene) copolymer, *Appl. Phys. Lett.* **2009**, 94. DOI: 10.1063/1.3123001.

71. Guan, F.; Pan, J.; Wang, J.; Wang, Q.; Zhu, L. Crystal orientation effect on electric energy storage in poly(vinylidene fluoride-co-hexafluoropropylene) copolymers, *Macromolecules* **2009**, 43, 384–392. DOI: 10.1021/ma901921h.

72. Zhou, X.; Chu, B.; Neese, B.; Lin, M.; Zhang, Q. M. Electrical energy density and discharge characteristics of a poly(vinylidene fluoride-chlorotrifluoroethylene)copolymer, *IEEE Trans. Dielectr. Electr. Insul.* **2007**, 14, 1133–1138. DOI: 10.1109/TDEI.2007.4339472.

73. Chu, B.; Zhou, X.; Ren, K.; Neese, B.; Lin, M.; Wang, Q.; Bauer, F.; Zhang, Q. M. A Dielectric polymer with high electric energy density and fast discharge speed, *Science* **2006**, 313, 334–336. DOI: 10.1126/science.1127798.

74. Lovinger, A. J.; Furukawa, T.; Davis, G. T.; Broadhurst, M. G. Crystallographic changes characterizing the Curie transition in three ferroelectric copolymers of vinylidene fluoride and trifluoroethylene: 2. Oriented or poled samples, *Polymer* **1983**, 24, 1233–1239. DOI: 10.1016/0032-3861(83)90051-4.

75. Lovinger, A. J.; Johnson, G. E.; Bair, H. E.; Anderson, E. W. Structural, dielectric, and thermal investigation of the Curie transition in a tetrafluoroethylene copolymer of vinylidene fluoride, *J. Appl. Phys.* **1984**, 56, 2412–2418. DOI: 10.1063/1.334303.

76. Ohigashi, H.; Koga, K. Ferroelectric copolymers of vinylidenefluoride and trifluoroethylene with a large electromechanical coupling factor, *Jpn. J. Appl. Phys.* **1982**, 21, L455–L457. DOI: 10.1143/jjap.21.l455.

77. Koga, K.; Ohigashi, H. Piezoelectricity and related properties of vinylidene fluoride and trifluoroethylene copolymers, *J. Appl. Phys.* **1986**, 59, 2142–2150. DOI: 10.1063/1.336351.

78. Xia, W.; Xu, Z.; Zhang, Q.; Zhang, Z.; Chen, Y. Dependence of dielectric, ferroelectric, and piezoelectric properties on crystalline properties of p(VDF-co-TrFE) copolymers, *J. Polym. Sci., Part B: Polym. Phys.* **2012**, 50, 1271–1276. DOI: 10.1002/polb.23125.

79. Xia, W.; Wang, Z.; Xing, J.; Cao, C.; Xu, Z. The dependence of dielectric and ferroelectric properties on crystal phase structures of the hydrogenized p(vdf-trfe) films with different thermal processing, *IEEE Trans. Ultrason. Ferroelectr. Freq. Control* **2016**, 63, 1674–1680. DOI: 10.1109/TUFFC.2016.2594140.

80. Bharti, V.; Zhao, X. Z.; Zhang, Q. M.; Romotowski, T.; Tito, F.; Ting, R. Ultrahigh field induced strain and polarization response in electron irradiated poly(vinylidene fluoride-trifluoroethylene) copolymer, *Mater. Res. Innovations* **2016**, 2, 57–63. DOI: 10.1007/s100190050063.

81. Forsythe, J. S.; Hill, D. J. T. The radiation chemistry of fluoropolymers, *Prog. Polym. Sci.* **2000**, 25, 101–136. DOI: 10.1016/S0079-6700(00)00008-3.

82. Lovinger, A. J. Polymorphic transformations in ferroelectric copolymers of vinylidene fluoride induced by electron irradiation, *Macromolecules* **1985**, 18, 910–918.

83. Zhang, Q. M.; Bharti, V.; Zhao, X. Giant electrostriction and relaxor ferroelectric behavior in electron-irradiated poly(vinylidene fluoride-trifluoroethylene) copolymer, *Science* **1998**, 280, 2101–2104. DOI: 10.1126/science.280.5372.2101.

84. Mabboux, P. Y.; Gleason, K. K. 19F NMR characterization of electron beam irradiated vinylidene fluoride–trifluoroethylene copolymers, *J. Fluorine Chem.* **2002**, 113, 27–35. DOI: 10.1016/S0022-1139(01)00440-7.

85. Cheng, Z. Y.; Zhang, Q. M.; Bateman, F. B. Dielectric relaxation behavior and its relation to microstructure in relaxor ferroelectric polymers: High-energy electron irradiated poly(vinylidene fluoride–trifluoroethylene) copolymers, *J. Appl. Phys.* **2002**, 92, 6749–6755. DOI: 10.1063/1.1518130.

86. Daudin, B.; Dubus, M.; Legrand, J. F. Effects of electron irradiation and annealing on ferroelectric vinylidene fluoride-trifluoroethylene copolymers, *J. Appl. Phys.* **1987**, 62, 994–997. DOI: 10.1063/1.339685.

87. Wang, L.; Zhao, X.; Feng, J. Effects of electron irradiation on poly(vinylidene fluoride–trifluoroethylene) copolymers studied by solid-state nuclear magnetic resonance spectroscopy, *J. Polym. Sci., Part B: Polym. Phys.* **2006**, 44, 1714–1724. DOI: 10.1002/polb.20830.

88. Li, Z. M.; Arbatti, M. D.; Cheng, Z. Y. Recrystallization study of high-energy electron-irradiated P(VDF–TrFE) 65/35 copolymer, *Macromolecules* **2004**, 37, 79–85. DOI: 10.1021/ma035251h.

89. Mijovic, J.; Sy, J. W.; Kwei, T. K. Reorientational dynamics of dipoles in poly(vinylidene fluoride)/poly(methyl methacrylate) (PVDF/PMMA) blends by dielectric spectroscopy, *Macromolecules* **1997**, 30, 3042–3050. DOI: 10.1021/ma961774w.

90. Zhu, L. Exploring strategies for high dielectric constant and low loss polymer dielectrics, *J. Phys. Chem. Lett.* **2014**, 5, 3677–3687. DOI: 10.1021/jz501831q.

91. Fried, J. R. Sub-T g transitions, In *Physical Properties of Polymers Handbook*. Springer, Mark, J. E. Ed.; **2007** pp. 217–232.

92. Zhao, B.; Zhu, L. Mixed polymer brush-grafted particles: A new class of environmentally responsive nanostructured materials, *Macromolecules* **2009**, 42, 9369–9383. DOI: 10.1021/ma902042x.

93. Bao, H. M.; Song, J. F.; Zhang, J.; Shen, Q. D.; Yang, C. Z.; Zhang, Q. M. Phase transitions and ferroelectric relaxor behavior in P(VDF–TrFE–CFE) terpolymers, *Macromolecules* **2007**, 40, 2371–2379. DOI: 10.1021/ma062800l.

94. Baojin, C.; Xin, Z.; Neese, B.; Zhang, Q. M.; Bauer, F. Relaxor ferroelectric poly (vinylidene fluoride-trifluoroethylene-chlorofluoroethylene) terpolymer for high-energy-density storage capacitors, *IEEE Trans. Dielectr. Electr. Insul.* **2006**, 13, 1162–1169. DOI: 10.1109/TDEI.2006.247845.

95. Yang, L.; Li, X.; Allahyarov, E.; Taylor, P. L.; Zhang, Q. M.; Zhu, L. Novel polymer ferroelectric behavior via crystal isomorphism and the nanoconfinement effect, *Polymer* **2013**, 54, 1709–1728. DOI: 10.1016/j.polymer.2013.01.035.

96. Bobnar, V.; Vodopivec, B.; Levstik, A.; Kosec, M.; Hilczer, B.; Zhang, Q. M. Dielectric properties of relaxor-like vinylidene fluoride–trifluoroethylene-based electroactive polymers, *Macromolecules* **2003**, 36, 4436–4442. DOI: 10.1021/ma034149h.

97. Li, J.; Sun, Z.; Yan, F. Solution processable low-voltage organic thin film transistors with high-k relaxor ferroelectric polymer as gate insulator, *Adv. Mater.* **2012**, 24, 88–93. DOI: 10.1002/adma.201103542.

98. Li, M.; Wondergem, H. J.; Spijkman, M. J.; Asadi, K.; Katsouras, I.; Blom, P. W.; De Leeuw, D. Revisiting the δ-phase of poly (vinylidene fluoride) for solution-processed ferroelectric thin films, *Nat. Mater.* **2013**, 12, 433–438.

99. Lu, Y.; Claude, J.; Zhang, Q.; Wang, Q. Microstructures and dielectric properties of the ferroelectric fluoropolymers synthesized via reductive dechlorination of poly(vinylidene fluoride-co-chlorotrifluoroethylene)s, *Macromolecules* **2006**, 39, 6962–6968. DOI: 10.1021/ma061311i.

100. Chung, T. C.; Petchsuk, A. Synthesis and properties of ferroelectric fluoroterpolymers with curie transition at ambient temperature, *Macromolecules* **2002**, 35, 7678–7684. DOI: 10.1021/ma020504c.

101. Yang, L.; Tyburski, B. A.; Dos Santos, F. D.; Endoh, M. K.; Koga, T.; Huang, D.; Wang, Y.; Zhu, L. Relaxor ferroelectric behavior from strong physical pinning in a poly(vinylidene fluoride-co-trifluoroethylene-co-chlorotrifluoroethylene) random terpolymer, *Macromolecules* **2014**, 47, 8119–8125. DOI: 10.1021/ma501852x.

102. Li, J.; Hu, X.; Gao, G.; Ding, S.; Li, H.; Yang, L.; Zhang, Z. Tuning phase transition and ferroelectric properties of poly(vinylidene fluoride-co-trifluoroethylene) via grafting with desired poly(methacrylic ester)s as side chains, *J. Mater. Chem. C* **2013**, 1, 1111–1121. DOI: 10.1039/c2tc00431c.

103. Zhang, Z.; Meng, Q.; Chung, T. C. M. Energy storage study of ferroelectric poly (vinylidene fluoride-trifluoroethylene-chlorotrifluoroethylene) terpolymers, *Polymer* **2009**, 50, 707–715. DOI: 10.1016/j.polymer.2008.11.005.

104. Zhang, Z.; Chung, T. M. Study of VDF/TrFE/CTFE terpolymers for high pulsed capacitor with high energy density and low energy loss, *Macromolecules* **2007**, 40, 783–785. DOI: 10.1021/ma0627119.

105. Xu, H.; Shen, D.; Zhang, Q. Structural and ferroelectric response in vinylidene fluoride/trifluoroethylene/hexafluoropropylene terpolymers, *Polymer* **2007**, 48, 2124–2129. DOI: 10.1016/j.polymer.2007.02.035.

106. Zhang, T.; Chen, X.; Thakur, Y.; Lu, B.; Zhang, Q.; Runt, J.; Zhang, Q. M. A highly scalable dielectric metamaterial with superior capacitor performance over a broad temperature, *Sci. Adv.* **2020**, 6, eaax6622. DOI: 10.1126/sciadv.aax6622.

107. Mao, H.; You, Y.; Tong, L.; Tang, X.; Wei, R.; Liu, X. Dielectric properties of diblock copolymers containing a polyarylene ether nitrile block and a polyarylene ether ketone block, *J. Mater. Sci. Mater. Electron.* **2017**, 29, 3127–3134. DOI: 10.1007/s10854-017-8245-z.

108. Chen, J.; Wang, Y.; Li, H.; Han, H.; Liao, X.; Sun, R.; Huang, X.; Xie, M. Rational design and modification of high-k bis(double-stranded) block copolymer for high electrical energy storage capability, *Chem. Mater.* **2018**, 30, 1102–1112. DOI: 10.1021/acs.chemmater.7b05042.

109. Li, Z.; Treich, G. M.; Tefferi, M.; Wu, C.; Nasreen, S.; Scheirey, S. K.; Ramprasad, R.; Sotzing, G. A.; Cao, Y. High energy density and high efficiency all-organic polymers with enhanced dipolar polarization, *J. Mater. Chem. A* **2019**, 7, 15026–15030. DOI: 10.1039/c9ta03601f.

110. Liu, W.; Chen, J.; Zhou, D.; Liao, X.; Xie, M.; Sun, R. A high-performance dielectric block copolymer with a self-assembled superhelical nanotube morphology, *Polym. Chem.* **2017**, 8, 725–734. DOI: 10.1039/c6py01571a.

111. Miao, B.; Liu, J.; Zhang, X.; Lu, J.; Tan, S.; Zhang, Z. Ferroelectric relaxation dependence of poly(vinylidene fluoride-co-trifluoroethylene) on frequency and temperature after grafting with poly(methyl methacrylate), *RSC Adv.* **2016**, 6, 84426–84438. DOI: 10.1039/c6ra17977k.

112. Zhou, D.; Zhang, H.; Ma, C.; Han, H.; Xie, M. Disubstituted pendant-functionalized insulating-conductive block copolymer with enhanced dielectric and energy storage performance, *React. Funct. Polym.* **2019**, 139, 50–59. DOI: 10.1016/j.reactfunctpolym.2019.03.013.

113. Gong, H.; Miao, B.; Zhang, X.; Lu, J.; Zhang, Z. High-field antiferroelectric-like behavior in uniaxially stretched poly(vinylidene fluoride-trifluoroethylene-chlorotrifluoroethylene)-grafted-poly(methyl methacrylate) films with high energy density, *RSC Adv.* **2016**, 6, 1/9–1599. DOI: 10.1039/c5ra22617a.

114. Li, J.; Gong, H.; Yang, Q.; Xie, Y.; Yang, L.; Zhang, Z. Linear-like dielectric behavior and low energy loss achieved in poly(ethyl methacrylate) modified poly(vinylidene-co-trifluoroethylene), *Appl. Phys. Lett.* **2014**, 104, 263901. DOI: 10.1063/1.4886391.

115. Guan, F.; Yang, L.; Wang, J.; Guan, B.; Han, K.; Wang, Q.; Zhu, L. Confined ferroelectric properties in poly(vinylidene fluoride-co-chlorotrifluoroethylene)-graft-polystyrene graft copolymers for electric energy storage applications, *Adv. Funct. Mater.* **2011**, 21, 3176–3188. DOI: 10.1002/adfm.201002015.

116. Guan, F.; Wang, J.; Yang, L.; Tseng, J. K.; Han, K.; Wang, Q.; Zhu, L. Confinement-induced high-field antiferroelectric-like behavior in a poly(vinylidene fluoride-co-trifluoroethylene-co-chlorotrifluoroethylene)-graft-polystyrene graft copolymer, *Macromolecules* **2011**, 44, 2190–2199. DOI: 10.1021/ma102910v.

117. Treufeld, I.; Wang, D. H.; Kurish, B. A.; Tan, L. S.; Zhu, L. Enhancing electrical energy storage using polar polyimides with nitrile groups directly attached to the main chain, *J. Mater. Chem. A* **2014**, 2, 20683–20696. DOI: 10.1039/c4ta03260h.

118. Yuan, X.; Matsuyama, Y.; Chung, T. C. M. Synthesis of functionalized isotactic polypropylene dielectrics for electric energy storage applications, *Macromolecules* **2010**, 43, 4011–4015. DOI: 10.1021/ma100209d.

119. Han, H.; Zhou, D.; Ren, Q.; Ma, F.; Ma, C.; Xie, M. High-performance all-polymer dielectric and electrical energy storage materials containing conjugated segment and multi-fluorinated pendants, *Eur. Polym. J.* **2020**, 122. DOI: 10.1016/j.eurpolymj.2019.109376.

120. Zhang, M.; Tan, S.; Xiong, J.; Chen, C.; Zhang, Y.; Wei, X.; Zhang, Z. Tailoring dielectric and energy storage performance of PVDF-based relaxor ferroelectrics with hydrogen bonds, *ACS Appl. Energy Mater.* **2021**, 4, 8454–8464. DOI: 10.1021/acsaem.1c01658.

121. Feng, Y.; Jiang, L.; Yang, A.; Liu, X.; Yang, L.; Lu, G.; Li, S. Interfacial effect on dielectric properties of self-assembled polythiourea-based copolymers for ultrahigh energy storage, *Macromol. Rapid Commun.* **2022**, 43, e2100700. DOI: 10.1002/marc.202100700.

122. Zhang, M.; Li, B.; Wang, J. J.; Huang, H. B.; Zhang, L.; Chen, L. Q. Polymer dielectrics with simultaneous ultrahigh energy density and low loss, *Adv. Mater.* **2021**, 33, e2008198. DOI: 10.1002/adma.202008198.

123. Meereboer, Niels L.; Terzić, I.; van der Steeg, P.; Portale, G.; Loos, K. Physical pinning and chemical crosslinking-induced relaxor ferroelectric behavior in P(VDF-ter-TrFE-ter-VA) terpolymers, *J. Mater. Chem. A* **2019**, 7, 2795–2803. DOI: 10.1039/c8ta11534f.

124. Tan, S.; Hu, X.; Ding, S.; Zhang, Z.; Li, H.; Yang, L. Significantly improving dielectric and energy storage properties via uniaxially stretching crosslinked P(VDF-co-TrFE) films, *J. Mater. Chem. A* **2013**, 1. DOI: 10.1039/c3ta11484h.

125. Yuan, M.; Zhang, G.; Li, B.; Chung, T. C. M.; Rajagopalan, R.; Lanagan, M. T. Thermally stable low-loss polymer dielectrics enabled by attaching cross-linkable anti-oxidant to polypropylene, *ACS Appl. Mater. Interfaces* **2020**, 12, 14154–14164. DOI: 10.1021/acsami.0c00453.

126. Chen, X. Z.; Li, Z. W.; Cheng, Z. X.; Zhang, J. Z.; Shen, Q. D.; Ge, H. X.; Li, H. T. Greatly enhanced energy density and patterned films induced by photo cross-linking of poly(vinylidene fluoride-chlorotrifluoroethylene), *Macromol Rapid Commun* **2011**, 32, 94–99. DOI: 10.1002/marc.201000478.

127. Khanchaitit, P.; Han, K.; Gadinski, M. R.; Li, Q.; Wang, Q. Ferroelectric polymer networks with high energy density and improved discharged efficiency for dielectric energy storage, *Nat. Commun.* **2013**, 4, 2845. DOI: 10.1038/ncomms3845.

128. Liu, Y.; Chen, J.; Jiang, X.; Jiang, P.; Huang, X. All-organic cross-linked polysiloxane-aromatic thiourea dielectric films for electrical energy storage application, *ACS Appl. Energy Mater.* **2020**, 3, 5198–5207. DOI: 10.1021/acsaem.9b02521.

129. Chen, X. Z.; Li, X.; Qian, X. S.; Wu, S.; Lu, S. G.; Gu, H. M.; Lin, M.; Shen, Q. D.; Zhang, Q. M. A polymer blend approach to tailor the ferroelectric responses in P(VDF–TrFE) based copolymers, *Polymer* **2013**, 54, 2373–2381. DOI: 10.1016/j.polymer.2013.02.041.

130. Shehzad, M.; Malik, T. Antiferroelectric behavior of P(VDF-TrFE) and P(VDF-TrFE-CTFE) ferroelectric domains for energy harvesting, *ACS Appl. Energy Mater.* **2018**, 1, 2832–2840. DOI: 10.1021/acsaem.8b00478.

131. Li, W.; Jiang, L.; Zhang, X.; Shen, Y.; Nan, C. W. High-energy-density dielectric films based on polyvinylidene fluoride and aromatic polythiourea for capacitors, *J. Mater. Chem. A* **2014**, 2, 15803–15807. DOI: 10.1039/c4ta03374d.

132. Liu, F.; Li, Z.; Wang, Q.; Xiong, C. High breakdown strength and low loss binary polymer blends of poly(vinylidene fluoride-trifluoroethylene-chlorofluoroethylene) and poly(methyl methacrylate), *Polym. Adv. Technol.* **2018**, 29, 1271–1277. DOI: 10.1002/pat.4238.

133. Luo, B.; Wang, X.; Wang, H.; Cai, Z.; Li, L. P(VDF-HFP)/PMMA flexible composite films with enhanced energy storage density and efficiency, *Compos. Sci. Technol.* **2017**, 151, 94–103. DOI: 10.1016/j.compscitech.2017.08.013.

134. Xia, W.; Zhang, Q.; Wang, X.; Zhang, Z. Electrical energy discharging performance of poly(vinylidene fluoride-co-trifluoroethylene) by tuning its ferroelectric relaxation with polymethyl methacrylate, *J. Appl. Polym. Sci.* **2014**, 131, n/a-n/a. DOI: 10.1002/app.40114.

135. Chi, Q.; Zhou, Y.; Yin, C.; Zhang, Y.; Zhang, C.; Zhang, T.; Feng, Y.; Zhang, Y.; Chen, Q. A blended binary composite of poly(vinylidene fluoride) and poly(methyl methacrylate) exhibiting excellent energy storage performances, *J. Mater. Chem. C* **2019**, 7, 14148–14158. DOI: 10.1039/c9tc04695j.

136. Ahmad, A.; Tong, H.; Fan, T.; Xu, J. All-organic polymer blend dielectrics of poly(arylene ether urea) and polyimide: Toward high energy density and high temperature applications, *J. Polym. Sci.* **2021**, 59, 1414–1423. DOI: 10.1002/pol.20200725.

137. Zhang, Q.; Chen, X.; Zhang, T.; Zhang, Q. M. Giant permittivity materials with low dielectric loss over a broad temperature range enabled by weakening intermolecular hydrogen bonds, *Nano Energy* **2019**, 64. DOI: 10.1016/j.nanoen.2019.103916.

138. Ahmad, A.; Tong, H.; Fan, T.; Xu, J. Binary polymer blend of ArPTU/PI with advanced comprehensive dielectric properties and ultra-high thermally stability, *J. Appl. Polym. Sci.* **2021**, 138. DOI: 10.1002/app.50997.

139. Rahimabady, M.; Yao, K.; Arabnejad, S.; Lu, L.; Shim, V. P. W.; Cheong Wun Chet, D. Intermolecular interactions and high dielectric energy storage density in poly(vinylidene fluoride-hexafluoropropylene)/poly(vinylidene fluoride) blend thin films, *Appl. Phys. Lett.* **2012**, 100. DOI: 10.1063/1.4730603.

140. Mao, P.; Wang, J.; Zhang, L.; Sun, Q.; Liu, X.; He, L.; Liu, S.; Zhang, S.; Gong, H. Tunable dielectric polarization and breakdown behavior for high energy storage capability in P(VDF-TrFE-CFE)/PVDF polymer blended composite films, *Phys. Chem. Chem. Phys.* **2020**, 22, 13143–13153. DOI: 10.1039/d0cp01071e.

141. Zhang, X.; Shen, Y.; Shen, Z.; Jiang, J.; Chen, L.; Nan, C. W. Achieving high energy density in pvdf-based polymer blends: Suppression of early polarization saturation and enhancement of breakdown strength, *ACS Appl. Mater. Interfaces* **2016**, 8, 27236–27242. DOI: 10.1021/acsami.6b10016.

142. Liu, K.; Liu, Y.; Ma, W.; Takesue, N.; Samart, C.; Tan, H.; Jiang, S.; Dou, Z.; Hu, Y.; Zhang, S.; Zhang, H. Realizing enhanced energy density in ternary polymer blends by intermolecular structure design, *Chem. Eng. J.* **2022**, 446. DOI: 10.1016/j. cej.2022.136980.

143. Zheng, M. S.; Zha, J. W.; Yang, Y.; Han, P.; Hu, C. H.; Dang, Z. M. Enhanced breakdown strength of poly(vinylidene fluoride) utilizing rubber nanoparticles for energy storage application, *Appl. Phys. Lett.* **2016**, 109. DOI: 10.1063/1.4961252.

144. Dang, Z. m.; Zhang, Y. x.; Liu, C.; Zhang, D. l.; Feng, Q. k. Preparation and Characterization of all-organic TPU/P(VDF-HFP) flexible composite films with high energy storage, *Acta Chim. Sinica* **2021**, 79. DOI: 10.6023/a21060273.

145. Wen, F.; Zhu, C.; Lv, W.; Wang, P.; Zhang, L.; Li, L.; Wang, G.; Wu, W.; Ying, Z.; Zheng, X.; Han, C.; Li, W.; Zu, H.; Yue, Z. Improving the energy density and efficiency of the linear polymer PMMA with a double-bond fluoropolymer at elevated temperatures, *ACS Omega* **2021**, 6, 35014–35022. DOI: 10.1021/acsomega.1c05676.

146. Zheng, M. S.; Zha, J. W.; Yang, Y.; Han, P.; Hu, C. H.; Wen, Y. Q.; Dang, Z. M. Polyurethane induced high breakdown strength and high energy storage density in polyurethane/poly(vinylidene fluoride) composite films, *Appl. Phys. Lett.* **2017**, 110. DOI: 10.1063/1.4989579.

147. Feng, Q. K.; Ping, J. B.; Zhu, J.; Pei, J. Y.; Huang, L.; Zhang, D. L.; Zhao, Y.; Zhong, S. L.; Dang, Z. M. All-organic dielectrics with high breakdown strength and energy storage density for high-power capacitors, *Macromol. Rapid Commun.* **2021**, 42, e2100116. DOI: 10.1002/marc.202100116.

148. Zhang, L.; Lu, X.; Zhang, X.; Jin, L.; Xu, Z.; Cheng, Z. Y. All-organic dielectric nanocomposites using conducting polypyrrole nanoclips as filler, *Compos. Sci. Technol.* **2018**, 167, 285–293. DOI: 10.1016/j.compscitech.2018.08.017.

149. Yuan, J. K.; Dang, Z. M.; Yao, S. H.; Zha, J. W.; Zhou, T.; Li, S. T.; Bai, J. Fabrication and dielectric properties of advanced high permittivity polyaniline/poly(vinylidene fluoride) nanohybrid films with high energy storage density, *J. Mater. Chem.* **2010**, 20. DOI: 10.1039/b923590f.

150. Zhou, Y.; Liu, Q.; Chen, F.; Zhao, Y.; Yang, W.; He, X.; Mao, X.; Yang, Y.; Xu, J. Enhanced breakdown strength and energy storage density in polyurea all-organic composite films, *Mater. Res. Express* **2019**, 6. DOI: 10.1088/2053-1591/ab493f.

151. Mao, X.; Yang, J.; Du, L.; Guo, W.; Li, C.; Tang, X. Effect of two synthetic methods of polyimide/poly(vinylidene fluoride) composites on their dielectric properties, *JOM* **2017**, 69, 2497–2500. DOI: 10.1007/s11837-017-2498-0.

152. Dang, Z. M.; Yan, W. T.; Xu, H. P. Novel high-dielectric-permittivity poly(vinylidene fluoride)/polypropylene blend composites: The influence of the poly(vinylidene fluoride) concentration and compatibilizer, *J. Appl. Polym. Sci.* **2007**, 105, 3649–3655. DOI: 10.1002/app.26447.

153. Cheng, Z.; Zhou, W.; Zhang, C.; Li, Q.; Sha, R.; Chen, X.; Chu, B.; Shen, Q. D. Composite of P(VDF-CTFE) and aromatic polythiourea for capacitors with high-capacity, high-efficiency, and fast response, *J. Polym. Sci., Part B: Polym. Phys.* **2018**, 56, 193–199. DOI: 10.1002/polb.24537.

154. Li, C.; Shi, L.; Yang, W.; Zhou, Y.; Li, X.; Zhang, C.; Yang, Y. All polymer dielectric films for achieving high energy density film capacitors by blending poly(vinylidene fluoride-trifluoroethylene-chlorofluoroethylene) with aromatic polythiourea, *Nanoscale Res. Lett.* **2020**, 15, 36. DOI: 10.1186/s11671-020-3270-x.

155. Zhu, H.; Liu, Z.; Wang, F. Improved dielectric properties and energy storage density of poly(vinylidene fluoride-co-trifluoroethylene-co-chlorotrifluoroethylene) composite films with aromatic polythiourea, *J. Mater. Sci.* **2017**, 52, 5048–5059. DOI: 10.1007/s10853-016-0742-6.

156. Yuan, C.; Zhou, Y.; Zhu, Y.; Liang, J.; Wang, S.; Peng, S.; Li, Y.; Cheng, S.; Yang, M.; Hu, J.; Zhang, B.; Zeng, R.; He, J.; Li, Q. Polymer/molecular semiconductor all-organic composites for high-temperature dielectric energy storage, *Nat Commun* **2020**, 11, 3919. DOI: 10.1038/s41467-020-17760-x.

157. Sun, S.; Shi, Z.; Sun, L.; Liang, L.; Dastan, D.; He, B.; Wang, H.; Huang, M.; Fan, R. Achieving concurrent high energy density and efficiency in all-polymer layered paraelectric/ferroelectric composites via introducing a moderate layer, *ACS Appl. Mater. Interfaces* **2021**, 13, 27522–27532. DOI: 10.1021/acsami.1c08063.

158. Wang, J.; Xie, Y.; Zhang, Y.; Peng, B.; Li, Q.; Ma, D.; Zhang, Z.; Huang, X. The ultrahigh discharge efficiency and energy density of P(VDF-HFP) via electrospinning-hot press with St-MMA copolymer, *Mater. Chem. Front.* **2021**, 5, 3646–3656. DOI: 10.1039/d1qm00105a.

159. Dan, Z.; Jiang, J.; Qian, J.; Shen, Z.; Li, M.; Nan, C.; Shen, Y. A Ferroconcrete-like all-organic nanocomposite exhibiting improved mechanical property, high breakdown strength, and high energy efficiency, *Macromol. Mater. Eng.* **2019**, 304. DOI: 10.1002/mame.201900433.

160. Zhou, Y.; Liu, Q.; Chen, F.; Zhao, Y.; Xu, J.; Yang, W.; He, X.; Mao, X.; Yang, Y. Improving breakdown strength and energy storage efficiency of poly(vinylidene fluoride-co-chlorotrifluoroethylene) and polyurea blend films by double layer structure design, *Polym. Test.* **2020**, 81. DOI: 10.1016/j.polymertesting.2019.106261.

161. Chen, C.; Xing, J.; Cui, Y.; Zhang, C.; Feng, Y.; Zhang, Y.; Zhang, T.; Chi, Q.; Wang, X.; Lei, Q. Designing of ferroelectric/linear dielectric bilayer films: An effective way to improve the energy storage performances of polymer-based capacitors, *J. Phys. Chem. C* **2020**, 124, 5920–5927. DOI: 10.1021/acs.jpcc.9b11486.

162. Feng, M.; Zhang, T.; Song, C.; Zhang, C.; Zhang, Y.; Feng, Y.; Chi, Q.; Chen, Q.; Lei, Q. Improved energy storage performance of all-organic composite dielectric via constructing sandwich structure, *Polymers (Basel)* **2020**, 12. DOI: 10.3390/polym12091972.

163. Liu, G.; Feng, Y.; Zhang, T.; Zhang, C.; Chi, Q.; Zhang, Y.; Zhang, Y.; Lei, Q. High-temperature all-organic energy storage dielectric with the performance of self-adjusting electric field distribution, *J. Mater. Chem. A* **2021**, 9, 16384–16394. DOI: 10.1039/d1ta02668b.

164. Wang, C.; He, G.; Chen, S.; Zhai, D.; Luo, H.; Zhang, D. Enhanced performance of all-organic sandwich structured dielectrics with linear dielectric and ferroelectric polymers, *J. Mater. Chem. A* **2021**, 9, 8674–8684. DOI: 10.1039/d1ta00974e.

165. Chen, J.; Wang, Y.; Dong, J.; Chen, W.; Wang, H. Ultrahigh energy storage density at low operating field strength achieved in multicomponent polymer dielectrics with hierarchical structure, *Compos. Sci. Technol.* **2021**, 201. DOI: 10.1016/j.compscitech.2020.108557.

166. Chen, J.; Wang, Y.; Xu, X.; Yuan, Q.; Niu, Y.; Wang, Q.; Wang, H. Sandwich structured poly(vinylidene fluoride)/polyacrylate elastomers with significantly enhanced electric displacement and energy density, *J. Mater. Chem. A* **2018**, 6, 24367–24377. DOI: 10.1039/c8ta09111k.

167. Wang, L.; Luo, H.; Zhou, X.; Yuan, X.; Zhou, K.; Zhang, D. Sandwich-structured all-organic composites with high breakdown strength and high dielectric constant for film capacitor, *Compos. Part A Appl. Sci. Manuf.* **2019**, 117, 369–376. DOI: 10.1016/j.compositesa.2018.12.007.

168. Yin, K.; Zhou, Z.; Schuele, D. E.; Wolak, M.; Zhu, L.; Baer, E. Effects of interphase modification and biaxial orientation on dielectric properties of poly(ethylene terephthalate)/Poly(vinylidene fluoride-co-hexafluoropropylene) multilayer films, *ACS Appl. Mater. Interfaces* **2016**, 8, 13555–13566. DOI: 10.1021/acsami.6b01287.

169. Jiang, J.; Shen, Z.; Qian, J.; Dan, Z.; Guo, M.; Lin, Y.; Nan, C. W.; Chen, L.; Shen, Y. Ultrahigh discharge efficiency in multilayered polymer nanocomposites of high energy density, *Energy Storage Mater.* **2019**, 18, 213–221. DOI: 10.1016/j.ensm.2018.09.013.

170. Li, Y.; Cheng, S.; Wang, S.; Yuan, C.; Luo, Z.; Zhu, Y.; Hu, J.; He, J.; Li, Q. Multilayered ferroelectric polymer composites with high energy density at elevated temperature, *Compos. Sci. Technol.* **2021**, 202. DOI: 10.1016/j.compscitech.2020.108594.

171. Jiang, Y.; Wang, J.; Yan, S.; Shen, Z.; Dong, L.; Zhang, S.; Zhang, X.; Nan, C. W. Ultrahigh energy density in continuously gradient-structured all-organic dielectric polymer films, *Adv. Funct. Mater.* **2022**, n/a, 2200848. DOI: 10.1002/adfm.202200848.

172. Tseng, J. K.; Tang, S.; Zhou, Z.; Mackey, M.; Carr, J. M.; Mu, R.; Flandin, L.; Schuele, D. E.; Baer, E.; Zhu, L. Interfacial polarization and layer thickness effect on electrical insulation in multilayered polysulfone/poly(vinylidene fluoride) films, *Polymer* **2014**, 55, 8–14. DOI: 10.1016/j.polymer.2013.11.042.

173. Mackey, M.; Schuele, D. E.; Zhu, L.; Flandin, L.; Wolak, M. A.; Shirk, J. S.; Hiltner, A.; Baer, E. Reduction of dielectric hysteresis in multilayered films via nanoconfinement, *Macromolecules* **2012**, 45, 1954–1962. DOI: 10.1021/ma202267r.

174. Sun, L.; Shi, Z.; He, B.; Wang, H.; Liu, S.; Huang, M.; Shi, J.; Dastan, D.; Wang, H. Asymmetric trilayer all-polymer dielectric composites with simultaneous high efficiency and high energy density: A novel design targeting advanced energy storage capacitors, *Adv. Funct. Mater.* **2021**, 31. DOI: 10.1002/adfm.202100280.

175. Luo, S.; Ansari, T. Q.; Yu, J.; Yu, S.; Xu, P.; Cao, L.; Huang, H.; Sun, R. Enhancement of dielectric breakdown strength and energy storage of all-polymer films by surface flattening, *Chem. Eng. J.* **2021**, 412. DOI: 10.1016/j.cej.2021.128476.

176. Pei, J. Y.; Zhong, S. L.; Zhao, Y.; Yin, L. J.; Feng, Q. K.; Huang, L.; Liu, D. F.; Zhang, Y. X.; Dang, Z. M. All-organic dielectric polymer films exhibiting superior electric breakdown strength and discharged energy density by adjusting the electrode–dielectric interface with an organic nano-interlayer, *Energy Environ. Sci.* **2021**, 14, 5513–5522. DOI: 10.1039/d1ee01960k.

177. Rahimabady, M.; Qun Xu, L.; Arabnejad, S.; Yao, K.; Lu, L.; Shim, V. P. W.; Gee Neoh, K.; Kang, E. T. Poly(vinylidene fluoride-co-hexafluoropropylene)-graft-poly(dopamine methacrylamide) copolymers: A nonlinear dielectric material for high energy density storage, *Appl. Phys. Lett.* **2013**, 103. DOI: 10.1063/1.4858397.

178. Wang, Z.; Liu, J.; Gong, H.; Zhang, X.; Lu, J.; Zhang, Z. Synthesis of poly(methyl methacrylate–methallyl alcohol) via controllable partial hydrogenation of poly(methyl methacrylate) towards high pulse energy storage capacitor application, *RSC Adv.* **2016**, 6, 34855–34865. DOI: 10.1039/c6ra03757g.

179. Chen, X. Z.; Cheng, Z. X.; Liu, L.; Yang, X. D.; Shen, Q. D.; Hu, W. B.; Li, H. T. Evolution of nanopolar phases, interfaces, and increased dielectric energy storage capacity in photoinitiated cross-linked poly(vinylidene fluoride)-based copolymers, *Colloid. Polym. Sci.* **2013**, 291, 1989–1997. DOI: 10.1007/s00396-013-2939-4.

180. Chen, Y.; Tang, X.; Shu, J.; Wang, X.; Hu, W.; Shen, Q. D. Crosslinked P(VDF-CTFE)/ PS-COOH nanocomposites for high-energy-density capacitor application, *J. Polym. Sci., Part B: Polym. Phys.* **2016**, 54, 1160–1169. DOI: 10.1002/polb.24023.

181. Li, H.; Gadinski, M. R.; Huang, Y.; Ren, L.; Zhou, Y.; Ai, D.; Han, Z.; Yao, B.; Wang, Q. Crosslinked fluoropolymers exhibiting superior high-temperature energy density and charge–discharge efficiency, *Energy Environ. Sci.* **2020**, 13, 1279–1286. DOI: 10.1039/ c9ee03603b.

182. Pei, J. Y.; Zha, J. W.; Zhou, W. Y.; Wang, S. J.; Zhong, S. L.; Yin, L. J.; Zheng, M. S.; Cai, H. W.; Dang, Z. M. Enhancement of breakdown strength of multilayer polymer film through electric field redistribution and defect modification, *Appl. Phys. Lett.* **2019**, 114. DOI: 10.1063/1.5088085.

4 Composite Dielectric Materials for Capacitive Energy Storage

Haibo Zhang, Hua Tan, and Bing Xie

1 INTRODUCTION

The primary shortcoming of dielectric capacitors is an insufficient energy density for more complex uses.[1] This kind of capacitor often has a ceramic or polymer dielectric. Among the properties shared by ceramic dielectrics are high stiffness, exceptional thermal stability, and a high dielectric constant.[2–4] However, their low breakdown strength, poor flexibility, large specific mass, and challenging manufacturing conditions restrict their applicability in energy storage capacitors. In contrast, polymer dielectrics include not only low dielectric loss and strong breakdown strength but also low specific mass, simple processing, applicability to large-scale production, and a unique self-healing capacity. A local dielectric breakdown clears away the material at the place of breakdown, thanks to the self-healing ability, preventing catastrophic failure of the whole capacitor. However, the poor dielectric constant is a fatal drawback of polymer dielectrics.[5–10]

Polymer-based dielectric composite films, which include ceramic fillers into a polymer matrix, are one logical option to overcome these drawbacks and achieve high-energy storage density in dielectric capacitors. Ceramic fillers are used to increase the dielectric constant and dielectric displacement of the polymer matrix, whereas the polymer matrix is included for its high breakdown strength. Micron-sized ceramic particle fillers have been the primary focus of research on polymer-based dielectric composites for energy storage applications.[11, 12] However, it was quickly discovered that these composites show deterioration in their dielectric characteristics owing to stress cracking and surface flaws. The size of the microparticles also sets a cap on the thickness of these composite films. However, dielectric coatings in the nanoscale range are needed to accommodate the ongoing structural downsizing trend in the electronics sector. Thinner films are better at storing the same amount of energy in less space. To reduce the size of these polymer-based composite dielectrics, nanofillers must be used instead of micro-fillers. Because of this, polymer nanocomposite film capacitors have been suggested and produced; they use composite dielectrics with negligible nanofiller loadings. Excellent flexibility, low density, and easy processability are maintained while yet achieving the requisite dielectric properties. Polymer nanocomposite film capacitors have been shown to have superior power density, operating voltage, and cycle lifespan.[13–17]

DOI: 10.1201/9781003454496-4

2 EFFECTS OF INTERFACES IN NANOCOMPOSITE

The incorporation of various types of nanofillers, i.e., conducting and nonconducting nanofillers, considerably influences the general characteristics of dielectrics via interface polarization.[18] Also, several constraints including chain mobility and conformation of polymers, crystallinity, Coulombic potential, etc. are reported in the literature for controlling the nanomaterials' interaction with the polymer matrix.[19, 20] Herein, two most famous theoretical models are hypothesized, i.e., Lewis's and Tanaka's models, in order to comprehend the impact of the interface in polymer nanocomposites. The detailed discussion of these theoretical models is described in the following subsections.

2.1 LEWIS'S MODEL

Several interfaces in nanoscale materials are resulted due to the nanofiller addition in the polymer matrix. The chemical potential or Fermi energy level of nanoparticles and polymer matrix may eventually charge the surface of nanofillers or at least part of it. In response, the polymer matrix develops counter charges close to nanofiller surfaces.[18] Suppose a nanoparticle having a positive charge is lying inside the polymer matrix, as depicted in Figure 4.1(a) and (c), while the surface of a nanoparticle is considered to be a planner, the Coulomb attraction develops between the newly introduced positive charge and the matrix charges, which leads to the redistribution of charges in a matrix.[18] As a result, it produces a double electrical layer, comprising a Gouy-Chapman diffused and a Stern layer. The formation of the Stern layer (also termed as Helmholtz double layer) occurs at the exterior of a nanoparticle, caused by

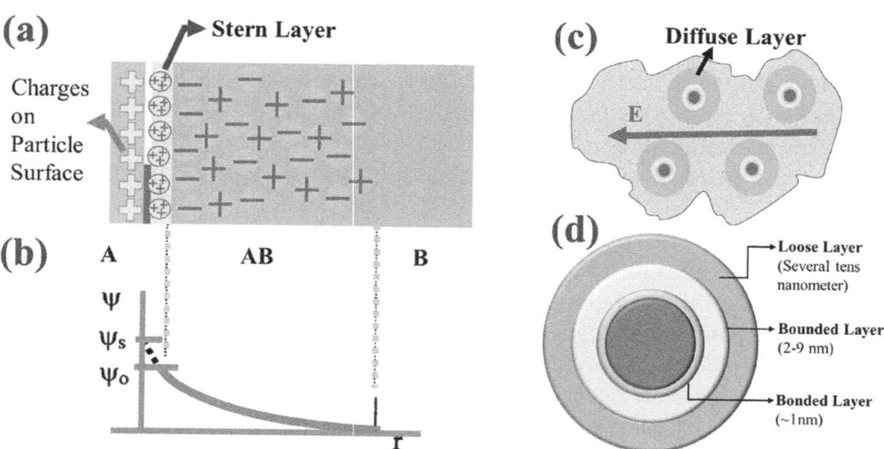

FIGURE 4.1 Schematic and graphic representation of Lewi's and Tanaka's theoretical model. The schematics of (a) charges in different layers, (b) co-ion and counterion concentrations at the AB interaction point, (c) the conduction in polymer nanocomposite by diffuse double layers. (d) The schematic of Tanaka's theoretical model containing three layers, i.e., bonded, bound, and loose layers; reproduced with permission.[18]

counterions adsorption, whereas the distribution of positive and negative ions forms a diffused layer around the Stern layer. This diffused layer is then considered as a controlling factor for the dielectric performance of the polymer-based nanocomposites; thus, adjacent to the nanofiller percolation threshold, it becomes more predominant.[21] The $\psi(r)$, termed as potential electrical distribution, represents the distance from the surface nanoparticle is shown in Figure 4.1(b), which is in agreement with the solution of Boltzmann and Poisson equations. The joint Poisson–Boltzmann relation is given as:

$$\nabla^2 \psi(r) = -e\varepsilon^{-1} \sum_i Z_i n_i(\infty) e^{-Z_i e \Psi(r)/kT} \tag{4.1}$$

Where r represents the distance from the nanoparticle surface, $\psi(r)$ denotes the r-dependent potential distribution function, k signifies the Boltzmann constant, ε indicates the permittivity of a material, and Z_i is the valency, while $n_i(\infty)$ is the concentration of ion species i inside the polymer matrix. The equation reduces to a simpler form called the Debye-Hückel approximation (Equation 4.2) when potential is small.

$$\frac{Z_i e \psi(r)}{kT} < 1 \tag{4.2}$$

Further solving results in Equation 4.3, i.e. more simplified form known Debye-Hückel equation which is represented as:

$$\psi(r) = \psi_0 e^{-\kappa T}; \kappa = \left[\frac{2e^2}{\varepsilon kT} \sum_i Z_i^2 n_i(\infty) \right]^{1/2} \tag{4.3}$$

Where κ represents a parameter known as the Debye-Hückel while κ^{-1} denotes Debye length. Further solving of the Debye-Hückel equation for the diffused double layer is simplified called the Gouy-Chapman equation, depicting the change of potential in the part of diffusion at the double layer and which starts from the layer known as the Stern layer.

$$\tan h \left[\frac{Z_i e \psi(r)}{4kT} \right] = \tanh \left[\frac{Z_i e \psi(o)}{4kT} \right] e^{-kT} \tag{4.4}$$

2.2 TANAKA'S MODEL

Tanaka et al. studied the development of interfaces, produced due to the electrical, physical, or chemical methods when sphere-shaped nanoparticles were embedded inside the matrix material and proposed a theoretical model called Tanaka's Model.[22] According to this model, "it is assumed that the interface of the nanoparticles consists of three layers, which are bonded layer, bounded layer, and loose layer", see Figure 4.1(d). The first layer or bonded layer corresponds to the transition layer with a thickness of ~1 nm and is strongly bonded to both inorganic and organic layers. The strong bonding can be attributed to either ionic bonds, covalent bonds, van der Waals force, or hydrogen bonds.[18] In an interfacial region or the bound layer, there is a strong interaction and/or bonding of polymer chains with bounded layers along with the surface of the inorganic particles. Even though its thickness ranges from 2 to 9 nm, it firmly relies upon the nanoparticles and polymer interaction. The thickness of the third (or loose layer) is approximately a few tens of nanometers and is loosely coupled with the bound layer.[22] At this loose layer, several physical parameters of the polymer matrix, such as chain conformation, crystallinity, free volume, and mobility are significantly different from its bulk. This model predicts numerous possibilities to tailor and/or improve the dielectric characteristics of the polymer nanocomposites for potential applications. The bound layer unfavorably affects the polar radical dipole orientation, while the loose layer is responsible for the decrease in free volume. Both effects result in a decreased dielectric constant. In this manner, the determination of a suitable coupling agent is one of the essential factors in achieving high-energy storage performance.[18]

2.3 Maxwell-Garnett Model

Maxwell-Garnett model is comparably easier for modeling because of its linearity; however, the major disadvantage of this type of model is that it can be applied to very small nanofiller loadings, i.e., below its percolation threshold. The model supposes that the interfiller distance of randomly oriented fillers is higher than their typical sizes.[23] In addition, it is not constrained by the matrix or filler's resistivity in nanocomposites. The theoretical effective permittivity (ε_{eff}) of 0D nanoparticles-filled nanocomposites is given as under:

$$\varepsilon_{\text{eff}} = \varepsilon_{\text{m}} \left[1 + \frac{3\varphi_{\text{f}}(\varepsilon_{\text{f}} - \varepsilon_{\text{m}})}{\varphi_{\text{m}}(\varepsilon_{\text{f}} - \varepsilon_{\text{m}}) + 3\varepsilon_{\text{m}}} \right] \tag{4.5}$$

Where the φ represents the volume fraction and its subscripts f and m indicate filler and matrix, respectively.

3 TYPES OF FILLERS

3.1 0D-Filled Nanocomposites

The insertion of 0D, i.e., spherical nanoparticles inside of the polymer matrix, is one of the solutions in polymer nanocomposites for energy storage applications that has been the subject of many studies and is very simple to produce. Adding a relatively small quantity of 0D nanoparticles to a polymer matrix has been observed to improve the material's dielectric breakdown strength. On the other hand, adding a greater quantity of nanoparticles can increase the dielectric constant almost linearly, but it will drastically lower the material's breakdown strength. It is important to point out that the research that has been published on the incorporation of simply 0D spherical nanoparticles into the polymer matrix without any interface engineering or its coupling with other dimensional nanoparticles, i.e., 1D and 2D, has become obsolete. Therefore, the research focuses on determining the optimal number of spherical nanoparticles in polymer matrices to produce a high breakdown strength and good energy storage capabilities. Some of the research about 0D polymer nanocomposites is summarized below.

The ternary composites of polyvinylidene fluoride (PVDF), barium titanate ($BaTiO_3$), and multi-walled carbon nanotubes (MWNTs) were made by Zhao et al.[24] using the solution casting technique. It is observed that increasing the dielectric constant and the breakdown field has resulted in a significant decrease in the percolation threshold (fraction of MWNTs) to below 0.4 vol%. The discharged energy density of up to 10.3 J/cm^3 was achieved with efficiencies of more than 77.2% when using a $BaTiO_3$/MWNTs/PVDF (11.5/0.35/88.15) composite with a dielectric constant of 59, a loss of less than 0.055, and a maximum working electric field of 324 MV/m. A higher fraction of the phases in polymers and more evenly distributed filler in composites are both responsible for the observed increases in polarization and dielectric constant in PVDF. To increase the composites' breakdown strength, spherical inorganic particles added during the third phase are used to stop the creation of conductive networks and make the local electric field more homogeneous.

Ding et al.[25] prepared PVDF/ZnS nanocomposites with tailorable filler loadings. Adjusting the filler content of polymer composites allowed for finer control of the materials' breakdown strength. Nanocomposites with 2 wt.% indicated a very high E_b of 4,416.5 kV/cm. Tuning the inter-particle distance led to a different distribution of charges and movement of polymer chains, which is why the breakdown failed to occur. Thus, the inter micro-electric field reduced the applied electric field at low filler loading and increased it at high filler loading. Therefore, polymer nanocomposites with 2 wt.% ZnS nanoparticles had the highest discharged energy density of 9.8 J/cm^3 at 4,500 kV/cm.

Synthesis of the novel polymer/ceramic nanocomposite by Huang et al.[26] included core-shell $BaTiO_3$@SiO_2 (BT@SO) structures with a diameter of less than 10 nm, see Figure 4.2. Since the interfaces in this ultrafine nanostructure are almost 10 times larger than those in conventional 100 nm fillers, the composite's breakdown strength and electrical displacement may be increased without sacrificing either. This is due to the fact that a composite's breakdown strength and electrical displacement may both benefit from high polarization. When tested at a 420 kV/mm electric field, the BT@SO/PVDF nanocomposite showed remarkable energy storage ability, with an U_d value of 11.5 J/cm^3.

To increase the discharged energy density U_d of PP film, Liu et al.[27] showed that a small loading of PMMA@$BaTiO_3$ (PMMA@BT) nanoparticles into the PP matrix at a volumetric concentration of 2.27% could significantly increase the U_d from 1.40 J/cm^3 of pure PP film to 3.86 J/cm^3 of PP nanocomposite film. The 3.86 J/cm^3 U_d value attained is the highest documented value for PP-based films manufactured using commercial PP resin. Meanwhile, a slight drop in charge-discharge efficiency is experienced from 99.5% to 94.1%.

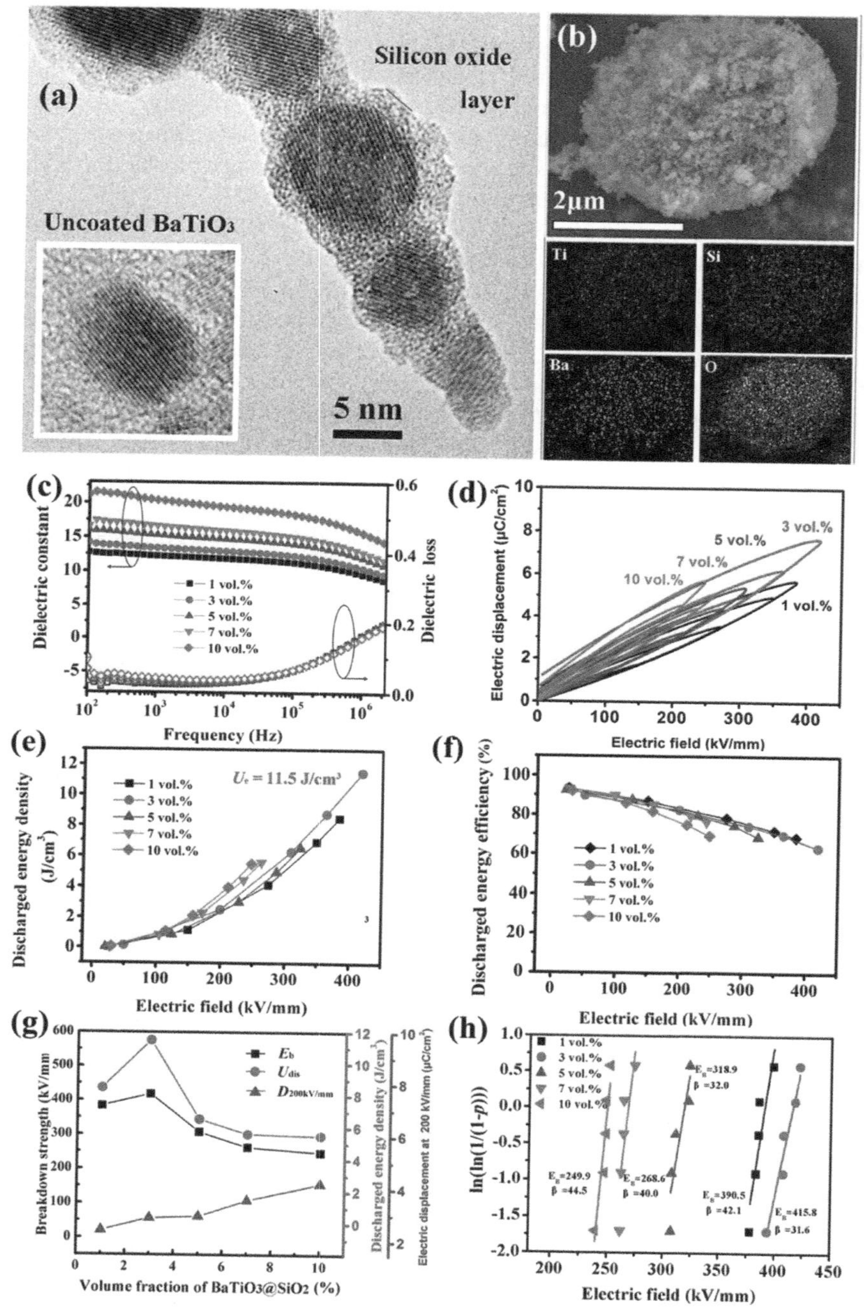

FIGURE 4.2 (a) TEM image; (b) SEM and EDS mapping images of BT@SO structure. (c) Dielectric constant and dielectric loss as a function of frequency, (d) D–E loops, (e) U_d, (f) charge-discharge efficiency, (g) E_b, U_d, and D variation at 2,000 kV/cm, and (h) Weibull's distribution of E_b with respect to the volume fraction of BT@SO.[26]

3.2 1D-Filled Nanocomposites

Recently, it has been suggested that 1D ceramic nanofillers with a bigger aspect rationed should be used instead of spherical NPs to solve the problems caused by a high NP volume fraction in the polymer matrix.[28] (1) One of the benefits of using 1D nanofillers is that, due to their strong dipole moments, they may boost the nanocomposites' permittivity at a lower nanofiller concentration.[29] (2) The reduced surface energy feature of polymer nanocomposites is another benefit of their one-dimensional structure. (3) 1D ceramic fillers cause less agglomeration inside the polymer matrix due to their low specific surface areas. The incorporation of high aspect ratio NWs into nanocomposites has been shown to enhance the dielectric constant compared to low aspect ratio NPs, as reported by Tang et al.[30]

Energy density may be increased by including several kinds of 1D nanofillers into polymers, such as linear (1D TiO_2 (TO)-based), ferroelectric (1D $BaTiO_3$(BT)-based), paraelectric (1D $SrTiO_3$ (ST)-based and $BaSrTiO_3$ (BST)-based), and relaxor-ferroelectric (1D $BaZrTiO_3$ (BZT)-based). High polarization and dielectric responses are used to justify the use of 1D ferroelectric nanofillers, in addition to the high aspect ratios of 1D nanofillers in general. To improve the dielectric properties of polymer-based nanocomposites, 1D linear nanofillers with high electric field-induced enhanced dielectric properties and a large area enclosed between $D–E$ loops and the vertical D-axis are added; 1D paraelectric nanofillers are incorporated because of their lower P_r values, and 1D relaxor-ferroelectric nanofillers are used because of their high dielectric constant and slim $D–E$ loops.

In order to increase the breakdown strength of the final nanocomposites, numerous studies have hypothesized that a smaller dielectric difference between the linear fillers, such as TiO_2, and polymers like PVDF is necessary. The leakage current densities and dielectric loss property of the anatase-phased TiO_2 nanowires (NWs) incorporated with P(VDF-HFP) ferroelectric polymer were greatly reduced in comparison to those of the raw TO NWs/P(VDF-HFP) nanocomposites. The surface of the NWs was modified with strong adhesion performance dopamine derivative (h-DA) containing elongated-chain tails with brush-like. Due to the excellent compatibility between the polymer and the filler, the dielectric breakdown strength of these polymer nanocomposites was 5,200 kV/cm for 2.5 vol% h-DA-modified TO NWs/ P(VDF-HFP). At an electric field of 5,200 kV/cm, 2.5 vol% h-DA-modified TO NWs/ P(VDF-HFP) discharged the ultrahigh energy density of around 11 J/cm^3. Moreover, at 5,000 kV/cm, the energy storage density of pure polymer only reached 8.8 J/cm^3. Interestingly yet, at a lower electric field of roughly 3,500 kV/cm, 15 vol% h-DA-modified TO NWs/ P(VDF-HFP) also yields the same energy density (8.6 J/cm^3).[31]

One of the most popular nanofillers employed in polymer nanocomposites is 1D ferroelectric nanofillers because of their strong polarization responses and, therefore, their high maximum dielectric displacements. BT is an example of a material with a high dielectric constant and a high D_{max}. Combining the high dielectric breakdown strength of the polymer with the large dielectric constant of the 1D ferroelectric has been shown to be possible via the use of ferroelectric nanofillers. Energy density in 1D ferroelectric-filled nanocomposites varied from 1 to 13 J/cm^3, which is below what is required for many potential applications. This is despite the high dielectric constant feature being one of the crucial characteristics in improving the nanocomposite's energy density. The high aspect ratio BT NWs were synthesized using a novel synthesis method proposed by Tang et al.[32] At an electric field of 3,000 kV/cm, the energy storage density of the P(VDF-TrFE-CFE) nanocomposite functionalized with 17.5 vol% BT NWs was about 45.3% higher than the pristine P(VDF-TrFE-CFE) polymer (7.2 J/cm³). Roughly 1.52 μs after the discharge began, the same nanocomposite showed a maximum power density of about 1,200 kW/cm³.

Although nanocomposites containing high dielectric constant 1D ferroelectric nanofillers incorporated with high dielectric breakdown strength polymers show promising results for pulsed power applications, the large remnant polarization of the ferroelectric nanofillers used limits the discharge energy density and charge-discharge efficiency. It follows that 1D nanofillers with bigger maximum polarization and smaller residual polarization are preferable since they may maximize discharge energy density and charge-discharge efficiency. Because of their greater maximum polarization and lower residual polarization, 1D paraelectric nanofillers are recommended.[33] Trilayer-structured polymer nanocomposites with excellent energy storage and power densities were reported by Liu et al. As a consequence of the BST-NWs-modified PVDF nanocomposite used in the sandwich layer, the produced nanocomposites had a high dielectric constant. This work presents a trilayer-structured polymer nanocomposite with an optimal filler content of 8 vol% At a dielectric breakdown strength of approximately 5,880 kV/cm, the BST NWs/PVDF nanocomposite demonstrated an ultrahigh-discharged energy density of around 20.5 J/cm³. A higher power density of roughly 910 kW/cm³ was also shown using the same trilayered nanocomposite, which is more than nine times that of the most advanced biaxially orientated polypropylene (BOPP).[34]

Despite the strong dielectric breakdown strength and high dielectric constant quality of nanocomposites, including 1D ferroelectric nanofillers, one of their high-energy densities compromising factors is their broad D–E loops (large energy loss, U_{loss}). So, 1D nanofillers are preferred because they provide considerably narrower D–E loops in the nanocomposite and increase the area between the D–E loop and the vertical dielectric displacement axis. Researchers mixed 1D relaxor-ferroelectrics nanofillers, which inherently have narrow D–E loops compared to 1D ferroelectric nanofillers, with diverse polymer matrices to achieve the desired features. Pan et al.[35] synthesized PATP-modified BCT-BZT nano fillers (NFs) by electrospinning, grafted them to the surface of the NFs, and then integrated the NFs into a PVDF polymer matrix. Chemical bonding and confinement of charge carrier motion at the filler-matrix interface area successfully improved NFs' distribution and compatibility in this way. Thus, small nanofiller loading, i.e., 2.1 vol%., increased the energy storage density to 8.2 J/cm³ at a relatively lower electric field of 3,800 kV/cm, see Figure 4.3.

3.3   2D-Filled Nanocomposites

In polymer nanocomposites, 2D fillers with a high aspect ratio and lateral dimensions may significantly improve ε_r and E_b at a low filling content by forming percolation systems or constructing effective conduction barriers. Adjustments to these characteristics are possible by using various filler stages and developing multilayer

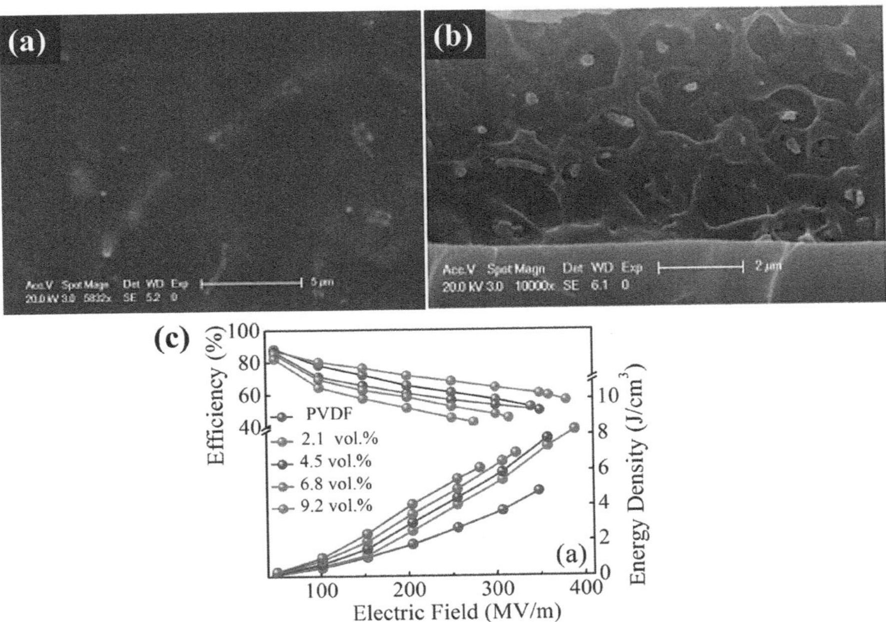

FIGURE 4.3 4.5 vol% BCZT@PATP NFs/PVDF nanocomposites' (a) top and (b) cross-section view. (c) BCZT NFs/PVDF nanocomposites' energy storage results as a function of electric field.[35]

designs.[6, 36, 37] PNDs reinforced with $BaTiO_3$ (BT) plates were described by Wen et al.[38] One weight percent PND loading results in a maximum U_d of 9.7 J/cm³, a 100% improvement over the U_d of unloaded PVDF.

$Ti_3C_2T_x$ nanosheet-reinforced p-n junctions were fabricated by Feng et al.[39] using a different tri-layer structure. As opposed to the symmetric sandwich construction, this one is asymmetric since the loading contents are dispersed along a gradient. As loading concentration rises from 0.1 to 2 vol%, the dielectric strength of single-layer PNDs drops while their maximum displacement (D_{max}) increases. The gradient-structure sample (2-1-0.1) has a computed average loading content slightly higher than 1 vol%, but its E_b and D_{max} are much larger than those of the single-layer sample with 1 vol% $Ti_3C_2T_x$ (MXene-1). Because the field strength is being redistributed across each sublayer in the 2-1-0.1 sample, the E_b is noticeably greater than in MXene-1. The interfacial barrier effect for suppressing the electric tree development over the full thickness of the gradient film is responsible for the abnormally high E_b of the 2-1-0.1 sample, which is even greater than that of pure PVDF. In addition to the barrier effect, the interface between neighboring layers also experiences the Maxwell-Wagner-Sillars (MWS) effect The gradient-structure sample achieves a high E_b and polarization by a combination of the $Ti_3C_2T_x$/PVDF MWS effect and the micro-capacitor mechanism. Therefore, compared to MXene-1, where the highest U_d is 4.57 J/cm³, and the efficiency at its E_b is 43.83%, the 2-1-0.1 sample has an U_d of 12.5 J/cm³ and an E_b efficiency of 64%.

Films with a BNNS layer in the center were the focus of an investigation by Zhu et al.[40] The current density in the PVDF matrix near the in-plane BNNSs is considerably lower than that near the through-plane BNNSs for the sample with 0.2 vol% randomly distributed BNNSs ($BNNS_{0.2}$). Thus, it is possible that high-current paths arise along the field direction, diminishing the effectiveness of the method used to reduce the leakage current throughout the composite material. By contrast, in PBP3 (a sandwich structure with 0.16 vol% BNNSs in the intermediate layer), all BNNSs lie in the in-plane direction and create a tight topological barrier to restrict the total leakage current. The dielectric characteristics of PBP3 are enhanced by the fact that its compact BNNS interlayer may serve as a line of effective scattering centers under strong electric fields, hence reducing the mobility of free charge carriers. To further understand the electrical breakdown process, a modified dielectric breakdown model was utilized to simulate the electrical tree propagation state of the two samples, taking into account the impact of space charge on the electric field distribution of composites. In PBP3, the BNNS barrier layer causes the electrical tree to be redirected or perhaps cut off entirely. Because BNNSs do not create a continuous topological barrier, the electrical tree may spread unhindered over the $BNNS_{0.2}$ sample. Consequently, PBP3's interlayer BNNS is more compact and continuous than $BNNS_{0.2}$'s, allowing it to limit the propagation of electric damage more effectively, resulting in a higher E_b. The reverse-sandwich-structured samples outperform neat PVDF and $BNNS_{0.2}$ in terms of energy storage due to the reduction in tan δ and the improvement in E_b.

3.4   CORE-SHELL-FILLED NANOCOMPOSITES

Enhancing the interfaces inside a nanocomposite is one of the most effective strategies to increase its energy storage capability, as was stated in the polarization section. One technique is to use core-shell nanostructures, whereby a low dielectric constant ceramic nano-layer (the shell) surrounds a high dielectric constant nanofillers (the core). The huge discrepancy between the polymer dielectric constant ($\varepsilon_r = 10$) and the ceramic nanofillers ($\varepsilon_r > 200$) is attenuated. This results in an increased dielectric breakdown strength and less internal distortion in polymer nanocomposites. This is also one of the methods used to increase the volume fraction of conductive fillers that can conduct electricity, such as metallic Al with a native oxide Al_2O_3 in the form of nanoparticles or nanowires, or semi-conductive Si with a coherent surface oxide SiO_2, and so on. Despite rapid increases in the dielectric constant, dielectric loss is kept to a minimum by the insulating sheath. So far, SiO_2 (SO), Al_2O_3 (AO), TiO_2 (TO), etc., have been the most effective nano-shells used in core-shell designs. Though the ε_r-mismatch is small, the large difference in electrical conductivity between ceramic nanofillers and polymers causes significant interfacial polarization due to the buildup of space charges at the interface. However, with an increase in interfacial polarization comes an increase in dielectric loss property and a decrease in dielectric breakdown strength in the polymer nanocomposites, which in turn leads to a decrease in the polymers' capacity to store energy. Due to polymer-nanofiller mismatch and distortions followed by poor breakdown strength, researchers have seldom studied the encapsulation of a relatively high dielectric constant shell on the surface of a core-nanofiller displaying a moderate ε_r.[41–46] To create high-energy storage nanocomposites, Wang et al.[43] used the high ε_r-core and low ε_r-shell method, preparing BT NWs encased by TO shells of varying thicknesses. At an electric field of 4,400 kV/cm, it was found that 5 wt.% TO@BT NWs-modified P(VDF-HFP), i.e., core-shell strategized nanocomposites, displayed ultrahigh-discharged energy density and high efficiency of around 9.5 J/cm^3 and 63%, respectively. The energy storage capability of the polymer nanocomposites is not linearly impacted by altering shell thickness, as shown by a comparison of core-shell strategized NWs with varied thicknesses. Using the finite element approach, we found that the higher energy storage capacity was mostly attributable to a reduction in the leakage current in the P(VDF-HFP) polymer matrix and an increase in the electric field concentration.

Meanwhile, Lin et al. investigated PVDF nanocomposites modified with BT@TO NFs as a potential strategy for increasing the maximum dielectric displacement of nanocomposites with less nanofiller concentration. The highly interfacial-engineered BT@TO NFs were prepared using a novel coaxial electrospinning method. The charge shift at the BT-TO interface was shown to significantly impact the dielectric constant and the maximum dielectric displacement. Extra polarization plays a crucial role in the increase when it comes to the maximum dielectric displacement of a nanocomposite. Because the charge shift is localized to the interface area, the dielectric breakdown strength of the polymer nanocomposite is maintained. Because of this, it was unable to create a percolation channel in the polymer matrix. At an electric field of 3,600 kV/cm, the BT@TO NFs-modified PVDF nanocomposite with 3 vol% nanofiller achieves a relatively greatest discharged energy density of around 10.9 J/cm^3. In addition, 2.5 vol% BT@Al$_2$O$_3$ NFs-modified PVDF nanocomposite at 3,800 kV/cm has a maximum discharged energy density and efficiency of around 7.1 J/cm^3 and 64%, respectively, compared to 2.5 vol% BT NFs-modified PVDF nanocomposite (5.6 J/cm^3 at 3,300 kV/cm). However, at an electric field over 1,000 kV/cm, the charge-discharge efficiency of 2.5 vol% BT NFs-modified PVDF nanocomposites dropped dramatically to 58.2%, whereas it stayed at 82.1% for 2.5 vol% BT@Al$_2$O$_3$ NFs-modified PVDF nanocomposite. According to the same scientists, a nanocomposite's dielectric breakdown strength is enhanced due to an AO-ability shell to drastically minimize leakage current by limiting contacts between neighboring BT NFs inside the nanocomposite. At the same time, the nanocomposites had greater D_{max} and lower P_r values under the same electric field compared to the pure PVDF polymer. The maximum discharged energy density of BT@Al$_2$O$_3$ NFs-modified PVDF nanocomposite was 7.1 J/cm^3 at an electric field of 3,800 kV/cm with a charge-discharge efficiency of around 62.5%.[41]

As an added bonus, with an electric field of 3,500 kV/cm, a core-shell structured PVDF nanocomposite modified with 2.5 vol% BT@SO NFs showed a comparatively high discharged energy density of around 6.6 J/cm^3 and a charge-discharge efficiency over 62.7%. The MWS interface polarization was suppressed by the SO-coating on the surface of the BT NFs, which led to a higher discharged energy density.[42] Pan et al.[45] investigated the single core-double shell (BT@TO@AO NFs)-modified PVDF nanocomposite. The filled PVDF nanocomposite displayed a maximum U_d of roughly 14.8 J/cm^3 and charge-discharge efficiency of 64.6% at 4,503 kV/cm, see Figure 4.4. In comparison to the current state-of-the-art BOPP (1.2 J/cm^3 at 6,400 kV/cm),

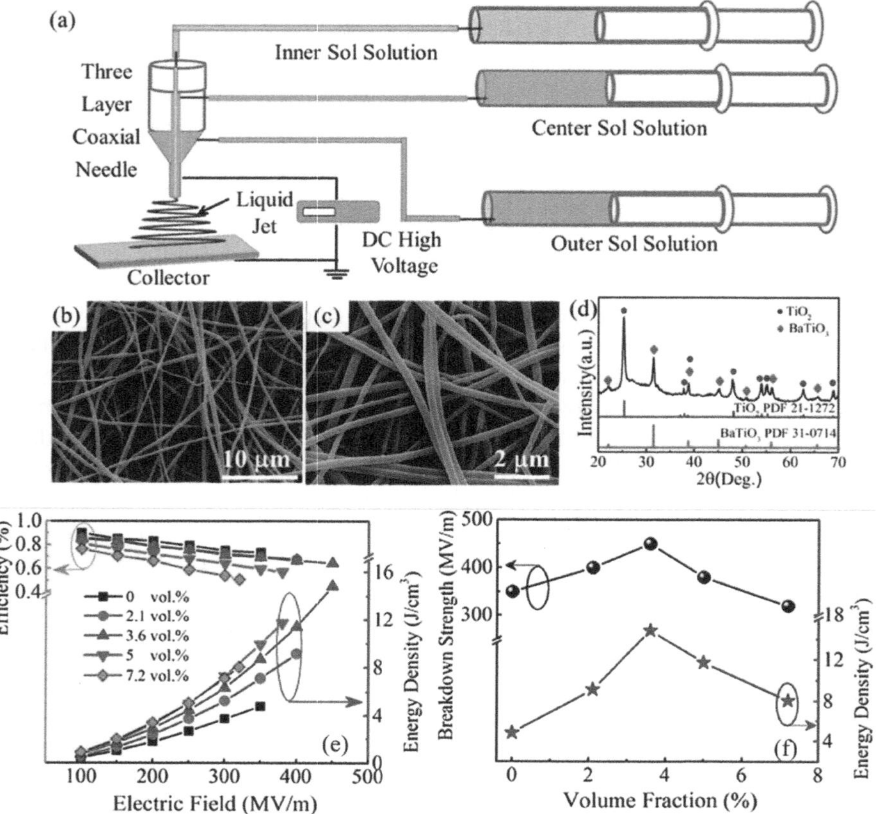

FIGURE 4.4 (a) Schematic of the preparation of BT@TO@AO/PVDF nanocomposites. (b) SEM images of the as-prepared and (c) heat-treated nanofibers. (d) XRD analysis of the heat-treated nanofibers. (e) Charge-discharge efficiency and energy density of various volume rationed BT@TO@AO/PVDF nanocomposites. (f) Energy density and breakdown strength of nanocomposites as a function of nanofillers volume fraction.[45]

this is almost 12 times greater. Maximum U_d is attributed to 3.6 vol% BT@TO@ AO NFs-modified PVDF nanocomposite due to its high relative E_b and D. Ultrafast discharge times on the order of microseconds were also seen under identical loading conditions. Comparing PVDF nanocomposites modified with 3.6 vol% BT NFs, BT@TO NFs, and BT@TO@AO NFs, the BT@TO NFs-modified PVDF nanocomposite showed the largest maximum dielectric displacement. Also, in the same electric field, the BT@TO@AO NFs-modified PVDF nanocomposite displayed the lowest P_r. This was due to three factors. (a) Interfacial polarization between the inner TO-layer and the outer AO-layer of this nanocomposite causes a slow but steady increase in the dielectric constant value and the conductivity. (b) the outer AO-layer insulation effectively inhibits free electrons mobility, increases current percolation, and minimizes structural flaws, resulting in lower dielectric loss and leakage current density. This is because (c) the DA-coating ensures excellent interfacial adhesion in the polymer matrix, and (d) the nanofillers are distributed evenly throughout the matrix. To improve dielectric breakdown strength, discharged energy density, and charge-discharge efficiency, BT@TO@AO NFs-modified PVDF nanocomposites remarkably owned the largest maximum dielectric displacement and yet retained a fairly lower P_r value. The ultrahigh-discharged energy density of around 10 J/cm^3 and of 65% at an electric field of 5,000 kV/cm was found in a P(VDF-HFP) polymer containing BT NPs encapsulated by TO NWs (BT@TO), which is more than 1.3 times higher than pure P(VDF-HFP) polymer (7.6 J/cm^3 at 4,400 kV/cm). However, the dielectric breakdown strength decreases as the nanofiller loading increases, which has a negative impact on energy storage ability. For instance, under a reduced electric field of 4,400 kV/cm, a P(VDF-HFP)-based nanocomposite containing 10 wt.% BT@TO NWs discharged 8.3 J/cm^3. When the nanofiller loading was increased to 15 and 20 wt.%, the results were 7 J/cm^3 at 4,000 kV/cm and 5.1 J/cm^3 at 3,400 kV/cm. The investigation of charge-discharge efficiency indicated that the P(VDF-HFP) nanocomposite was modified with 5 wt.% BT@TO NWs and had the comparatively greatest performance.[46]

3.5 SURFACE-DECORATED-FILLED NANOCOMPOSITES

As we've shown above, improving interfaces is crucial for increasing polymer nanocomposites' total energy-storage efficiency. Creating hierarchical internal and external interfaces, or interfaces on the inside and outside of each 1D nanofiller is one of the most successful methods for improving these interfaces. Here, 0D nanoparticles are tethered to 1D nanofillers from inside or outside. To take advantage of the synergy between the internal interfaces of the NFs and the high aspect ratio of the NFs, Zhang et al.[47] created TiO$_2$ NFs embedded with BaTiO$_3$ NPs (BT@TO) within P(VDF-HFP) polymers. Dielectric displacement and dielectric breakdown strength characteristics in flexible P(VDF-HFP) nanocomposite films were decoupled due to the presence of high aspect ratio composite NFs with internal hierarchical interconnections. Dielectric polymer nanocomposites with a modest nanofiller loading (3 vol% BT@TO nanofibers) produced the highest energy density to date (31.2 J/cm^3). Because the NFs were aligned perpendicular to the electric field, the dielectric breakdown strength of the polymer nanocomposites was significantly improved. Results from a thorough high-resolution transmission electron microscopy (HRTEM) study also demonstrated the coexistence of Ba^{2+} and Ti^{4+} ions. The density functional theory (DFT) simulations of the dielectric polarization behavior of the individual BTO@TO NFs were similarly in agreement with the actual data. In addition, the dielectric breakdown strength and maximum dielectric displacement of the BT@TO NFs/P(VDF-HFP) nanocomposite simultaneously increased to 7,977 kV/cm and 10.5 C/cm^2, respectively, leading to a large energy density of 31.2 J/cm^3, see Figure 4.5. When compared to current-generation BOPP, the energy density of these nanocomposites is greater by more than a factor of 1,400, making them the highest-density polymer-based energy storage materials to date. Equally impressive, even with an electric field of 8,000 kV/cm, the charge-discharge efficiency remained over 78%. These findings show that surface-decorated nanofiller has great promise

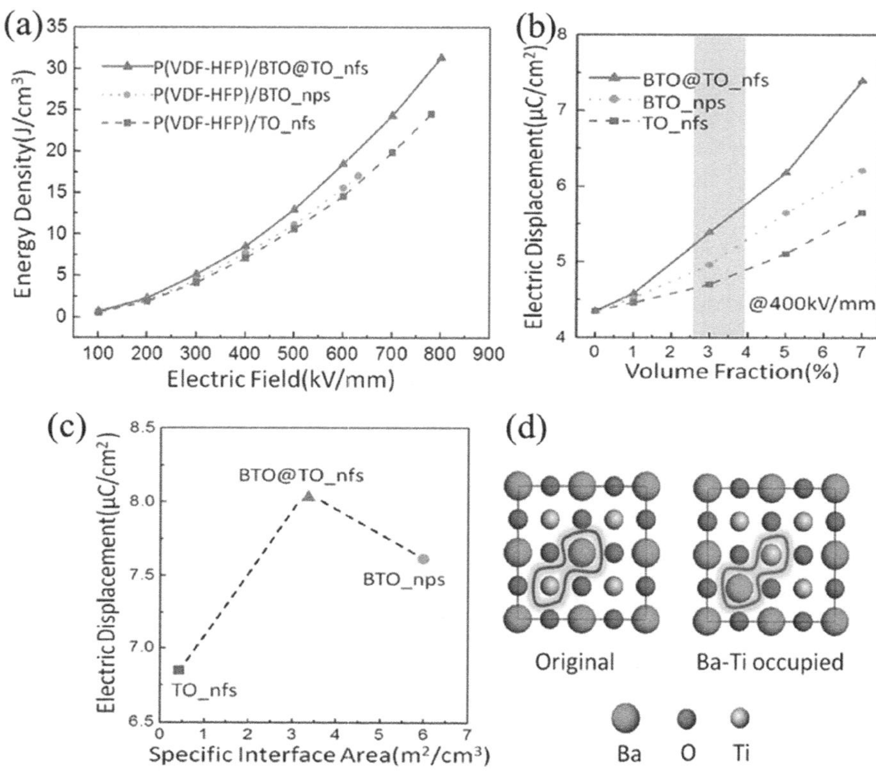

FIGURE 4.5 (a) Energy density of BTO@TO/PVDF nanocomposites as a function of electric field. (b) Electric displacement of TO, BTO, and BTO@TO with respect to various volume fractions. (c) Electric displacement of TO, BTO, and BTO@TO as a function of surface interface area.[47]

for resolving the seeming contradiction between improved dielectric performance and lower breakdown strength. It has also provided a new approach for atomic-scale interface engineering of nanofillers to improve polymer nanocomposites' energy density.[47]

Similar authors used PVDF to prepare BT@TO NFs. Incorporating the same BT@TO NFs with PVDF resulted in a significantly better ε_r (>41) at a relatively lower nanofiller volume fraction, i.e., near 10 vol%. However, modifying the polymer matrix brought the E_b value down to 6,467 kV/cm, which is a substantial improvement. Since then both the ultrahigh energy density and the charge-discharge efficiency have dropped to about 20 J/cm³.[48]

4 SYNTHESIS OF NANOCOMPOSITE FILMS

Many techniques may be used to get polymer nanocomposites ready for use. These include the Solution Cast Method, the Spin Coating Method, the Hot Press, and

Dip Coating Route, the Melt Intercalation Method, and the In Situ Polymerization Method.[49]

When it comes to producing polymer films of varying thicknesses, the simplest, most time-tested way is the solution cast technique. To begin, you'll need to make your own nanoparticle solution and polymer salt. The polymer salt may be prepared by dissolving the necessary quantity of the polymer host into the designated solvent with constant stirring. The homogeneous polymer matrix is then saline to the correct concentration and agitated until a clear, uniform solution is obtained. Solution nanoparticles are introduced to a solvent and dispersed via ultra-sonication as part of the nanoparticle synthesis process. Now, the nanoparticle solution is mixed with the polymer salt solution until a homogeneous mixture is achieved; this is the polymer nanocomposite. Solvent evaporation was accomplished by leaving the viscous solution in the petri dish at room temperature. After removing the film from the petri dish, the solvent is placed in a vacuum oven to evaporate any leftover solvent.[50]

The spin coating technology may easily and quickly create uniform coatings anywhere from a few nanometers to a few microns in thickness. This technique employs a controlled spin coater speed to evenly distribute the solvent across the substrate. Using centrifugal acceleration, the heat evaporates the solvent after the mixture has been distributed over a substrate. It is crucial that the substrate be held perpendicular to the rotating axis when coating the thin layer. Four variables affect film thickness: (i) the mixture's viscosity, (ii) the mixture's concentration, (iii) the rotational speed, and (iv) the spin duration. This technique works well with low-viscosity solution combinations. It is not relevant for mixtures of higher-order viscosities. There is no way for the spin coater to thin down a gel-type mixture.[51]

The hot press process is a quick and inexpensive option. This approach eliminates the need for solvents, leading to superior thin films. This film is made from a thick substance. The heating chamber in this method is paired with a temperature controller, and the sample weighing cylinder is housed in the base of the chamber. This method is often used for polymers. The polymer, nanoparticle, and solvent are mixed together when using an agate pestle. After the requisite amount of time has passed, the resulting slurry and polymer combination is heated to a temperature just below their respective melting points. The collected material is placed in the center of two stainless steel plates. The apparatus applies pressure to the slurry, which results in the film.[50]

Dip Coating is a method for coating a substrate on both sides with a high-quality film. This approach is one of a kind because of its low price. Immersion, deposition, and drainage are the three stages of the process. When this happens, the solvent evaporates. First, the substrate is submerged in the solution at a steady pace, allowing for enough coating time. Disposal and drainage come next. The substrate is then maintained immersed in the solvent. For the substrate to interact, the solvent must be present. As a result, the substrate is being slowly removed at a steady speed in order to facilitate the formation of a thin coating on the substrate. Solvent evaporation is the last stage. After this, the substrate is heated to drive out any remaining solvent. The speed of the removal and the density of the solution determine the film quality and thickness.[52]

The melt intercalation process is significant since it is inexpensive, does not need a solvent, and is thus beneficial to the environment. This procedure improves nanoparticle dispersion and enhances thermal characteristics, which are two of its primary benefits. Extreme heat has the potential to destroy nanoparticles by changing their surface characteristics. First, the polymer matrix is annealed at high temperatures, then the nanoparticles are added, and last, the mixture is vigorously stirred for a certain amount of time. Surface modification of the nanofiller, compatibility with the host polymer, and processing conditions may all be managed by dispersion.[53, 54]

The in-situ polymerization technique is a generic approach for creating polymer nanocomposites. This technique combines the monomer and nanoparticles in an appropriate solvent. To create a polymer nanocomposite, monomers are intercalated with nanoparticles and then polymerized using a suitable reagent or free radical initiator. It takes a bottom-up strategy. With the aid of this technique, well-defined structures in several dimensions may be synthesized, each with unique characteristics that cannot be attributed to the initial building blocks. Generally, homogenous dispersion is difficult in the polymeric matrix. This method is used to do this. Metal or metal oxide nanoparticles are formed within the polymer matrix by the metal precursor. The in-situ technique is useful for regulating particle size and shape. In order to convert metal nanoparticles into the organic phase, the most often used technique is the sol-gel process, which is based on a chemical pathway.[55]

5   CONSIDERATIONS FOR FILMS' QUALITY OF NANOCOMPOSITES

Due to the dielectric film's low surface energy, the metal electrode cannot establish intimate contact with it. Corona or plasma treatment is often used to boost the surface energy of polymer films before they are coated with a metal electrode. The energy of the particles in plasma is typically several to tens of electron volts, which is comparable to or more than the chemical bond energy of plastic molecules. Nanocomposites containing polyethylene (PE), polypropylene (PP), polyvinyl chloride (PVC), polyamide (PA), polyester (PET), polycarbonate (PC), and different fluoropolymers (and associated copolymers), all benefit greatly from the corona treatment approach. More care must be taken during procedures using thinner polymer sheets. Fortunately, the corona treatment only affects the topmost few nanometers of a film's surface. Thus, it has no negative impact on the film's mechanical qualities and does not lead to the release of harmful substances or pollutants.

It is also crucial to carefully consider the methodology used to evaluate dielectric strength. Common techniques include the use of a ball electrode, a metalized film electrode, or a metal electrode deposited directly onto the film. For thin films and soft films, in particular, the actual value of the breakdown will be impacted by the mechanical force applied by the conventional ball-plane test fixture. Many labs have started using the metalized film electrode approach for their thin-film research since it eliminates this limitation.[56] Roughness, wrinkles, and pits on the surface of thinner films significantly lower their breakdown strength. Important factors in determining charge behaviors at interfaces include specimen surface roughness, electrode material, surface pollution, and gases absorbed by electrodes and materials. The importance of the metal-polymer contact problem cannot be overstated. The effect of electrode material on the dielectric breakdown field of P(VDF-TrFE-CFE) was studied by Chen et al.[57] According to their findings, the interfacial characteristics of the polymer have a significant role in determining the breakdown field of the relaxor terpolymer, and an electrode with a lower charge injection would result in a stronger breakdown field. The interface barrier layer and the variation in Schottky barrier height account for the observed discrepancy between the charge injection and breakdown fields.[58]

Moreover, the dispersibility of nanoparticles within the polymer matrix material is one of the significant challenges for large-scale production. If the nanoparticle concentration in the polymer matrix is higher than the threshold value, it will agglomerate and lead to the early breakdown of the nanocomposite. Thus, optimal nanoparticle content and novel processing techniques are required to remedy this problem. Furthermore, intrinsic,[59] electromechanical,[60, 61] thermal,[62, 63] partial discharge,[64–67] bulk-limited conduction and electrode-limited conduction needs serious consideration in order to enhance the electric breakdown strength of polymer nanocomposites and in turn its energy storage properties.[68–71]

6 HIGH-TEMPERATURE NANOCOMPOSITES

BOPP is the best dielectric polymer currently on the market; however, it can only function at temperatures below 105°C. As a result, dielectric polymers can never be used in high-temperature applications without first implementing some kind of heat control system. For instance, cooling systems must be used to reduce the ambient temperature from about 140°C to about 70°C in order to accommodate BOPP film capacitors in the power inverters of hybrid and electric vehicles, which are used to control and convert direct current from batteries into the alternating current required to power the motor. The integrated power system loses efficiency and reliability as a result of the additional mass and size. Due to the exponential growth in the amount of heat produced by electronic devices and circuitry as they get smaller and more useful, the demand for high-temperature dielectric polymers has skyrocketed with the popularity of lightweight and flexible electronic devices.

High-temperature uses of polymer-based capacitors are constrained by the polymer's low glass transition temperature and higher losses with increasing temperature. Li et al.[63] evaluated the dielectric and energy storage characteristics of the composite crosslinked divinyl tetramethyl disiloxane-bis(benzocyclobutene)/boron nitride nanosheets (c-BCB/BNNS) to those of several high-temperature polymer films, they discovered that the c-BCB/BNNS demonstrated superior energy density and efficiency with excellent temperature stability. Due to the increased thermal conductivity provided by the BNNS, superior temperature stability can be ascribed to its usage in nanocomposites. Also, the insertion of BNNS results in a boost to Young's modulus. Due to the increased Young's modulus, electromechanical failure occurs less often. For this reason, expanding the use of ceramic/polymer composites in energy storage may be as simple as combining the two types of materials. Next, Li et al. analyzed the high-field capacitive energy storage characteristics at high temperatures. When comparing the discharged energy density (U_d) and the charge-discharge efficiency (η), the c-BCB/BNNS is obviously superior to all of the high-T_g polymer dielectrics throughout a temperature range of 150°C to 250°C. For instance, at 150°C, c-BCB/BNNS may discharge an U_d of more than 2.2 J/cm³ with an of more than 90% under 400 MV/m. At 200°C, the U_d of c-BCB/BNNS is 2 J/cm³, under 400 MV/m, which is double that of PEI, and the η value is more than 1.5 times higher than PEI. At 250°C, when all other high-Tg polymer dielectrics fail at voltages over 150 MV/m, c-BCB/BNNS continues to function up to 400 MV/m with an U_d of 1.8 J/cm³. Amazingly, the value for c-BCB/BNNS at 150°C (97%) is the same as that of BOPP at 70°C (200 MV/m), which is the operational state of BOPP film capacitors in electric cars. This suggests that the complicated cooling system for power inverters in electric cars may be avoided if c-BCB/BNNS were used instead of BOPP. Furthermore, under these circumstances, the U_d of c-BCB/BNNS is approximately 40% greater than that of BOPP due to its higher K; that is, 3.1, compared to 2.2 for BOPP, by replacing BOPP with c-BCB/BNNS, the complex cooling system for power inverters in electric vehicles could be eliminated. Furthermore, under these conditions, the U_d of c-BCB/BNNS is over 40% higher than that of BOPP owing to its higher K; that is, 3.1 versus 2.2 for BOPP.

Zhu et al.[72] presented a novel category of high-temperature dipolar polymers that are founded on sulfonylated poly(2,6-dimethyl-1,4-phenylene oxide) (SO_2-PPO), which was produced by post-polymer functionalization. As a result of the effective rotation of highly polar methyl sulfonyl side groups below the glass transition temperature (T_g and 220°C), the dipolar polarization of these SO_2-PPOs was boosted, and as a consequence, the dielectric constant was high. As a direct result of this, the discharge achieved an energy density of up to 22 J/cm³. The SO_2-PPO$_{25}$ sample displayed a low dielectric loss in addition to its high T_g, which was due to the fact that its T_g was so high. For instance, the dissipation factor, expressed as tan, was 0.003, and the charge-discharge efficiency, expressed as a percentage, was 92% at 800 MV/m. Because of this, the use of these dipolar glass polymers for the storage of electrical energy in applications that require high temperatures, a high-energy density, and minimal loss is considered to be very promising.

Zhang et al.[73] showed that poly(arylene ether urea), a high-temperature (high glass transition temperature) semicrystalline dipolar polymer, can be modified by the incorporation of nanofillers at very low-volume content to significantly decrease conduction losses at high electric fields and across a wide temperature range. This modification is possible because poly(arylene ether urea) has a high glass transition temperature. As a consequence of this, the polymer with a modest nanofiller loading (0.2 vol%) provides a high discharged energy density of around 5 J/cm³ even when the temperature is set to 150°C. The experimental data uncovered changes in the microstructure of the nanocomposites, and these changes, which occurred at a nanofiller loading of 0.2 vol%, shortened the mean free path for mobile charges, increased the deep trap level, and loosened constraints on dipole motions locally in the glassy state of the polymer. All of these effects occurred in the glassy state of the polymer.

According to the findings of Chi et al.,[74] the matrix material is high-temperature resistant polyimide (PI), and the filling phase is composed of $0.5Ba(Zr_{0.2}Ti_{0.8})O_3$–$0.5(Ba_{0.7}Ca_{0.3})TiO_3$ (BZT-BCT) nanofibers. When the doping content of BZT-BCT nanofibers is more than 1 vol%, the dielectric strength of the composites drops sharply when the temperature increases from 25°C to 150°C, which results in serious deterioration of energy storage properties. This can be found out by analyzing the energy storage behaviors of BZT-BCT/PI composites at different temperatures. On the basis of this, a composite film with a sandwich structure has been designed. The intermediate layer is comprised of BZT-BCT/PI, whose volume fractions vary, and hexagonal boron nitride (h BN), which possesses excellent thermal and insulating properties, is introduced in the top and bottom layers with a content of 5 vol%. As a consequence of this, the findings have shown that the energy storage capabilities of the sandwiched dielectric composite films that were created demonstrate remarkable temperature stability. The maximum field strength of the composite film with a BZT-BCT content of 1 vol% in the intermediate layer is 360 kV/mm and 350 kV/mm at temperatures of 25°C and 150°C, while the storage density is 2.3 J/cm^3 and 1.83 J/cm^3, respectively. These values are based on the fact that the composite film has a BZT-BCT content of 1 vol% in the intermediate layer.

Yue et al.[75] showed polyimide (PI) matrices, reinforced with reduced BaTiO$_3$ (rBT) particles sintered in a reducing environment (95N$_2$/5H$_2$) with a high defect density. The dielectric constant and energy storage density of the rBT/PI composite films made via in-situ polymerization were greatly improved. Compared to pure PI (ε_r = 4.1), the rBT/PI composite with 30 wt.% rBT achieved a dielectric constant of up to 31.6 while preserving reduced loss (tan δ = 0.031@1,000 kHz). Its energy storage density (9.7 J/cm^3 at 2,628 kV/cm) was more than the energy density of the finest commercial BOPP (1.2 J/cm^3 at 6,400 kV/cm). This was an improvement of nearly 400% over the energy storage density of pure PI (1.9 J/cm^3 at 3,251 kV/cm). As a result of the linear dielectric performance of the rBT/PI composite films, the energy storage efficiency was close to 90%. Interface contact between two phases and surface imperfections of rBT caused by the reducing environment may both contribute to the enhanced dielectric constant and energy storage density. Given their high dielectric constant, energy storage density, and storage efficiency, rBT/PI composite films may find use in the fabrication of embedded capacitors.

Further, a discharged energy density of 2.92 J/cm^3 and Weibull breakdown strength of 547 MV/m at 150°C were reported for the ternary polymer nanocomposite by Li et al., which is 83% and 25% higher, respectively, than those of the virgin polymer.[76] In order to understand the energy-storing capabilities of ternary polymer nanocomposites, researchers have looked at their conduction characteristics, such as the interfacial barrier and the carrier transport mechanism.

Using a solution casting method, Marwat et al.[77] produced a sandwich structure of barium titanate/poly(ether imide) (BT/PEI) nanocomposites, with two insulation layers (ILs) on each side for high breakdown strength and a polarization layer (PL) in the center for high dielectric constant. Thus, the sandwich-structured BT/PEI nanocomposites with optimal BT NPs concentration in ILs and PL demonstrated tremendously enhanced discharged energy density (U_d) of 5.7 J/cm^3, which was 256% and 307% higher than the pristine PEI (with 1.6 J/cm^3) and its single-layered counterpart, i.e., 9 wt.% BT/PEI (with 1.4 J/cm^3). With a comparable sandwich configuration, the discharge efficiency was 62% at a high electric field. High-temperature hysteresis loops further showed that the best sandwich structure nanocomposites could withstand temperatures up to 150 °C while maintaining a mechanical stress of 200 MV/m. In addition, heterogeneous sandwich-structured nanocomposites, composed of pure PEI on the outside and xDA@Ag@BT/PVDF with x = 1–11 wt.%, were also developed by Marwat et al.[78] Based on the results, PEI helped insulate charge injection from electrodes, decreased polymer free volume, increased electrical insulation, and enhanced thermal and mechanical stability. Using Ag-grafts on top of BT NPs helped DA@Ag@BT/PVDF achieve more polarizability, and the well-known Coulomb blockade effect synergistically increased E_b. Finite element modeling supported a decrease in the local electric field concentration inside the polymer matrices of the outer and inner layers, indicating improved E_b and U_d performance. The ideal heterogeneous sandwich-structured nanocomposites showed the ultrahigh U_e of 21.03 J/cm³ at 592.1 MV/m, which is the greatest U_d recorded to date when the comparable NP content and an analogous electric field are taken into account. Moreover, essential heterogeneous sandwich-structured nanocomposites demonstrated remarkable thermal stability up to 170°C in high-temperature energy storage experiments.

In conclusion, a lot of research is still being done to improve the thermal stability of polymer nanocomposites for energy storage applications. For this, various strategies are incorporated, including using high-T_g polymers, introducing thermally conductive nanoparticles, and engineering the interface between polymer and nanoparticles. All these strategies facilitate the high-energy storage capability in polymer nanocomposites at high temperatures. However, to create an innovative, industrially feasible solution that can replace BOPP, it is still necessary to develop novel engineered polymers, nanoparticles, interface engineering, and innovative methodologies to obtain thermally stable high-energy storage dielectrics.

7 MULTILAYER STRUCTURE NANOCOMPOSITES

In order to obtain polymer-based dielectric nanocomposite dielectrics with U_d increasing the dielectric constant of polymer-based dielectric nanocomposite dielectrics is a very effective method. Numerous studies have been conducted to improve ε_r of polymer-based dielectric nanocomposite dielectrics by increasing the filler content. However, the breakdown strength of polymer-based dielectric nanocomposite dramatically decreases with a large increase in filler content.[79–89] Despite the increment in the dielectric constant of polymer-based nanocomposites, the decrease in the breakdown strength causes a very minor enhancement in energy storage density. It is vital to increase the dielectric constant of polymer-based dielectric nanocomposite while improving or maintaining breakdown strength to obtain a high-energy storage density of polymer-based dielectric nanocomposite. Therefore, some research indicates that only a small amount of filler is needed while E_b of polymer-based dielectric nanocomposites is desired to be increased. Nonetheless, the increment in polymer-based dielectric nanocomposites is negligible. Therefore, it is necessary to improve the structure of polymer-based dielectric nanocomposites to obtain simultaneously enhanced dielectric constant and breakdown strength of low filler content. Recently, MSPBDNs have drawn great attention to improving dielectric constant and breakdown strength, thus promoting discharged energy density simultaneously. This section first summarizes MSPBDNs and then focuses on multilayer-structured models, breakdown mechanisms, and finite element simulations, including a lot of examples to obtain detailed explanations. There are still many opportunities and challenges in this field.

7.1 TYPES OF MULTILAYER-STRUCTURED POLYMER-BASED DIELECTRIC NANOCOMPOSITES

A lot of research indicates that MSPBDNs can fully utilize the physical characteristics of each layer.[5, 26, 50, 90–95] Such MSPBDNs exhibit outperformed breakdown strength and suppressed dielectric loss compared to the single-layered polymer-based dielectric nanocomposites with randomly distributed nanofillers.[3, 11, 20, 90, 91, 96–106] According to the nature of the interlayer of MSPBDNs, MSPBDNs can generally be classified into MSPBDNs with high insulation interlayer, MSPBDNs with high dielectric interlayer, and MSPBDNs with gradient structure. In these three structures, each layer serves as a high insulation layer or high dielectric layer to achieve high dielectric constant and high breakdown strength simultaneously. The MSPBDNs construct interfaces between the adjacent layers on a micro-scale. The heterogeneous mesoscopic interfaces not only largely enhance Maxwell-Wagner interfacial polarization but also impede electric breakdown path by distorting electrical trees along the horizontal direction of interfaces in MSPBDNs. These interfacial polarization and barrier effects contribute to improving the dielectric constant and breakdown strength of MSPBDNs, thus helping to promote energy storage performance systematically.

7.1.1 High Insulation Interlayer

The MSPBDNs with high insulation interlayer are composed of the polymer film with insulation as an interlayer and the polymer/ceramic nanofillers with high dielectric constant as outer layers.[101, 107–111] The polymer film as interlayer with high insulation in MSPBDNs can take up most of the applied electric field. This is because the huge difference of dielectric constant between the interlayer and outer layers leads to the redistribution of electric filed under applied electric filed. As well as, the high insulation interlayer of the MSPBDNs prevents further migration of charge to effectively improve the breakdown strength.

The role played by high insulation interlayers in MSPBDNs can be better explained by the following examples. Yu and co-workers first prepared hollow porous carbon (HPC) spheres via a one-pot surfactant-free method, then opened pores and formed internal decoration of $BaTiO_3$ (BT) to synthesize BT@HPC hybrids.[109] Finally, they prepared MSPBDNs, which contained pure poly(vinylidene fluoride) (PVDF) as the insulation interlayer and the BT@HPCs/PVDF as the outer layers, described in Figure 4.6(a). The dielectric constant of all specimens with various BT@HPC loadings (1 wt.%, 3wt.%, 5 wt.%, 7 wt.%, and 9 wt.%) is higher than that of pure PVDF film due to the appearance of space charge polarizations at the interfacial region between different layers of MSPBDNs. All specimens maintain low dielectric loss due to the interfaces between adjacent layers impeding carrier transportation, shown in Figure 4.6(b). Compared to other works, MSPBDNs are filled with BT@HPC with their own preferable dielectric properties as shown in Figure 4.6(c). These can be interpreted as plentiful space charge polarizations that are generated at the interfaces between the BT@HPCs/PVDF and the pure PVDF as well as between BT@HPC and PVDF producing deep traps to enhance the dielectric constant of MSPBDNs. Notably, these demonstrate that filling BT@HPC hybrid and building multilayer

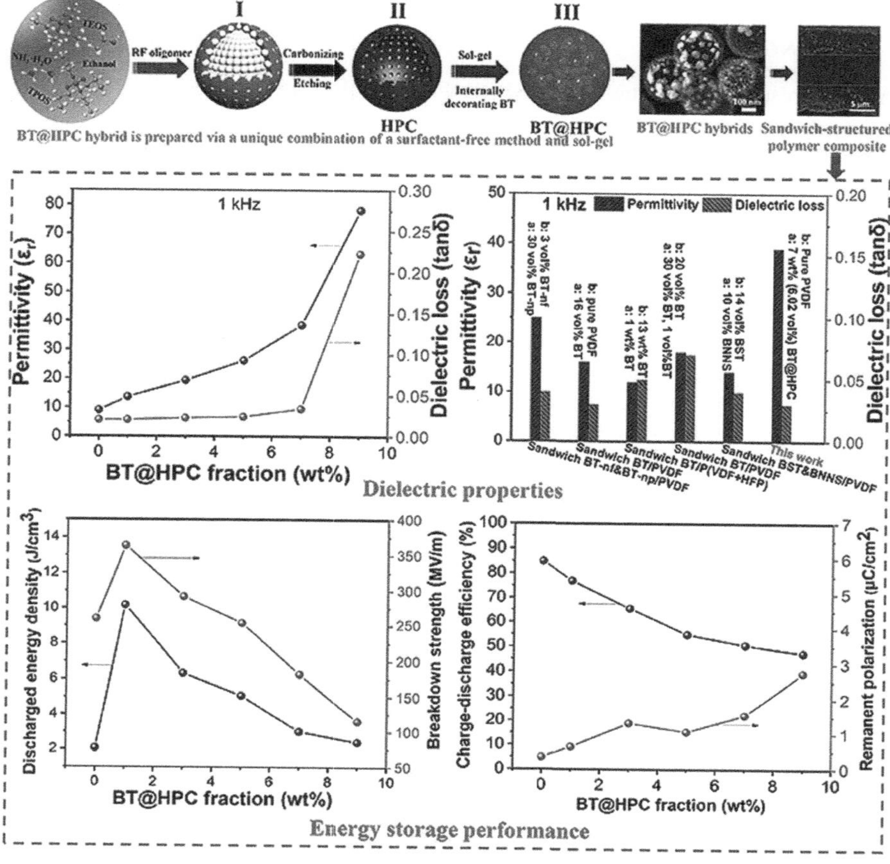

FIGURE 4.6   (a) The schematic diagram of the preparation process of BT@HPC hybrids. (b) Permittivity and dielectric loss of the MSPBDNs filled with different BT@HPC content under 1 kHz. (c) The comparison of dielectric properties of the MSPBDNs filled with BT@HPC and reported works. (d) Discharged energy density and breakdown strength; (e) charge-discharge efficiency and remnant polarization of the MSPBDNs filled with BT@HPC.[109]

structures can effectively enhance dielectric properties. Besides, E_b is also a key factor of MSPBDNs for energy storage performance. The relationship between the content of BT@HPC and E_b and U_e of MSPBDNs filled with BT@HPC is presented in Figure 4.6(d). It is observed that E_b of MSPBDNs filled with BT@HPC is higher than that of pure PVDF. It is attributed to the significant difference of dielectric constant between outer layers and interlayers inducing redistribution of the local electric field which can make the electric tree breakdown path winding for enhancing E_b. From Figure 4.6(d), the U_e of MSPBDNs filled with 1.0 wt.% BT@HPC hybrid can reach the highest value ~10.2 J/cm³ among all specimens under 360 MV/m. The charge-discharge efficiency and the remnant polarization of MSPBDNs filled with BT@HPC are described in Figure 4.6(e). In general, MSPBDNs with high insulation interlayers own superior energy storage performance attributed to bearing more electric field and preventing the breakdown of insulation interlayer.

7.1.2 High Dielectric Interlayer

In addition to the MSPBDNs with high insulation interlayer mentioned in Section 2.1, the MSPBDNs with high dielectric interlayer are also widely studied, which efficaciously redistribute the electric field between the adjacent layers and restrain the growth of electric trees due to interfacial barrier effects. More applied electric fields would concentrate on the insulation outer layers due to the dielectric differences between the interlayer and outer layers to enhance the E_b of MSPBDNs.

In our previous work,[50] we also present a series of MSPBDNs to certificate the multilayer structure promoting the energy storage performance. First, we put Ag nanoparticles attached to the surface of boron nitride nanosheets (BNNs) (Ag@ BNN) as the filler of MSPBDNs. Then, we made the MSPBDNs with Ag@ BNN. The Ag@BNN as the filler of interlayer can improve the dielectric constant as well as E_b, while the PEI as the polymer of the interlayer reduces the dielectric losses due to its linear characteristic. Besides, the pure P(VDF-HFP) film as the outer layer enhances the dielectric constant. As a result, the discharged energy density of 11.3 J/cm^3 and efficiency of 80% under 510 MV/m can be achieved in MSPBDNs with 5 wt.% Ag@BNN (Figure 4.7).

Zhang et al.[51] designed a series of single-layered BaTiO$_3$/PVDF nanocomposites and MSPBDNs with BaTiO$_3$ nanowires. Two kinds of MSPBDNs with BaTiO$_3$ nanowires were prepared, which are MSPBDNs with the BaTiO$_3$/PVDF as interlayer and MSPBDNs with PVDF as the interlayer. The MSPBDNs loaded with 3 wt.% BaTiO$_3$ nanowires as outer layers and the pure PVDF as interlayer is called "3-0-3" and the MSPBDNs loaded with 3 wt.% BaTiO$_3$ nanowires as interlayer and the pure PVDF as outer layers is called "0-3-0". The dielectric constant and dielectric loss with the increment of frequency for the single-layered BaTiO$_3$/PVDF and the MSPBDNs with BaTiO$_3$ nanowires are shown in Figure 4.8(a) and (b). It is worth noticing that from 1 kHz to 10 MHz, the dielectric constant of all specimens gradually decreases with the enhancement of frequency while the dielectric loss gradually increases. This can

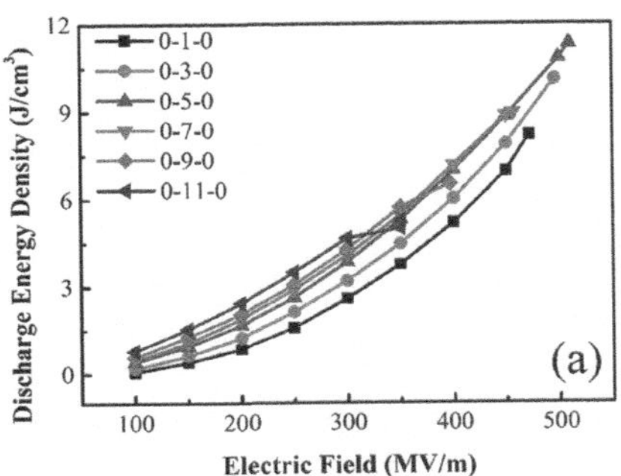
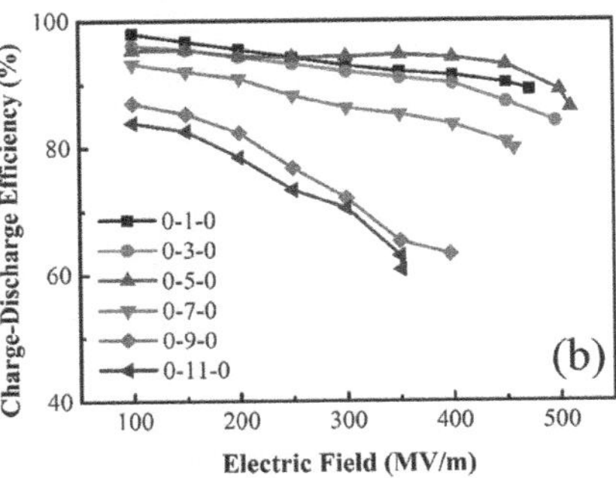

FIGURE 4.7   (a) Discharge energy density and (b) charge-discharge efficiency of various heterogeneous sandwich-structured P(VDF-HFP)-Ag@BNN/PEI-P(VDF-HFP).[50]

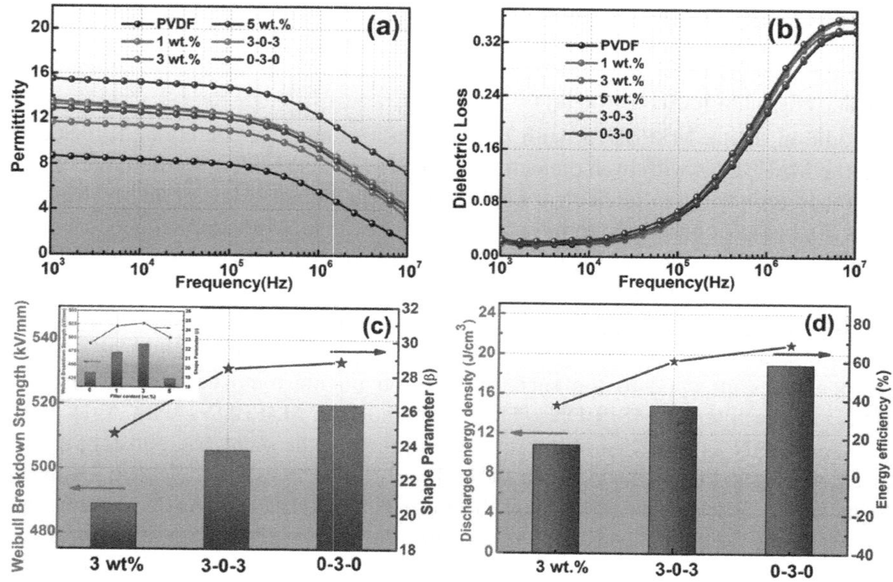

FIGURE 4.8 (a) Dielectric constant and (b) dielectric loss of the single-layered $BaTiO_3$/ PVDF nanocomposites and MSPBDNs with $BaTiO_3$ nanowires. (c) Characteristic E_b and shape parameter, as well as (d) discharged energy density and energy efficiency of 3 wt.% single-layered $BaTiO_3$/PVDF nanocomposites and 3-0-3, 0-3-0 MSPBDNs with $BaTiO_3$ nanowires under their E_b.[5]

be explained that the mobility of the dipole moment is not sufficient to contribute to the dielectric constant when the frequency of the applied electric field exceeds the relaxation frequency. It can be observed that E_b of 3-0-3 and 0-3-0 MSPBDNs are 505.6 MV/m and 519.7 MV/m in Figure 4.8(c). Notably, they are higher than that of the single-layered nanocomposites loaded with 3 wt.% $BaTiO_3$ nanowires. It helps us to reveal that multilayer structures can facilitate electrical trees to propagate along the in-plane directions and thus extend the breakdown path to enhance E_b. As Figure 4.8(c) displayed, the E_b of the 0-3-0 MSPBDNs is higher than that of the 3-0-3 MSPBDNs, attributing to PVDF as outer layers of 0-3-0 MSPBDNs with high insulation and lower electron mobility. The U_e of 0-3-0 MSPBDNs is 19.1 J/cm³ (under 519.7 MV/m), which is 76.3% and 28.4% higher than that of the single-layered nanocomposite with 3 wt.% $BaTiO_3$ nanowires and the 3-0-3 MSPBDNs in Figure 4.8(d). The 0-3-0 MSPBDNs exhibit the highest U_e of 19.1 J/cm³ (under 519.7 MV/m) attributed to (1) the $BaTiO_3$ nanowires having a high aspect ratio to enhance dielectric constant and polarization; (2) the Maxwell-Wagner interfacial polarization and blocking effect in multilayer structures causing an enhancement in E_b and max dielectric displacement; (3) more insulation outer layers helping to improve E_b and U_e. In conclusion, compared with the 3-0-3 MSPBDNs, the 0-3-0 sandwich-structured nanocomposite with high dielectric interlayer has a better advantage in the energy storage performance of MSPBDNs, which can effectively inhibit the charge injection from the electrode region and prevent the electrical tree from extending.

Compared with the MSPBDNs with high insulation interlayer, the MSPBDNs with high dielectric interlayer have better energy storage performance. It is attributed to the higher insulation and lower electron mobility of polymer as outer layer than that of nanofillers in the interlayer.

7.1.3 GRADIENT STRUCTURE

In Sections 7.1.1 and 7.1.2, we have introduced two common MSPBDNs by giving some examples. They all outperform well in E_b than that of single-layered composites. In addition to the above two traditional MSPBDNs, there is another MSPBDNs called gradient structure, which generate a dielectric distribution via regulating the distribution of nanofillers is a gradient in MSPBDNs. Gradient-structured nanocomposites facilitate the regulation of electric field distribution, and they can control the variation of the electric field between adjacent layers to a very low level. This prevents the generation of conductive pathways and the occurrence of electric breakdown.

Zhang et al.[112] constructed gradient-structured MSPBDNs via the hot-pressed method, in which poly(vinylidene fluoride-co-hexafluoropropylene) (P(VDF-HFP)) was used as the polymer matrix. Spherical $BaTiO_3$ (BT) nanoparticles were dispersed in P(VDF-HFP) generating the gradient-structured P(VDF-HFP)/BT MSPBDNs. The schematic illustration of BT nanoparticles dispersed gradient in P(VDF-HFP) is shown in Figure 4.9(a). The number of grades signifies BT concentration intervals between adjacent layers. They are allocated a fixed content of BT nanoparticles into corresponding numbers of layers. The dielectric constant and dielectric loss of MSPBDNs with different BT volume fractions and distributions are shown in Figure 4.9(b). The dielectric constant of all specimens decreases with increasing frequency, while increase with the enhancing incorporation of BT nanoparticles. Dielectric losses of all specimens still remain at low levels under whole frequency. As illustrated in Figure 4.9(c), the dielectric constant of P(VDF-HFP)/BT MSPBDNs enhances with the promotion of the volume fraction of BT nanoparticles. However, these are almost unvaried with increment in the numbers of grades. The E_b of single-layered nanocomposites (the grade is one) declines with an increment of the amount of BT nanoparticles in Figure 4.9(d). By regulating the amount of grades from one to three, five, and even seven, the E_b of MSPBDNs starts to rise gradually. It is worth noting that, as indicated in Figure 4.9(d), the E_b of MSPBDNs increases with an enhancement of the number of grades exceeding the P(VDF-HFP) film baseline. The local dielectric constant of the single-layered nanocomposite enhances attributed to the loading of BT nanoparticles with a high dielectric constant in Figure 4.9(e). The amount of charge is injected from the electrode area to single-layered nanocomposites to accelerate the generation of electrical breakdown. Compared to single-layered nanocomposites, the outer layers of gradient-structured MSPBDNs nearing the electrode area are pure PVDF films with insulating properties inhibiting charge inject from the electrode area to gradient-structured MSPBDNs. Moreover, the macroscopic

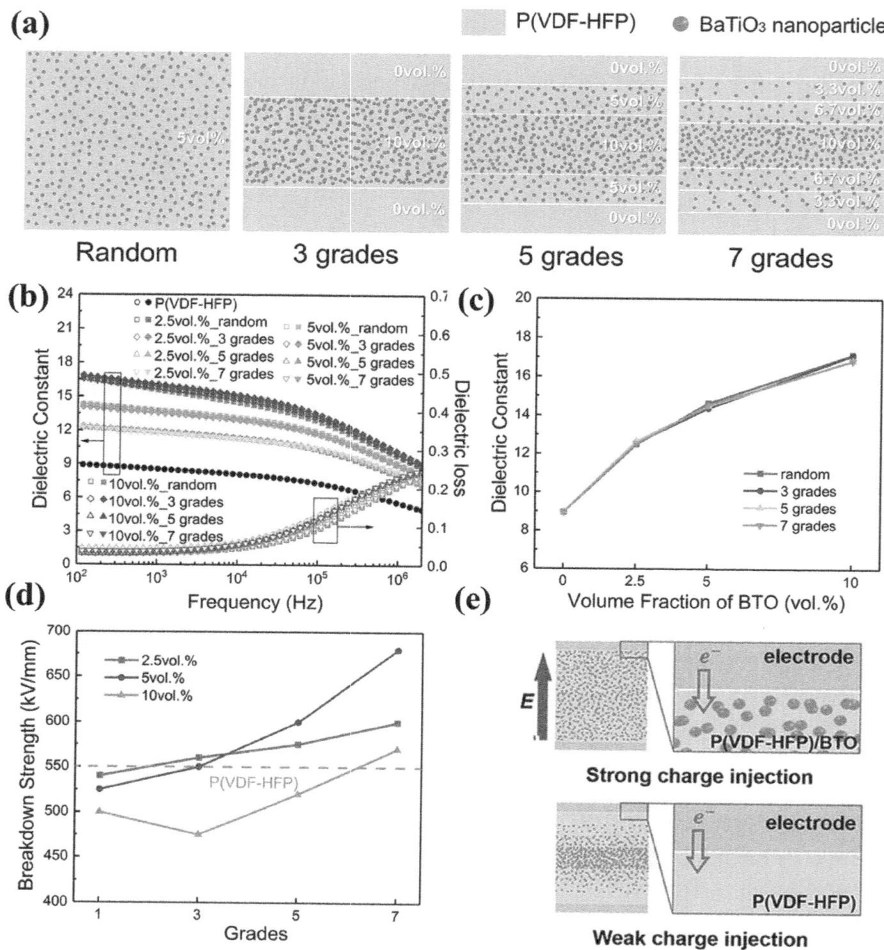

FIGURE 4.9　(a) Schematic illustration of BT nanoparticles gradient distributions in P(VDF-HFP). (b) Dependencies of dielectric constant and dielectric loss on frequency for P(VDF-HFP)/BT MSPBDNs with different BT volume fractions and distributions. (c) Dielectric constant of BT nanoparticles for P(VDF-HFP)/BT nanocomposites with different nanoparticle distributions and volume fractions. (d) Breakdown strength of P(VDF-HFP)/BT MSPBDNs with different BT volume fractions and grade numbers. (e) Schematic illustration of interfaces for nanocomposites with random distribution and gradient distribution of BT nanoparticles.[112]

interfaces between adjacent layers create barriers for the charge migration improving the E_b in the gradient-structured MSPBDNs. The E_b and U_e of gradient-structured MSPBDNs can be enhanced via regulating the number of grades. Moreover, the positive effect on energy storage performance can be enlarged by enhancing the grade number of gradient structures. Therefore, the gradient-structured MSPBDNs integrate the advantages of each layer and supply a high breakdown strength.

7.2 Multilayer-Structured Models, Breakdown Mechanisms, and Simulations

7.2.1 Multilayer-Structured Models

To further understand the mechanisms of breakdown strength in the MSPBDNs, models of MSPBDNs are constructed in much research.[113, 114] Building multilayer-structured models facilitates the understanding of the distribution of the electric field of MSPBDNs easily and quickly. The distribution of electric field in outer layers and interlayers is determined with a capacitive voltage divider in MSPBDNs. The distribution of local electric fields in interlayers and outer layers is rewritten as $E_i = V / 2d_i + \dfrac{\varepsilon_i}{\varepsilon_o} d_o$ and $E_o = V / d_o + 2 \dfrac{\varepsilon_o}{\varepsilon_i} d_i$, respectively. The V presents the applied voltage of MSPBDNs. The d_i and d_o are the thickness of the interlayer and outer layers. The ε_i and ε_o are the dielectric constant of the interlayers and outer layers.

For example, in order to understand MSPBDNs filled with high dielectric interlayers, Pan et al.[115] construct models of MSPBDNs. Notably, the process of building NBT/PVDF MSPBDNs is demonstrated. The pure PVDF is adopted as the outer layers enhancing the breakdown strength of MSPBDNs and the NBT/PVDF is adopted as the interlayers improving the dielectric constant of MSPBDNs, prompting the energy storage performance of NBT/PVDF MSPBDNs. Wang et al.[101] designed the model of BT/PVDF MSPBDNs. The BT/PVDF MSPBDNs consist of pure PVDF as the interlayers and BT/PVDF composites as the outer layers, which can simultaneously improve E_b and inhibit dielectric loss.

7.2.2 Breakdown Mechanism

According to the analysis in the previous sections, constructing multilayer-structured models can enhance the E_b for further improving the E_b of MSPBDNs. Therefore, we need to understand the breakdown mechanism of the MSPBDNs. The breakdown mechanisms of MSPBDNs are mainly divided into electrical breakdown, thermal breakdown, and partial discharge breakdown. The occurrence of electrical breakdown is attributed that the presence of a small number of electrons in MSPBDNs on the conduction band energy state collides with the atoms on the lattice nodes under the acceleration of the electric field, causing the ionization of lattice atoms to produce electron avalanches further generating electrical breakdown. According to the different conditions for the occurrence of electrical breakdown of MSPBDNs, the theory of electric breakdown can be divided into two main categories: (1) the start of collisional ionization as a criterion for electric breakdown is called collisional ionization theory. (2) After ionization, the number of electrons multiplying to a certain value can be enough to destroy the insulation state of MSPBDNs as a basis for judgment which is called the avalanche breakdown theory. The electrical breakdown has many characteristics, for example, short action time, high breakdown voltage, affected by ambient temperature, and closely related to the electric field.

With the increment of the electric field, the accumulation of heat leads to an increment of the temperature in MSPBDNs because the heat generated by the dielectric loss in polymer-based nanocomposites cannot be dissipated in time. When the temperature rises, the conductivity of MSPBDNs increases sharply which makes the conductivity loss of MSPBDNs generate heat to further increase the temperature, forming a vicious circle. This leads polymer-based nanocomposites to undergo oxidation, fusion, and scorch, or even breakdown. It is worth noticing that the thermal breakdown voltage of MSPBDNs is closely related to the ambient temperature and heat dissipation conditions. When the temperature increases, the breakdown voltage of MSPBDNs decreases according to the exponential law. Also, the smaller the heat dissipation coefficient, the lower will be the breakdown voltage of MSPBDNs.

Partial discharge is one of the main factors of breakdown in MSPBDNs. If the presence of an air gap in MSPBDNs, there is a very high conductivity presenting in MSPBDNs. For instance, the original E_b of many polymer materials is generally high, however, is lower in practice due to various factors, one of the important reasons being partial discharge.

In order to explain the breakdown process of the MSPBDNs with high insulation interlayer, Wang et al.[111] made $Ba(Fe_{0.5}Ta_{0.5})O_3$ (BFT) powder modified by Dopamine (DA) generated BFT@DA, then designed BFT@DA/PVDF MSPBDNs via the solution-casting method. As depicted in Figure 4.10, the breakdown schematic diagram illustrates the conductive pathways in single-layered BFT@DA/PVDF nanocomposites and BFT@DA/PVDF MSPBDNs. When an electric field is applied to single-layered BFT@DA/PVDF nanocomposites, conductive pathways are easily formed without obstacles, as shown in Figure 4.10(a). However, when an electric field is applied to the BFT@DA/PVDF MSPBDNs, it is difficult to generate conductive pathways due to pure PVDF film as interlayer inhibiting the motion of charge, as revealed in Figure 4.10(b).

In order to understand the breakdown mechanism, the schematic of the breakdown pathways of gradient-structured MSPBDNs has been demonstrated in

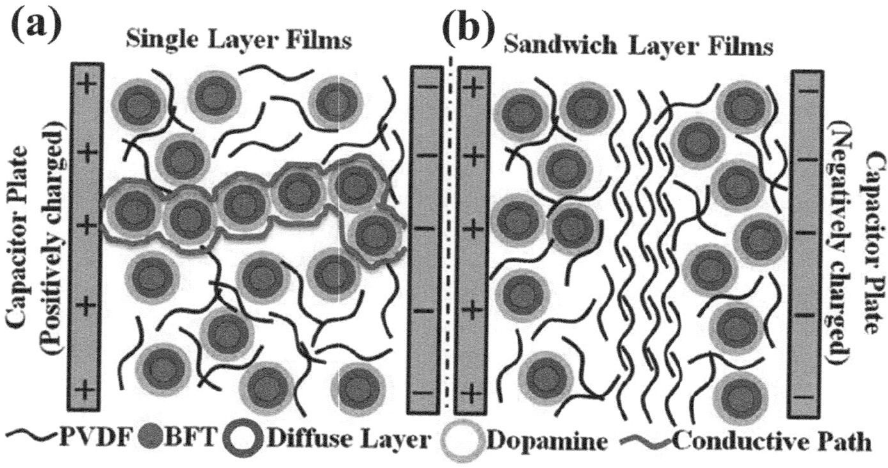

FIGURE 4.10　Breakdown mechanism diagram of conductive pathways in single-layered BFT@DA/PVDF nanocomposites and BFT@DA/PVDF MSPBDNs.[111]

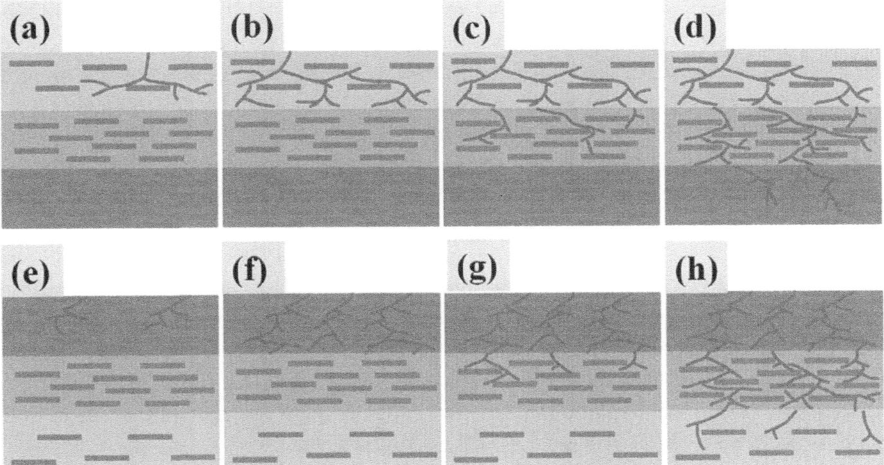

FIGURE 4.11 The schematic of the electric tree pathway of gradient-structured MSPBDNs.[114]

Figure 4.11. Wang et al.[114] designed two kinds of electric tree direction growth of gradient-structured $Na_{0.5}Bi_{4.5}Ti_4O_{15}$ (NBT)/PVDF nanocomposites. One is that electric tree direction growth of gradient-structured NBT/PVDF nanocomposites is from low to high-volume fractions of NBT particles. The other is that electric tree direction growth of gradient-structured NBT/PVDF nanocomposites is from high to low-volume fractions of NBT particles. The NBT/PVDF nanocomposites with low-volume fractions of NBT particles are called high breakdown strength layers. On the contrary, the NBT/PVDF nanocomposites with high-volume fractions of NBT particles are called low breakdown strength layers. Assuming E_1, E_2, and E_3 to represent the breakdown strength of the top layer, interlayer, and bottom layer, it can assist in analyzing the breakdown of each layer. It is easy to see the film does not break down because the electric field of the NBT/PVDF nanocomposites was less than E_1 (Figure 4.11(a)). Then, increasing the electric field to E_1, each layer does not occur from the electric tree, suggesting there is no breakdown in whole MSPBDNs, as shown in Figure 4.11(b). Further increasing the electric field over E_1, the electric tree only crosses the top layer in Figure 4.11(c). When the electric field further increases, the electric tree finally crosses the interlayer of MSPBDNs in Figure 4.11(d). The difference in dielectric constants between the three layers weakens the electric tree when it crosses layer by layer. The same mechanism can be summarized in Figure 4.11(e)–(h), in which the direction of the electric tree pathway is from high to low-volume fractions of NBT particles.

7.3 FINITE ELEMENT SIMULATIONS

As the traditional trial-and-error method has low efficiency and a long development cycle, it can no longer meet the current requirements for rapid design and preparation of materials. The multi-scale simulation technology can be used to guide the material design and elucidate the microscopic mechanism by establishing the quantitative relationship between microstructure and macrostructure properties, which becomes an effective way for research of MSPBDNs. With the continuous development of accurate and effective theoretical calculations and simulation levels and numerical computing capabilities, as well as the introduction of high-throughput screening and material genome engineering, the study of structure-property relationships of composite materials is rapidly developing toward digital, numerical, virtualization, and modularity. In nature, the physical laws of space and time-related problems can usually be described by partial differential equations (PDEs), which also become boundary value problems. When the model geometry, load structure, and material properties are complex, it is difficult to obtain the exact solution of the boundary value problem by analytical methods. The finite element method (FEM) is a numerical calculation method to solve the approximate solution of the PDE. Currently, finite element methods have become mature, and commercial FEM software are available, such as ANSYS, ABAQUS, and COMSOL Multiphysics.

Finite element simulations (FEM) can be used to obtain the distribution of electric field and thus select a superior structure and design of polymer-based dielectric nanocomposite. For example, applying a voltage at the upper electrode and keeping the lower electrode grounded while imposing a type second Neumann boundary condition on the lateral structure, which is rewritten as $\partial\varphi/\partial\hat{n} = 0$. In the above equation, φ is the magnitude of the potential waiting to be solved; $\hat{n}$ is the unit outer normal vector of the side. Then the solution region can be regarded as satisfying the electrostatic field conditions and the potential φ satisfies the Laplace equation, which is rewritten as $\nabla^2\varphi = 0$. The potential distribution in the solution region can be obtained by solving this Laplace equation by the finite element method, and the electric field strength and potential shift vector in the region satisfy the following relationship with the potential, which is rewritten as, $E = -\nabla\varphi$ and $D = \varepsilon_r\varepsilon_0 E$. The negative gradient of the potential is the electric field strength, and then the regional electric field distribution map can be solved according to the potential shift vector, the electric field strength, and the intrinsic relationship of the polymer-based dielectric nanocomposite. FEM is widely used in many different MSPBDNs (high insulation interlayer, high insulation interlayer, and gradient structure), allowing a clearer view of the electric field distribution.

In order to have a deep exploration of the influence on the E_b of in MSPBDNs with high insulation interlayer, Pan et al. analyzed the electric field distribution of MSPBDNs with NaNbO$_3$ (NN) platelets through FEM by using MATLAB and COMSOL Multiphysics.[108] The MSPBDNs consist of the high dielectric outer layers (NN/PVDF) and the high insulation interlayer (PVDF). As is illustrated in Figure 4.12(a), NN is dispersed randomly in PVDF generated via MATLAB. There is the electric field distribution of the 3-0-3 MSPBDNs. Due to the large dielectric mismatch between the interlayer and outer layers, the outer layer is subjected to a smaller electric field, while the middle layer only needs to be subjected to a larger electric field and the local field inside the 2D NN platelets is close to zero. The developed electric field channel is easily truncated passing through the outer layer to the interlayer. The electric field distribution of cross-section images of the 3-0-3 MSPBDNs are illustrated in Figure 4.12(b) and (c). The adjacent fillers of outer layers in 3-0-3

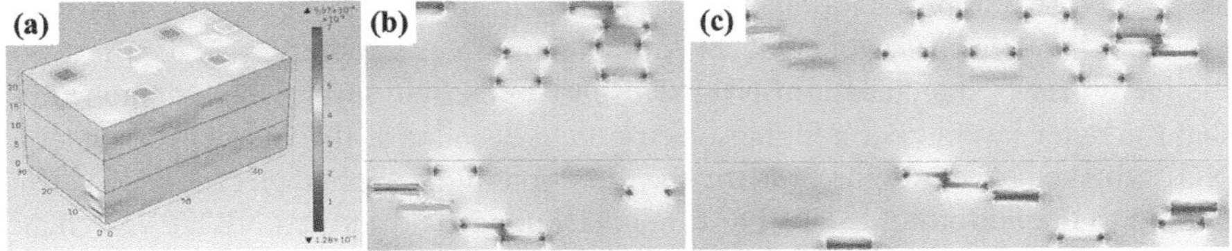

FIGURE 4.12 Three-dimensional images of electric field strength of (a) 3-0-3 MSPBDNs. (b), (c) The electric field distribution of cross-section images of 3-0-3 MSPBDNs.[108]

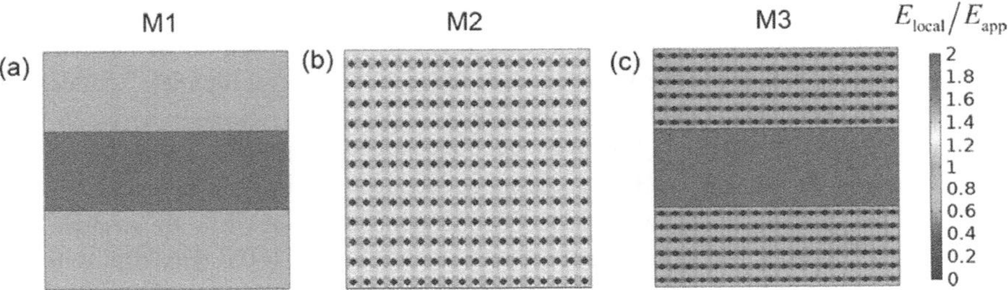

FIGURE 4.13 The local electric fields of (a) P(VDF-HFP)/PMMA/P(VDF-HFP) without BT particles; (b) P(VDF-HFP)/BT particles; and (c) sandwich-structured P(VDF-HFP)/PMMA/P(VDF-HFP) with BT particles.[107]

MSPBDNs easily generate channels along the electric field, which is potentially dangerous for breakdown. It is demonstrated that PVDF as an interlayer not only plays an important role in supporting the breakdown performance of 3-0-3 MSPBDNs but also, can alleviate the electric field stress for improving the breakdown strength.

To investigate the effect on the structure of MSPBDNs, Li et al.[107] designed MSPBDNs with a high insulation interlayer and simulated the electric field distribution of each layer of MSPBDNs, which contain PMMA as the interlayer and PVDF-HFP loaded with $BaTiO_3$ nanoparticles as the outer layers. From Figure 4.13(a), the local electric fields are redistributed because of the differences in dielectric constants between P(VDF-HFP) and PMMA. The P(VDF-HFP) as outer layers take a lower electric field, while the PMMA as interlayer takes a higher electric field. The local electric field is distorted and concentrates at the interfaces in Figure 4.13(b). Notably, the MSPBDNs combining M_1 and M_2 exhibit good voltage resistance attributed to the homogenized P(VDF-HFP) layer and better insulating PMMA layer in Figure 4.13(c). Comparing the M_2, the local field of the outer layers in MSPBDNs is slightly lower. The interlayer of MSPBDNs in M_3 avoids an excessive voltage applied onto the outer layers resulting in a smaller probability of electric breakdown.[107]

Lin et al.[115] used MATLAB and COMSOL Multiphysics to achieve utilization of FEM. The electric potential distribution in sandwich-structured NBT/PVDF nanocomposites is demonstrated in Figure 4.14(a). It is worth mentioning that the electric field of the interlayer distorts because of the introduction of NBT whiskers. As well as, the redistribution of the applied electric field of each layer occurs. It is worth noticing that the spacing of adjacent equipotential lines in outer layers declines with the increment of the content of NBT whiskers in the interlayer, which demonstrates that the outer layers bear a higher electric field because of the smaller resistance of NBT whiskers. From Figure 4.14(b), the direction of the applied electric field is from the top to the bottom of the PVDF film and sandwich-structured NBT/PVDF nanocomposites, which demonstrates that the electric field strength changes from low to high marked as from blue to red color. In order to explore the breakdown mechanism, the electric field distribution and the growth of electrical trees of MSPBDNs are simulated. The electrical trees easily grow through the PVDF film because there are no nanofillers in PVDF film as obstacles to block the growth of electric trees in Figure 4.14(b1). Notably, the distortion of the electric field of NBT whiskers tip (marked as yellow) is significantly increased while the distortion of the electric field of two sides of NBT whiskers (marked as light blue) is significantly reduced, as shown in Figure 4.14(b2–b4). There are large dielectric differences between NBT whiskers and the PVDF, which lead to the local electric field distortion. The distortion of the electric field is used to weaken the electric field concentration to enhance E_b. It is obvious that the pathway of electrical trees is blocked while they reach NBT whiskers in Figure 4.14(b2). However, excessive loading of NBT whiskers will lead to a connection of electric field around the NBT whiskers tip, thus making electrical trees pass quickly in the inner layer of sandwich-structured NBT/PVDF nanocomposites, as shown in Figure 4.14(b4). Notably, the 0-3-0 sandwich-structured NBT/PVDF nanocomposites have a modest loading of NBT whiskers, which owns the highest breakdown strength. It is worth noticing that FEM of composites can figure out how electrical trees cross composite, thus explaining what structure and how many fillers in composites should be used.

In our previous work,[50] we made MSPBDNs filled with Ag@BNN and BNN nanofillers and then simulated the electric field distributions via FEM. There are distinct electric field distributions of each layer in MSPBDNs. Moreover, the concentration of the local electric field of interlayer in MSPBDNs is less, which are filled with Ag@BNN and BNN nanofillers attributed to Ag with positive effects. Furthermore, the enlarged local electric field distribution of a single BNN and Ag@BNN nanoparticles demonstrates the positive effect of Ag on energy storage performance.

Comparing the MSPBDNs with high insulation interlayer and with high dielectric interlayer, it is found that sandwich-structured nanocomposites with high dielectric interlayer have higher breakdown field strength. This conclusion is also confirmed in the simulation section of some research.[5, 116] Wang et al.[116] designed a series of sandwich-structured BT/P(VDF-HFP) nanocomposites. The 1-9-1 MSPBDNs consist of the 9 wt.% BT nanoparticles dispersed in P(VDF-HFP) as interlayer and 1 wt.% BT nanoparticles dispersed in P(VDF-HFP) as outer layers. Correspondingly, 9-1-9 MSPBDNs consists of the 1 wt.% BT nanoparticles dispersed in P(VDF-HFP) as interlayer and 9 wt.% BT nanoparticles dispersed in P(VDF-HFP) as outer layers. Because the outer layer of 9-1-9 MSPBDNs is less insulating than the outer layer of 1-9-1 MSPBDNs, 9-1-9 MSPBDNs nanocomposites easily generate breakdown. Thus, the E_b of the 1-9-1 MSPBDNs is more than that of the 9-1-9 MSPBDNs. In 1-9-1

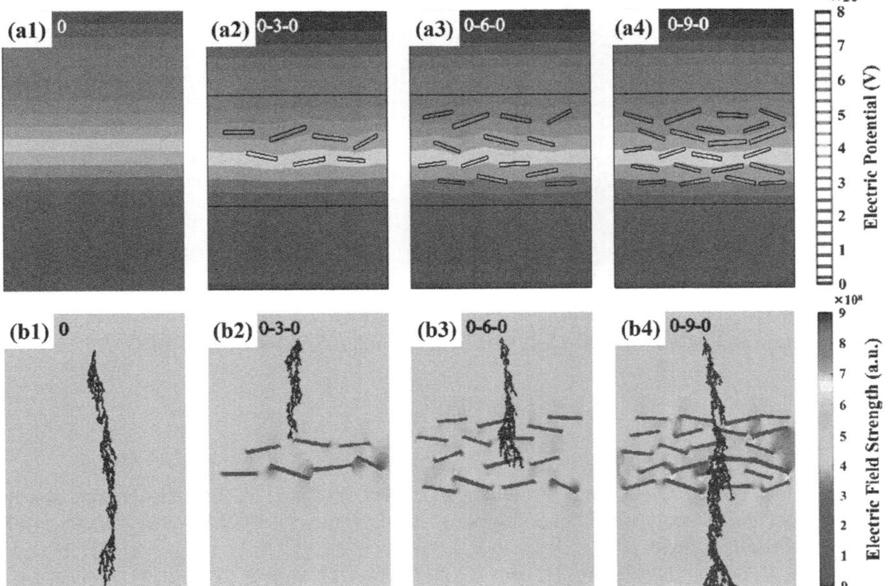

FIGURE 4.14 (a) Electric potential distribution and (b) electric field distribution and electrical trees of PVDF film and sandwich-structured NBT/PVDF nanocomposites loaded with 3 vol%, 6 vol%, and 9 vol% NBT whiskers by E_b.[115]

MSPBDNs, simulated results demonstrate that the strong barrier interfaces attributed to the electric field gradient between adjacent layers efficiently inhibit the development of electrical trees during the breakdown process and protect the composites from a complete breakdown. Through the analysis of the above experimental data, when the interlayer and the outer layer are filled with a certain content of nanofillers, the MSPBDNs with less fillers in the interlayer than the outer layer obtain greater E_b. It can be interpreted that the ILs with high E_b as the outer layer can effectively impede the charge injection and introduce deep traps for charge carriers.

Wang et al.[98] designed a group of sandwich-structured PVDF-based nanocomposites with the content of BT nanoparticles via layer by layer. It shows that the electric field distribution and shape characteristics of the electrical tree pathway in MSPBDNs in Figure 4.15. Except for 20-10-1 MSPBDNs, all samples broke down under 360 MV/m, which indicates the electrical trees growing through the three layers and finally arriving at the electrode. The favorable insulation characteristics of 20-10-1 MSPBDNs are attributable to the rationally adjusted electric distribution achieved by tailoring the BT contents.

According to the above examples, the electric field distribution can be analyzed with the FEM to better understand the breakdown strength-enhancing mechanism in MSPBDNs.

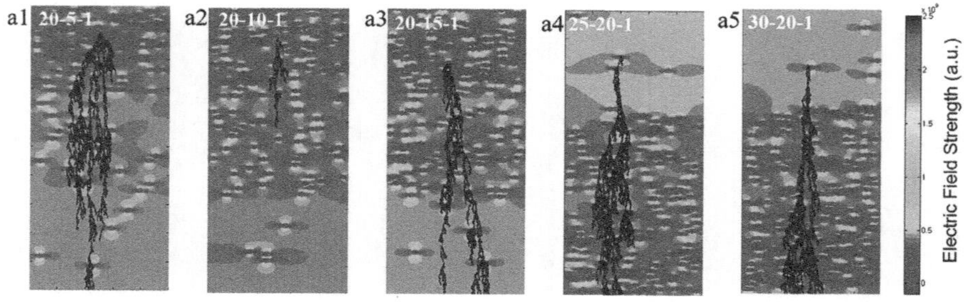

FIGURE 4.15 The electric field distribution and growth of breakdown channels in GLNs (a1) 20-5-1, (a2) 20-10-1, (a3) 20-15-1, (a4) 25-20-1, and (a5) 30-20-1 MSPBDNs.[98]

REFERENCES

1. Li, W.; Meng, Q.; Zheng, Y.; Zhang, Z.; Xia, W.; Xu, Z. Electric energy storage properties of poly(vinylidene fluoride), *Appl. Phys. Lett.* **2010**, 96, 192905. DOI: 10.1063/1.3428656.

2. Ren, X.; Meng, N.; Zhang, H.; Wu, J.; Abrahams, I.; Yan, H.; Bilotti, E.; Reece, M. J. Giant energy storage density in PVDF with internal stress engineered polar nanostructures, *Nano Energy* **2020**, 72, 104662. DOI: 10.1016/j.nanoen.2020.104662.

3. Xie, H.; Wang, L.; Gao, X.; Luo, H.; Liu, L.; Zhang, D. High breakdown strength and energy density in multilayer-structured ferroelectric composite, *ACS Omega* **2020**, 5, 32660–32666. DOI: 10.1021/acsomega.0c05031.

4. Lovinger, A. J.; Furukawa, T.; Davis, G. T.; Broadhurst, M. G. Crystallographic changes characterizing the curie transition in three ferroelectric copolymers of vinylidine fluoride and trifluoroethylene: 2. Oriented or poled samples, *Polymer* **1983**, 24, 1733–1739.

5. Guo, R.; Luo, H.; Yan, M.; Zhou, X.; Zhou, K.; Zhang, D. Significantly enhanced breakdown strength and energy density in sandwich-structured nanocomposites with low-level BaTiO$_3$ nanowires, *Nano Energy* **2021**, 79. DOI: 10.1016/j.nanoen.2020.105412.

6. Hu, J.; Zhang, S.; Tang, B. 2D filler-reinforced polymer nanocomposite dielectrics for high-k dielectric and energy storage applications, *Energy Storage Mater.* **2021**, 34, 260–281. DOI: 10.1016/j.ensm.2020.10.001.

7. Li, L.; Cheng, J.; Cheng, Y.; Han, T.; Liu, Y.; Zhou, Y.; Han, Z.; Zhao, G.; Zhao, Y.; Xiong, C.; Dong, L.; Wang, Q. Significantly enhancing the dielectric constant and breakdown strength of linear dielectric polymers by utilizing ultralow loadings of nanofillers, *J. Mater. Chem. A* **2021**, 9, 23028–23036. DOI: 10.1039/d1ta05408b.

8. Wang, C.; He, G.; Chen, S.; Zhai, D.; Luo, H.; Zhang, D. Enhanced performance of all-organic sandwich structured dielectrics with linear dielectric and ferroelectric polymers, *J. Mater. Chem. A* **2021**, 9, 8674–8684. DOI: 10.1039/d1ta00974e.

9. Zhang, Y.; Zhang, C.; Feng, Y.; Zhang, T.; Chen, Q.; Chi, Q.; Liu, L.; Wang, X.; Lei, Q. Energy storage enhancement of P(VDF-TrFE-CFE)-based composites with double-shell structured BZCT nanofibers of parallel and orthogonal configurations, *Nano Energy* **2019**, 66, 104195. DOI: 10.1016/j.nanoen.2019.104195.

10. Qiao, R.; Wang, C.; Chen, S.; He, G.; Liu, Z.; Luo, H.; Zhang, D. High-temperature dielectric polymers with high breakdown strength and energy density via constructing the electron traps in blends, *Compos. Part A: Appl. Sci. Manuf.* **2022**, 152, 106679. DOI: 10.1016/j.compositesa.2021.106679.

11. Sun, Q.; Wang, J.; Sun, H.; He, L.; Zhang, L.; Mao, P.; zhang, X.; Kang, F.; Wang, Z.; Kang, R.; Zhang, L. Simultaneously enhanced energy density and discharge efficiency of layer-structured nanocomposites by reasonably designing dielectric differences between $BaTiO_3@SiO_2$/PVDF layers and BNNSs/PVDF-PMMA layers, *Compos. Part A: Appl. Sci. Manuf.* **2021**, 149. DOI: 10.1016/j.compositesa.2021.106546.

12. Chi, Q.; Zhou, Y.; Feng, Y.; Cui, Y.; Zhang, Y.; Zhang, T.; Chen, Q. Excellent energy storage performance of polyetherimide filled by oriented nanofibers with optimized diameters, *Mater. Today Energy* **2020**, 18, 100516. DOI: 10.1016/j.mtener.2020.100516.

13. Sun, Q.; Wang, J.; Zhang, L.; Mao, P.; Liu, S.; He, L.; Kang, F.; Xue, R. Achieving high energy density and discharge efficiency in multi-layered PVDF-PMMA nanocomposites composed of 0D BaTiO3and 1D $NaNbO_3@SiO_2$, *J. Mater. Chem. C* **2020**, 8, 7211–7220. DOI: 10.1039/d0tc00838a.

14. Liu, Y.; Hou, Y.; Ji, Q.; Wei, S.; Du, P.; Luo, L.; Li, W. Modulation of individual-layer properties results in excellent discharged energy density of sandwich-structured composite films, *J. Mater. Sci.: Mater. Electron.* **2020**, 31, 7663–7671. DOI: 10.1007/s10854-020-03302-0.

15. Li, L.; Zhou, Y.; Liu, Y.; Chen, X.; Han, Z.; Wang, Q. Perspective on scalable high-energy-density polymer dielectrics with ultralow loadings of inorganic nanofillers, *Appl. Phys. Lett.* **2022**, 120, 050502. DOI: 10.1063/5.0080825.

16. Zhang, Y.; Li, S.; Cong, X.; Zhang, C.; Feng, Y.; Zhang, Y.; Zhang, T.; Chi, Q.; Wang, X.; Lei, Q. Interesting influence of different inorganic particles on the energy storage performance of a polyethersulfone-based dielectric composite, *ACS Appl. Energ. Mater.* **2022**, 5, 3545–3557. DOI: 10.1021/acsaem.1c04075.

17. Su, Y.; Chen, C.; Pan, H.; Yang, Y.; Chen, G.; Zhao, X.; Li, W.; Gong, Q.; Xie, G.; Zhou, Y.; Zhang, S.; Tai, H.; Jiang, Y.; Chen, J. Muscle fibers inspired high-performance piezoelectric textiles for wearable physiological monitoring, *Adv. Funct. Mater.* **2021**, 31, 2010962. DOI: 10.1002/adfm.202010962.

18. Zhong, S. L.; Dang, Z. M.; Zhou, W. Y.; Cai, H. W. J. I. N. Past and future on nanodielectrics, *IET Nanodielectrics* **2018**, 1, 41–47. DOI: 10.1049/iet-nde.2018.0004.

19. Luo, H.; Zhou, X. F.; Ellingford, C.; Zhang, Y.; Chen, S.; Zhou, K. C.; Zhang, D.; Bowen, C. R.; Wan, C. Y. Interface design for high energy density polymer nanocomposites, *Chem. Soc. Rev.* **2019**, 48, 4424–4465. DOI: 10.1039/c9cs00043g.

20. Prateek; Bhunia, R.; Siddiqui, S.; Garg, A.; Gupta, R. K. Significantly enhanced energy density by tailoring the interface in hierarchically structured TiO_2-$BaTiO_3$-TiO_2 nanofillers in PVDF-based thin-film polymer nanocomposites, *ACS Appl. Mater. Interfaces* **2019**, 11, 14329–14339. DOI: 10.1021/acsami.9b01359.

21. Lewis, T. J. Interfaces are the dominant feature of dielectrics at the nanometric level, *IEEE Trans. Dielectr. Electr. Insul.* **2004**, 11, 739–753. DOI: 10.1109/Tdei.2004.1349779.

22. Tanaka, T. Dielectric nanocomposites with insulating properties, *IEEE Trans. Dielectr. Electr. Insul.* **2005**, 12, 914–928. DOI: 10.1109/Tdei.2005.1522186.

23. Ren, Z. H.; Yu, J.; Li, Y. J.; Zhi, C. Y. Tunable free-standing ultrathin porous nickel film for high performance flexible nickel-metal hydride batteries, *Adv. Energy Mater.* **2018**, 8. DOI: 10.1002/aenm.201702467.

24. Zhao, M.; Fu, Q.; Hou, Y.; Luo, L.; Li, W. $BaTiO_3$/MWNTs/Polyvinylidene fluoride ternary dielectric composites with excellent dielectric property, high breakdown strength, and high-energy storage density, *ACS Omega* **2019**, 4, 1000–1006. DOI: 10.1021/acsomega.8b02504.

25. Ding, S.; Kong, Y. Physical mechanism of breakdown failure in polymer composites with various filler loading, *ChemistrySelect* **2019**, 4, 12043–12048. DOI: 10.1002/slct.201902422.

26. Bi, K.; Bi, M.; Hao, Y.; Luo, W.; Cai, Z.; Wang, X.; Huang, Y. Ultrafine core-shell $BaTiO_3@SiO_2$ structures for nanocomposite capacitors with high energy density, *Nano Energy* **2018**, 51, 513–523. DOI: 10.1016/j.nanoen.2018.07.006.

27. Liu, B.; Yang, M.; Zhou, W.-Y.; Cai, H.-W.; Zhong, S.-L.; Zheng, M.-S.; Dang, Z.-M. High energy density and discharge efficiency polypropylene nanocomposites for potential high-power capacitor, *Energy Storage Mater.* **2020**, 27, 443–452. DOI: 10.1016/j.ensm.2019.12.006.

28. Song, Y.; Shen, Y.; Liu, H.; Lin, Y.; Li, M.; Nan, C. W. Enhanced dielectric and ferroelectric properties induced by dopamine-modified $BaTiO_3$ nanofibers in flexible poly(vinylidene fluoride-trifluoroethylene) nanocomposites, *J. Mater. Chem.* **2012**, 22, 8063–8068. DOI: 10.1039/c2jm30297g.

29. Guo, N.; DiBenedetto, S. A.; Tewari, P.; Lanagan, M. T.; Ratner, M. A.; Marks, T. J. Nanoparticle, size, shape, and interfacial effects on leakage current density, permittivity, and breakdown strength of metal oxide-polyolefin nanocomposites: Experiment and theory, Chem. Mater. **2010**, 22, 1567–1578. DOI: 10.1021/cm902852h.

30. Tang, H.; Sodano, H. A. Ultra high energy density nanocomposite capacitors with fast discharge using $Ba_{0.2}Sr_{0.8}TiO_3$ nanowires, *Nano Letters* **2013**, 13, 1373–1379. DOI: 10.1021/nl3037273.

31. Wang, G.; Huang, X.; Jiang, P. Bio-inspired polydopamine coating as a facile approach to constructing polymer nanocomposites for energy storage, *J. Mater. Chem. C* **2017**, 5, 3112–3120. DOI: 10.1039/c7tc00387k.

32. Tang, H.; Lin, Y.; Sodano, H. A. Synthesis of high aspect ratio BaTiO3 nanowires for high energy density nanocomposite capacitors, *Adv. Energy Mater.* **2013**, 3, 451–456. DOI: 10.1002/aenm.201200808.

33. Pan, Z. B.; Yao, L. M.; Zhai, J. W.; Liu, S. H.; Yang, K.; Wang, H. T.; Liu, J. H. Fast discharge and high energy density of nanocomposite capacitors using $Ba_{0.6}Sr_{0.4}TiO_3$ nanofibers, *Ceram. Int.* **2016**, 42, 14667–14674. DOI: 10.1016/j.ceramint.2016.06.090.

34. Liu, F.; Li, Q.; Cui, J.; Li, Z.; Yang, G.; Liu, Y.; Dong, L.; Xiong, C.; Wang, H.; Wang, Q. High-energy-density dielectric polymer nanocomposites with trilayered architecture, *Adv. Funct. Mater.* **2017**, 27, 1606292. DOI: 10.1002/adfm.201606292.

35. Pan, Z.; Yao, L.; Zhai, J.; Wang, H.; Shen, B. Ultrafast discharge and enhanced energy density of polymer nanocomposites loaded with $_{0.5}(Ba_{0.7}Ca_{0.3})TiO_3$-$_{0.5}Ba(Zr_{0.2}Ti_{0.8})O_3$ one-dimensional nanofibers, *ACS Appl. Mater. Interfaces* **2017**, 9, 14337–14346. DOI: 10.1021/acsami.7b01381.

36. Pan, Z.; Liu, J.; Xu, J.; Cheng, Y.; Bai, H.; Chen, H.; Yao, L.; Ding, X.; Shi, S.; Xie, Z.; Zhai, J. Two-dimensional fillers induced superior electrostatic energy storage performance in trilayered architecture nanocomposites, *ACS Appl. Mater. Interfaces* **2022**, 14, 8448–8457. DOI: 10.1021/acsami.1c23086.

37. Chen, J.; Shen, Z.; Kang, Q.; Qian, X.; Li, S.; Jiang, P.; Huang, X. Chemical adsorption on 2D dielectric nanosheets for matrix free nanocomposites with ultrahigh electrical energy storage, *Sci. Bull.* **2022**, 67, 609–618. DOI: 10.1016/j.scib.2021.10.011.

38. Wen, F.; Lou, H.; Ye, J.; Bai, W.; Wang, L.; Li, L.; Wu, W.; Xu, Z.; Wang, G.; Zhang, Z.; Zhang, L. Preparation and energy storage performance of transparent dielectric films with two-dimensional platelets, *Compos. Sci. Technol.* **2019**, 182, 107759. DOI: 10.1016/j.compscitech.2019.107759.

39. Feng, Y.; Deng, Q.; Peng, C.; Hu, J.; Li, Y.; Wu, Q.; Xu, Z. An ultrahigh discharged energy density achieved in an inhomogeneous PVDF dielectric composite filled with 2D MXene nanosheets: Via interface engineering, *J. Mater. Chem. C* **2018**, 6, 13283–13292. DOI: 10.1039/c8tc05180a.

40. Zhu, Y.; Zhu, Y.; Huang, X.; Chen, J.; Li, Q.; He, J.; Jiang, P. High energy density polymer dielectrics interlayered by assembled boron nitride nanosheets, *Adv. Energy Mater.* **2019**, 9, 1901826. DOI: 10.1002/aenm.201901826.

41. Liu, S.; Wang, J.; Shen, B.; Zhai, J.; Hao, H.; Zhao, L. Poly(vinylidene fluoride) nanocomposites with a small loading of core-shell structured $BaTiO_3@Al_2O_3$ nanofibers exhibiting high discharged energy density and efficiency, *J. Alloys Compd.* **2017**, 696, 136–142. DOI: 10.1016/j.jallcom.2016.11.186.

42. Liu, S.; Wang, J.; Wang, J.; Shen, B.; Zhai, J.; Guo, C.; Zhou, J. Core-shell structured $BaTiO_3@SiO_2$ nanofibers for poly(vinylidene fluoride) nanocomposites with high discharged energy, *Mater. Lett.* **2017**, 189, 176–179. DOI: 10.1016/j.matlet.2016.12.008.

43. Wang, G.; Huang, Y.; Wang, Y.; Jiang, P.; Huang, X. Substantial enhancement of energy storage capability in polymer nanocomposites by encapsulation of $BaTiO_3$ NWs with variable shell thickness, *Phys. Chem. Chem. Phys.* **2017**, 19, 21058–21068. DOI: 10.1039/c7cp04096b.

44. Liu, S.; Wang, J.; Shen, B.; Zhai, J. Enhanced discharged energy density and efficiency of poly(vinylidene fluoride) nanocomposites through a small loading of core-shell structured $BaTiO_3@Al_2O_3$ nanofibers, *Ceram. Int.* **2017**, 43, 585–589. DOI: 10.1016/j.ceramint.2016.09.198.

45. Pan, Z.; Zhai, J.; Shen, B. Multilayer hierarchical interfaces with high energy density in polymer nanocomposites composed of $BaTiO_3@TiO_2@Al_2O_3$ nanofibers, *J. Mater. Chem. A* **2017**, 5, 15217–15226. DOI: 10.1039/c7ta03846a.

46. Kang, D.; Wang, G.; Huang, Y.; Jiang, P.; Huang, X. Decorating TiO_2 nanowires with $BaTiO_3$ nanoparticles: A new approach leading to substantially enhanced energy storage capability of high-k polymer nanocomposites, *ACS Appl. Mater. Interfaces* **2018**, 10, 4077–4085. DOI: 10.1021/acsami.7b16409.

47. Zhang, X.; Shen, Y.; Xu, B.; Zhang, Q.; Gu, L.; Jiang, J.; Ma, J.; Lin, Y.; Nan, C. W. Giant energy density and improved discharge efficiency of solution-processed polymer nanocomposites for dielectric energy storage, *Adv. Mater.* **2016**, 28, 2055–2061. DOI: 10.1002/adma.201503881.

48. Zhang, X.; Shen, Y.; Zhang, Q.; Gu, L.; Hu, Y.; Du, J.; Lin, Y.; Nan, C. W. Ultrahigh energy density of polymer nanocomposites containing $BaTiO_3@TiO_2$ nanofibers by atomic-scale interface engineering, *Adv. Mater.* **2015**, 27, 819–824. DOI: 10.1002/adma.201404101.

49. Shameem, M. M.; Sasikanth, S. M.; Annamalai, R.; Raman, R. G. A brief review on polymer nanocomposites and its applications, *Mater. Today: Proc.* **2021**, 45, 2536–2539. DOI: 10.1016/j.matpr.2020.11.254.

50. Marwat, M. A.; Yasar, M.; Ma, W. G.; Fan, P. Y.; Liu, K.; Lu, D. J.; Tian, Y.; Samart, C.; Ye, B. H.; Zhang, H. B. Significant energy density of discharge and charge-discharge efficiency in Ag@BNN nanofillers-modified heterogeneous sandwich structure nanocomposites, *ACS Appl. Energ. Mater.* **2020**, 3, 6591–6601. DOI: 10.1021/acsaem.0c00770.

51. Nunes-Pereira, J.; Sencadas, V.; Correia, V.; Cardoso, V. F.; Han, W.; Rocha, J. G.; Lanceros-Méndez, S. Energy harvesting performance of $BaTiO_3$/poly(vinylidene fluoride-trifluoroethylene) spin coated nanocomposites, *Compos. Part B: Eng.* **2015**, 72, 130–136. DOI: 10.1016/j.compositesb.2014.12.001.

52. Brinker, C. J.; Frye, G. C.; Hurd, A. J.; Ashley, C. S. Fundamentals of sol-gel dip coating, *Thin Solid Films* **1991**, 201, 97–108. DOI: 10.1016/0040-6090(91)90158-T.

53. Pavlidou, S.; Papaspyrides, C. D. A review on polymer-layered silicate nanocomposites, *Prog. Polym. Sci.* (Oxford) **2008**, 33, 1119–1198. DOI: 10.1016/j.progpolymsci.2008.07.008.

54. Erceg, M.; Jozić, D.; Banovac, I.; Perinović, S.; Bernstorff, S. Preparation and characterization of melt intercalated poly(ethylene oxide)/lithium montmorillonite nanocomposites, *Thermochim. Acta* **2014**, 579, 86–92. DOI: 10.1016/j.tca.2014.01.024.

55. Marroquin, J. B.; Rhee, K. Y.; Park, S. J. Chitosan nanocomposite films: Enhanced electrical conductivity, thermal stability, and mechanical properties, *Carbohydr. Polym.* **2013**, 92, 1783–1791. DOI: 10.1016/j.carbpol.2012.11.042.

56. Grabowski, C. A.; Fillery, S. P.; Westing, N. M.; Chi, C.; Meth, J. S.; Durstock, M. F.; Vaia, R. A. Dielectric breakdown in silica-amorphous polymer nanocomposite films: The role of the polymer matrix, *ACS Appl. Mater. Interfaces* **2013**, 5, 5486–5492. DOI: 10.1021/am4005623.

57. Chen, Q.; Chu, B.; Zhou, X.; Zhang, Q. M. Effect of metal-polymer interface on the breakdown electric field of poly(vinylidene fluoride-trifluoroethylene-chlorofluoroethylene) terpolymer, *Appl. Phys. Lett.* **2007**, 91, 062907. DOI: 10.1063/1.2768205.

58. Zhou, P.; Wang, Y.; Nasreen, S.; Cao, Y. Barrier heights of polymer-electrode interfaces measured via photo injection current method, *Surf. Interfaces* **2021**, 24, 101070. DOI: 10.1016/j.surfin.2021.101070.

59. Gerson, R.; Marshall, T. C. Dielectric breakdown of porous ceramics, *J. Appl. Phys* **1959**, 30, 1650–1653. DOI: 10.1063/1.1735030.

60. Tan, D. Q. The search for enhanced dielectric strength of polymer-based dielectrics: A focused review on polymer nanocomposites, *J. Appl. Polym. Sci.* **2020**, 137, 49379. DOI: 10.1002/app.49379.

61. Neusel, C.; Jelitto, H.; Schmidt, D.; Janssen, R.; Felten, F.; Schneider, G. A. Thickness-dependence of the breakdown strength: Analysis of the dielectric and mechanical failure, *J. Eur. Ceram. Soc.* **2015**, 35, 113–123. DOI: 10.1016/j.jeurceramsoc.2014.08.028.

62. Claude, J.; Lu, Y.; Wang, Q. Effect of molecular weight on the dielectric breakdown strength of ferroelectric poly(vinylidene fluoride-chlorotrifluoroethylene)s, *Appl. Phys. Lett.* **2007**, 91, 212904. DOI: 10.1063/1.2816327.

63. Li, Q.; Chen, L.; Gadinski, M. R.; Zhang, S.; Zhang, G.; Li, H.; Haque, A.; Chen, L. Q.; Jackson, T.; Wang, Q. Flexible higherature dielectric materials from polymer nanocomposites, *Nature* **2015**, 523, 576–579. DOI: 10.1038/nature14647.

64. Huang, X.; Li, Y.; Liu, F.; Jiang, P.; Iizuka, T.; Tatsumi, K.; Tanaka, T. Electrical properties of epoxy/POSS composites with homogeneous nanostructure, *IEEE Trans. Dielectr. Electr. Insul.* **2014**, 21, 1516–1528. DOI: 10.1109/TDEI.2014.004314.

65. Li, J.; Zhang, Y.; Xia, Z.; Qin, X.; Peng, Z. Action of space charge on aging and breakdown of polymers, *Chin. Sci. Bull.* **2001**, 46, 796–800. DOI: 10.1007/BF02900426.

66. Li, S.; Yin, G.; Chen, G.; Li, J.; Bai, S.; Zhong, L.; Zhang, Y.; Lei, Q. Short-term breakdown and long-term failure in nanodielectrics: A review, *IEEE Trans. Dielectr. Electr. Insul.* **2010**, 17, 1523–1535. DOI: 10.1109/TDEI.2010.5595554.

67. Li, Z.; Sheng, G.; Jiang, X.; Tanaka, T. Effects of inorganic fillers on withstanding short-time breakdown and long-time electrical aging of epoxy composites, *IEEJ Trans. Electr. Electron. Eng.* **2017**, 12, S10–S15. DOI: 10.1002/tee.22548.

68. Angle, R. L.; Talley, H. E. Electrical and charge storage characteristics of the tantalum oxide-silicon dioxide device, *IEEE Trans. Electron Devices* **1978**, 25, 1277–1283. DOI: 10.1109/T-ED.1978.19266.

69. Chiu, F. C. A review on conduction mechanisms in dielectric films, *Adv. Mater. Sci. Eng.* **2014**, 2014, 1–18. DOI: 10.1155/2014/578168.

70. Liu, A.; Zhu, H.; Sun, H.; Xu, Y.; Noh, Y. Y. Solution processed metal oxide high-κ dielectrics for emerging transistors and circuits, *Adv. Mater.* **2018**, 30, 1706364. DOI: 10.1002/adma.201706364.

71. Vecchio, M. A.; Meddeb, A. B.; Lanagan, M. T.; Ounaies, Z.; Shallenberger, J. R. Plasma surface modification of P(VDF-TrFE): Influence of surface chemistry and structure on electronic charge injection, *J. Appl. Phys.* **2018**, 124, 114102. DOI: 10.1063/1.5042751.

72. Zhang, Z.; Wang, D. H.; Litt, M. H.; Tan, L. S.; Zhu, L. High-temperature and high-energy-density dipolar glass polymers based on sulfonylated poly(2,6-dimethy l-1,4-phenylene oxide), *Angew. Chem - Int. Ed.* **2018**, 57, 1528–1531. DOI: 10.1002/anie.201710474.

73. Zhang, T.; Chen, X.; Thakur, Y.; Lu, B.; Zhang, Q.; Runt, J.; Zhang, Q. M. A highly scalable dielectric metamaterial with superior capacitor performance over a broad temperature, *Sci. Adv.* **2020**, 6. DOI: 10.1126/sciadv.aax6622.

74. Chi, Q.; Gao, Z.; Zhang, T.; Zhang, C.; Zhang, Y.; Chen, Q.; Wang, X.; Lei, Q. Excellent energy storage properties with high-temperature stability in sandwich-structured polyimide-based composite films, *ACS Sustainable Chem. Eng.* **2019**, 7, 748–757. DOI: 10.1021/acssuschemeng.8b04370.

75. Yue, S.; Wan, B.; Liu, Y.; Zhang, Q. Significantly enhanced dielectric constant and energy storage properties in polyimide/reduced BaTiO3 composite films with excellent thermal stability, *RSC Adv.* **2019**, 9, 7706–7717. DOI: 10.1039/c8ra10434d.

76. Li, H.; Ren, L.; Ai, D.; Han, Z.; Liu, Y.; Yao, B.; Wang, Q. Ternary polymer nanocomposites with concurrently enhanced dielectric constant and breakdown strength for high-temperature electrostatic capacitors, *InfoMat* **2020**, 2, 389–400. DOI: 10.1002/inf2.12043.

77. Marwat, M. A.; Xie, B.; Zhu, Y.; Fan, P.; Liu, K.; Shen, M.; Ashtar, M.; Kongparakul, S.; Samart, C.; Zhang, H. Sandwich structure-assisted significantly improved discharge energy density in linear polymer nanocomposites with high thermal stability, *Colloids Surf. A: Physicochem. Eng. Asp.* **2019**, 581, 123802. DOI: 10.1016/j.colsurfa.2019.123802.

78. Marwat, M. A.; Ma, W.; Fan, P.; Elahi, H.; Samart, C.; Nan, B.; Tan, H.; Salamon, D.; Ye, B.; Zhang, H. Ultrahigh energy density and thermal stability in sandwich-structured nanocomposites with dopamine@Ag@BaTiO3, *Energy Storage Mater.* **2020**, 31, 492–504. DOI: 10.1016/j.ensm.2020.06.030.

79. Zhang, Z.; Meng, Q.; Chung, T. C. M. Energy storage study of ferroelectric poly (vinylidene fluoride-trifluoroethylene-chlorotrifluoroethylene) terpolymers, *Polymer* **2009**, 50, 707–715. DOI: 10.1016/j.polymer.2008.11.005.

80. Zhang, Y.; Liu, P.; Shen, M.; Li, W.; Ma, W.; Qin, Y.; Zhang, H.; Zhang, G.; Wang, Q.; Jiang, S. High energy storage density of tetragonal PBLZST antiferroelectric ceramics with enhanced dielectric breakdown strength, *Ceram. Int.* **2020**, 46, 3921–3926. DOI: 10.1016/j.ceramint.2019.10.120.

81. Xu, J.; Lu, X.; Yang, L.; Zhou, C.; Zhao, Y.; Zhang, H.; Zhang, X.; Qiu, W.; Wang, H. Enhanced electrical energy storage properties in La-doped (Bi0.5Na0.5)0.93Ba0.07 TiO3 lead-free ceramics by addition of La2O3 and La(NO3)3, *J. Mater. Sci.* **2017**, 52, 10062–10072. DOI: 10.1007/s10853-017-1209-0.

82. Xie, L.; Huang, X.; Li, B. W.; Zhi, C.; Tanaka, T.; Jiang, P. Core-satellite Ag@BaTiO3 nanoassemblies for fabrication of polymer nanocomposites with high discharged energy density, high breakdown strength and low dielectric loss, *Phys. Chem. Chem. Phys.* **2013**, 15, 17560–17569. DOI: 10.1039/c3cp52799a.

83. Wu, Y.; Fan, Y.; Liu, N.; Peng, P.; Zhou, M.; Yan, S.; Cao, F.; Dong, X.; Wang, G. Enhanced energy storage properties in sodium bismuth titanate-based ceramics for dielectric capacitor applications, *J. Mater. Chem. C* **2019**, 7, 6222–6230. DOI: 10.1039/c9tc01239g.

84. Wu, L.; Wu, K.; Liu, D.; Huang, R.; Huo, J.; Chen, F.; Fu, Q. Largely enhanced energy storage density of poly(vinylidene fluoride) nanocomposites based on surface hydroxylation of boron nitride nanosheets, *J. Mater. Chem. A* **2018**, 6, 7573–7584. DOI: 10.1039/c8ta01294f.

85. Wu, L.; Cai, Z.; Zhu, C.; Feng, P.; Li, L.; Wang, X. Significantly enhanced dielectric breakdown strength of ferroelectric energy-storage ceramics via grain size uniformity control: Phase-field simulation and experimental realization, *Appl. Phys. Lett.* **2020**, 117. DOI: 10.1063/5.0027405.

86. Li, J.; Chen, G.; Lin, X.; Huang, S.; Cheng, X. Enhanced energy density in poly(vinylidene fluoride) nanocomposites with dopamine-modified BNT nanoparticles, *J. Mater. Sci.* **2019**, 55, 2503–2515. DOI: 10.1007/s10853-019-04132-0.

87. Huang, H.; Chen, X.; Li, R.; Fukuto, M.; Schuele, D. E.; Ponting, M.; Langhe, D.; Baer, E.; Zhu, L. Flat-on secondary crystals as effective blocks to reduce ionic conduction loss in polysulfone/poly(vinylidene fluoride) multilayer dielectric films, *Macromolecules* **2018**, 51, 5019–5026. DOI: 10.1021/acs.macromol.8b01037.

88. Chen, J.; Yu, X.; Yang, F.; Fan, Y.; Jiang, Y.; Zhou, Y.; Duan, Z. Enhanced energy density of polymer composites filled with BaTiO3@Ag nanofibers for pulse power application, *J. Mater. Sci. Mater. Electron.* **2017**, 28, 8043–8050. DOI: 10.1007/s10854-017-6510-9.

89. Chen, J.; Yu, X.; Fan, Y.; Duan, Z.; Jiang, Y.; Yang, F. Enhanced the breakdown strength and energy density in flexible composite films via optimizing electric field distribution, *J. Mater. Sci. Mater. Electron.* **2017**, 28, 18200.

90. Zhang, X.; Jiang, J.; Shen, Z.; Dan, Z.; Li, M.; Lin, Y.; Nan, C. W.; Chen, L.; Shen, Y. Polymer nanocomposites with ultrahigh energy density and high discharge efficiency by modulating their nanostructures in three dimensions, *Adv. Mater.* **2018**, 30, e1707269. DOI: 10.1002/adma.201707269.

91. Cui, Y.; Zhang, T.; Feng, Y.; Zhang, C.; Chi, Q.; Zhang, Y.; Chen, Q.; Wang, X.; Lei, Q. Excellent energy storage density and efficiency in blend polymer-based composites by design of core-shell structured inorganic fibers and sandwich structured films, *Compos. Part B: Eng.* **2019**, 177. DOI: 10.1016/j.compositesb.2019.107429.

92. Chen, J.; Wang, Y.; Dong, J.; Chen, W.; Wang, H. Ultrahigh energy storage density at low operating field strength achieved in multicomponent polymer dielectrics with hierarchical structure, *Compos. Sci. Technol.* **2021**, 201. DOI: 10.1016/j.compscitech.2020.108557.

93. Yao, Z.; Song, Z.; Hao, H.; Yu, Z.; Cao, M.; Zhang, S.; Lanagan, M. T.; Liu, H. Homogeneous/inhomogeneous-structured dielectrics and their energy-storage performances, *Adv. Mater.* **2017**, 29. DOI: 10.1002/adma.201601727.

94. Cao, T.; Li, Y.; Wang, C.; Shao, C.; Liu, Y. A facile in situ hydrothermal method to SrTiO3/TiO2 nanofiber heterostructures with high photocatalytic activity, *Langmuir* **2011**, 27, 2946–2952. DOI: 10.1021/la104195v.

95. Bouharras, F. E.; Raihane, M.; Ameduri, B. Recent progress on core-shell structured BaTiO3@polymer/fluorinated polymers nanocomposites for high energy storage: Synthesis, dielectric properties and applications, *Progr. Mater. Sci.* **2020**, 113. DOI: 10.1016/j.pmatsci.2020.100670.

96. Sun, L.; Shi, Z.; Liang, L.; Wei, S.; Wang, H.; Dastan, D.; Sun, K.; Fan, R. Layer-structured BaTiO3/P(VDF–HFP) composites with concurrently improved dielectric permittivity and breakdown strength toward capacitive energy-storage applications, *J. Mater. Chem. C* **2020**, 8, 10257–10265. DOI: 10.1039/d0tc01801e.

97. Li, Z.; Liu, F.; Li, H.; Ren, L.; Dong, L.; Xiong, C.; Wang, Q. Largely enhanced energy storage performance of sandwich-structured polymer nanocomposites with synergistic inorganic nanowires, *Ceram. Int.* **2019**, 45, 8216–8221. DOI: 10.1016/j.ceramint.2019.01.124.

98. Wang, Y.; Wang, L.; Yuan, Q.; Niu, Y.; Chen, J.; Wang, Q.; Wang, H. Ultrahigh electric displacement and energy density in gradient layer-structured BaTiO3/PVDF nanocomposites with an interfacial barrier effect, *J. Mater. Chem. A* **2017**, 5, 10849–10855. DOI: 10.1039/c7ta01522d.

99. Lin, Y.; Sun, C.; Zhan, S.; Zhang, Y.; Yuan, Q. Ultrahigh discharge efficiency and high energy density in sandwich structure K0.5Na0.5NbO3 nanofibers/poly(vinylidene fluoride) composites, *Adv. Mater. Inter.* **2020**, 7. DOI: 10.1002/admi.202000033.

100. Li, W.; Song, Z.; Zhong, J.; Qian, J.; Tan, Z.; Wu, X.; Chu, H.; Nie, W.; Ran, X. Multilayer-structured transparent MXene/PVDF film with excellent dielectric and energy storage performance, *J. Mater. Chem. C* **2019**, 7, 10371–10378. DOI: 10.1039/c9tc02715g.

101. Wang, Y.; Cui, J.; Wang, L.; Yuan, Q.; Niu, Y.; Chen, J.; Wang, Q.; Wang, H. Compositional tailoring effect on electric field distribution for significantly enhanced breakdown strength and restrained conductive loss in sandwich-structured ceramic/polymer nanocomposites, *J. Mater. Chem. A* **2017**, 5, 4710–4718. DOI: 10.1039/c6ta10709e.

102. Marwat, M. A.; Xie, B.; Zhu, Y. W.; Fan, P. Y.; Ma, W. G.; Liu, H. M.; Ashtar, M.; Xiao, J. Z.; Salamon, D.; Samart, C.; Zhang, H. B. Largely enhanced discharge energy density in linear polymer nanocomposites by designing a sandwich structure, *Compos. Part A: Appl. Sci. Manuf.* **2019**, 121, 115–122. DOI: 10.1016/j.compositesa.2019.03.016.

103. Zhou, Y.; Liu, Q.; Chen, F.; Li, X.; Sun, S.; Guo, J.; Zhao, Y.; Yang, Y.; Xu, J. Gradient dielectric constant sandwich-structured BaTiO3/PMMA nanocomposites with strengthened energy density and ultralow-energy loss, *Ceram. Int.* **2021**, 47, 5112–5122. DOI: 10.1016/j.ceramint.2020.10.089.

104. Zhang, Y.; Zhang, T.; Liu, L.; Chi, Q.; Zhang, C.; Chen, Q.; Cui, Y.; Wang, X.; Lei, Q. Sandwich-structured PVDF-based composite incorporated with hybrid Fe3O4@BN nanosheets for excellent dielectric properties and energy storage performance, *J. Phys. Chem. C* **2018**, 122, 1500–1512. DOI: 10.1021/acs.jpcc.7b10838.

105. Chen, J.; Wang, Y.; Xu, X.; Yuan, Q.; Niu, Y.; Wang, Q.; Wang, H. Ultrahigh discharge efficiency and energy density achieved at low electric fields in sandwich-structured polymer films containing dielectric elastomers, *J. Mater. Chem. A* **2019**, 7, 3729–3736. DOI: 10.1039/c8ta11790j.

106. Park, S. H.; Hwang, J.; Park, G. S.; Ha, J. H.; Zhang, M.; Kim, D.; Yun, D. J.; Lee, S.; Lee, S. H. Modeling the electrical resistivity of polymer composites with segregated structures, *Nat. Commun.* **2019**, 10, 2537. DOI: 10.1038/s41467-019-10514-4.

107. Li, Z.; Shen, Z.; Yang, X.; Zhu, X.; Zhou, Y.; Dong, L.; Xiong, C.; Wang, Q. Ultrahigh charge-discharge efficiency and enhanced energy density of the sandwiched polymer nanocomposites with poly(methyl methacrylate) layer, *Compos. Sci. Technol.* **2020**, 202. DOI: 10.1016/j.compscitech.2020.108591.

108. Pan, Z.; Liu, B.; Zhai, J.; Yao, L.; Yang, K.; Shen, B. NaNbO3 two-dimensional platelets induced highly energy storage density in trilayered architecture composites, *Nano Energy* **2017**, 40, 587–595. DOI: 10.1016/j.nanoen.2017.09.004.

109. Liang, X. W.; Yu, X. C.; Lv, L. L.; Zhao, T.; Luo, S. B.; Yu, S. H.; Sun, R.; Wong, C. P.; Zhu, P. L. BaTiO3 internally decorated hollow porous carbon hybrids as fillers enhancing dielectric and energy storage performance of sandwich-structured polymer composite, *Nano Energy* **2020**, 68. DOI: 10.1016/j.nanoen.2019.104351.

110. Liu, G.; Zhang, T. D.; Feng, Y.; Zhang, Y. Q.; Zhang, C. H.; Zhang, Y.; Wang, X. B.; Chi, Q. G.; Chen, Q. G.; Lei, Q. Q. Sandwich-structured polymers with electrospun boron nitrides layers as high-temperature energy storage dielectrics, *Chem. Eng. J.* **2020**, 389. DOI: 10.1016/j.cej.2020.124443.

111. Wang, Z.; Li, Y.; Wang, X.; Fan, J.; Yi, Z.; Kong, M.; Xu, N. Reduced conductivity in sandwich-structured BFT@DA/PVDF flexible nanocomposites for high energy storage density in lower electric field, *J. Materiomics* **2020**, 6, 315–320. DOI: 10.1016/j.jmat.2019.12.009.

112. Jiang, Y.; Zhang, X.; Shen, Z.; Li, X.; Yan, J.; Li, B. W.; Nan, C. W. Ultrahigh breakdown strength and improved energy density of polymer nanocomposites with gradient distribution of ceramic nanoparticles, *Adv. Funct. Mater.* **2019**, 30. DOI: 10.1002/adfm.201906112.

113. Wang, Z.; Wang, X.; Sun, Z.; Li, Y.; Li, J.; Shi, Y. Enhanced energy storage property of plate-like Na0.5Bi4.5Ti4O15/poly(vinylidene fluoride) composites through texture arrangement, *Ceram. Int.* **2019**, 45, 18356–18362. DOI: 10.1016/j.ceramint.2019.06.050.

114. Wang, Z.; Kong, M.; Wang, X.; Li, Y.; Yi, Z.; Sun, Z. Ultrahigh energy storage performance in gradient textured composites of plate-like $Na_{0.5}Bi_{4.5}Ti_4O_{15}$/PVDF through interface engineering, *Ceram. Int.* **2020**, 47, 8787–8794. DOI: 10.1016/j.ceramint.2020.11.244.

115. Lin, Y.; Zhang, Y. J.; Zhan, S. L.; Sun, C.; Hu, G. L.; Yang, H. B.; Yuan, Q. B. Synergistically ultrahigh energy storage density and efficiency in designed sandwich-structured poly(vinylidene fluoride)-based flexible composite films induced by doping $Na_{0.5}Bi_{0.5}TiO_3$ whiskers, *J. Mater. Chem. A* **2020**, 8, 23427–23435. DOI: 10.1039/d0ta07937e.

116. Wang, D.; Dang, Z. M. Processing of polymeric dielectrics for high energy density capacitors, In *Dielectric Polymer Materials for High-Density Energy Storage*; Elsevier, **2018** pp. 429–446.

5 Applications and Prospects of Dielectric Materials for Capacitive Energy Storage

Haibo Zhang, Hua Tan, and Mohsin Ali Marwat

1 INTRODUCTION

In recent years, with the rapid development of clean energy, electric vehicles, high-end medical equipment, and smart grids, the requirements of pulse power systems for high power output are getting higher and higher.[1] As the core part of the whole extensive system, the pulse power system, in practical application, is required to release all the stored energy to the load in the shortest time and to release a large current in an instant, which is the key to affect the output of electromagnetic energy equipment directly.[2] The energy storage module in pulse power can be divided into inductive energy storage, capacitive energy storage, inertial energy storage, and chemical energy storage according to the principle of energy storage. Compared with several commonly used pulse power energy storage components, pulse energy storage capacitors, which can realize instantaneous high power conversion between electromagnetic energy and kinetic energy, have the highest releasable power density (MW level), the highest operating voltage (kV level), fast charging and discharging rate (μs~ns level), and long cycle life (more than 10^4 times), and have an irreplaceable role in the field of pulse power systems.[3] It can be used for defense as a power source for fast start-up devices of military vehicles such as armored vehicles and tanks and as a pulse power source for new concept laser weapons and electromagnetic rail guns.[1] It can be used for civil applications in artificial intelligence, offshore oil and gas and geological exploration, medical technology, excimer lasers, and other fields. In addition, as a kind of capacitor with a special purpose, a pulse energy storage capacitor is also a key component planned for development by the Ministry of Industry and Information Technology of China. Therefore, high-energy density pulse dielectric energy storage devices have received wide attention from academia and industry in recent years (Figure 5.1).[4, 5]

In the new energy market, according to the State Council's "Energy-saving and New Energy Vehicle Industry Development Plan (2012–2020)," by 2015, the cumulative production and sales volume of pure electric vehicles and plug-in hybrid vehicles will reach 500,000 units; by 2020, the production capacity of pure electric vehicles and plug-in hybrid vehicles will reach 2 million units, and the cumulative production

DOI: 10.1201/9781003454496-5

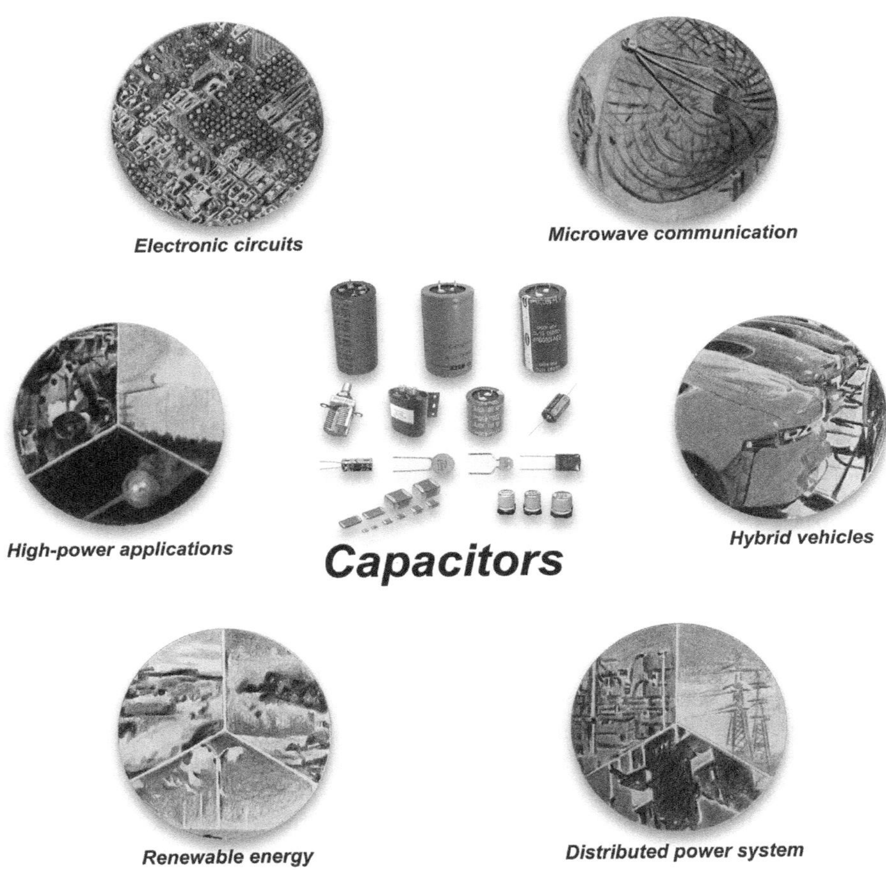

FIGURE 5.1 Application fields of capacitors.[6]

and sales volume will exceed 5 million units. According to the test standard, the power consumption of electric vehicles is about 0.15 KWh/km. Based on the 3 μm ultra-thin polypropylene films calculation, the cumulative demand for new energy films in China will exceed 6,000 tons and 33,000 tons in 2015 and 2020, respectively.

As for the power grid construction market, the construction of ultra-high voltage smart grid has been included in the 12th Five-Year Plan for National Economic and Social Development, which has become the future development trend of the power grid. To solve the imbalance of energy resources and load distribution in China, the State Grid will invest 620 billion RMB to build 20 UHV lines from 2013 to 2017, which is called "three vertical, three horizontal and one ring network plus 13 UHV DC lines" to transmit hydropower from the southwest and wind power from the northwest to the east of China. The new round of UHV power grid construction also brings huge development space for energy storage materials for capacitors.

In terms of rail transportation, by the end of the 12th Five-Year Plan, China's railroad mileage will reach about 120,000 km, and the electrification rate will reach more than 60%. As the process of railroad electrification accelerates, the market for energy storage capacitors will grow rapidly accordingly.

In the national defense construction, the electromagnetic catapult on the aircraft carrier is about 1,000 cubic meters in volume and weighs more than 500 tons; China's ultra-high power pulsed device "Jurong No.1" and Shenguang III laser fusion large scientific device also needs to use many large volume pulsed dielectric capacitors to store large-scale electrical energy to In the civil field, the newly developed off-grid power supply system for islands, mobile power supply system for islands, large power testing system, and high-voltage AC/DC conversion system, the dielectric capacitor that converts DC power to AC power to drive motors occupies more than 30% of the inverter volume. Therefore, with the urgent need for miniaturization and lightweight development of pulsed power devices, developing high-performance dielectric energy storage materials with high-energy storage density, high-energy storage efficiency, and can work under very high electric field is increasingly urgent.

2 PRINCIPLES OF OPERATION

Capacitors, as one of the most efficient energy storage devices, are also the most used passive components in electronic devices.[7] Capacitors have the following basic properties: (1) The voltage across the capacitor cannot be changed abruptly; (2) the capacitor through the AC DC isolation; (3) capacitor through the high-frequency, blocking low-frequency; (4) capacitor voltage lags behind the current; (5) capacitor just energized moment, equivalent to a short circuit; (6) capacitor capacitive resistance with the signal frequency increases and decreases, with the signal frequency decreases and increases. People use the basic characteristics of capacitors, design many practical applications of the circuit, and realize the function of many products; capacitors in the circuit play the main role of coupling, decoupling (bypass), filtering, energy storage, delay (timing), buck, resonance, buffer absorber (RC Snubber), the waveform changes (integral, differential, shaping), etc. (Note: Sometimes, the same circuit in the same capacitor, play the role of a variety of).[8]

Capacitors are useful in the following ways:

1. Energy Storage: Capacitors can store electrical energy. When a voltage is applied to the ends of a capacitor, charge accumulates between the two extreme plates of the capacitor. In this way, when a supply of electrical energy is needed, the capacitor can release the stored charge to provide energy to meet the needs of the circuit.[9]
2. Voltage Filtering: Capacitors have the property of low impedance, and connecting a capacitor in series in a circuit can help smooth out voltage fluctuations.[10] When there is a transient change in the supply voltage, the capacitor is able to provide a stable voltage output by absorbing or releasing charge, which makes the voltage change in the circuit smaller.
3. Improvement of Power Factor: Capacitors can improve the power factor of a circuit by compensating for the reactive power of the circuit.[11] In electrical equipment, some components need to consume reactive power (such as

inductive components), and capacitors can compensate for this part of the reactive power to make the whole circuit more efficient.

4. Timing Control: capacitor charging and discharging time constants are related to the capacitor and other components of the resistance value. Based on this characteristic, capacitors can be used as timing control elements for generating specific time delays, pulse signals, or oscillations.
5. Coupling: Capacitor signaling transfers the energy of an AC signal from one circuit to another.
6. Decoupling: Capacitors can eliminate mutual interference and influence between power sources, thus ensuring stable operation of the circuit.
7. Bypassing: Capacitors can bypass the AC component of the input signal so that the DC component of the signal can fully enter the subsequent circuit, thus improving the stability of the circuit.

According to the main role played by capacitors in the actual circuit and people's naming conventions, we divided into coupling capacitors, decoupling capacitors, bypass capacitors, filtering capacitors, energy storage capacitors, delay (timing) capacitors, buck capacitors, resonance capacitors, absorbing capacitors, integral capacitors, differential capacitors and so on. A capacitor in the circuit can play different roles at the same time. In the following section, we will specifically introduce the most commonly used capacitors in the circuit principle of operation.

2.1 Coupling Capacitor

Coupling, in electronics, denotes the transfer (passing) of energy from one circuit to another. Such as, through the light-emitting diode can be passed to the light-sensitive transistor, through an inductor can be passed to the magnetic energy of another inductor, through a capacitor can be passed from one end of the capacitor to the other end of the capacitor, the transfer of the above energy, can be called coupling. Capacitive coupling refers to transferring AC signal energy from one circuit to another through capacitance. This is a kind of energy coupling method; other energy coupling methods include inductive, optical, wire, and others.

Capacitors are used to block DC and allow AC to pass through. When capacitors are connected in series in a circuit, they block low-frequency signals and allow high-frequency signals to pass. In coupling capacitance, they are connected between the front and back stages of the circuit to transmit AC signals without attenuation while isolating DC signals.[12]

$$X_c = \frac{1}{2\pi fC} \tag{5.1}$$

where X_c is the capacitive reactance in Ω; π is the circumference; f is the signal frequency in Hz; C is the capacitive capacity in F.

According to the capacitance formula, in the case of a certain signal frequency f, the larger the value of capacitance, the smaller the capacitance; the smaller the obstruction attenuation of the signal, the better the coupling effect; the smaller the value of capacitance, the larger the capacitance, the larger the obstruction attenuation of the signal, the worse the coupling effect.

If the unwanted AC signal (noise signal) is superimposed on the desired AC signal (useful signal), the coupling capacitor with proper parameters can hinder the attenuation of low-frequency noise below the frequency of the useful signal and reduce the interference (cannot be eliminated); for high-frequency noise above the frequency of the useful signal, the coupling capacitor cannot be effective in hindering and attenuating the noise; to eliminate the interference, you need to use the coupling capacitor. In order to eliminate the interference, it is necessary to add appropriate filtering circuits (such as high-pass, low-pass, and band-pass, etc.) in the coupling capacitor's pre-stage circuit or post-stage circuit.

2.2 BYPASS CAPACITOR

This section summarizes the bypass capacitor and decoupling capacitor together due to their similarity. Bypass means to provide a new path of lower impedance than the original transmission path, allowing energy to bypass the original high impedance path and be transmitted from the new low impedance path. Decoupling, in electronics, means not allowing energy to pass from one circuit to another. In fact, bypass and decoupling have similar meanings, both of which mean filtering out undesired signals. Therefore, much domestic and foreign literature on the concept of bypass and decoupling is not strictly differentiated; decoupling and bypass designation can be interchangeable; decoupling can be called bypass, and bypass can be called decoupling. Because their essence is "not let the energy through a circuit to another circuit."[13] For the same circuit, the bypass capacitor filters out the high-frequency noise in the input signal and filters out the high-frequency clutter carried by the pre-stage, while the decoupling capacitor filters out the interference in the output signal.

Capacitors are used for AC and DC applications, as well as for high and low-frequency resistance. When placed in parallel with the rear circuit, this parallel capacitor serves as a bypass (decoupling) capacitor, with the following main roles:

1. Bypass the input of high-frequency AC signals (noise), hindering its transmission to the rear circuit so that DC or low-frequency signals go through to play the decoupling noise, filtering out the purpose of interference.
2. Bypass the high-frequency AC noise (power supply and ground noise) from the reverse output of the rear-stage circuit, impede its transmission to the front-stage circuit, play a role in decoupling the power supply and ground noise, and filter out the purpose of interference. The principle is as follows:

   If the ESR and ESL of the front-stage power supply path are large, when the power consumption current of the back-stage load circuit changes in a fast time and with a large amplitude, i.e., the larger the result of $\Delta i/\Delta t$ is, the larger the equivalent impedance Z of the front-stage power supply path is, the larger the power demand of the back-stage load's high-frequency sudden current cannot be met, which results in the generation of track collapse at the power supply input terminal of the load (power supply noise) as well as ground bounce (ground noise) at the ground output terminal of the load.[14] This results in track collapse (power noise) at the power input of the load and ground bounce (ground noise) at the ground output of the load, which not only causes the load itself to fail to operate normally but also causes the power noise and ground noise of the load to be transmitted to the front-stage network through the rear-stage circuit in reverse, causing electromagnetic interference to the entire circuit.

3. Immediately adjacent to the load and connected in parallel with the decoupling capacitor, shorten the path between the power supply and ground with the load, reducing the ESR and ESL, the equivalent impedance Z will be reduced, the decoupling capacitor charge stored in real-time to meet the load's high-frequency mutation current demand, there will be no power supply noise and ground noise, thereby improving the integrity of the power supply to ensure that the load works properly, and inhibit electromagnetic interference.

    for the latter stage of the circuit energy storage voltage stabilization. When the former circuit voltage dips, short-term interruptions occur, and there is a voltage gradient, the charge stored on the capacitor continues to power the latter circuit to play a role in stabilizing the voltage.[15] on the other hand, the capacitor has the role of energy storage to meet the demand for power to meet the latter circuit of the instantaneous current, the principle of and the above (2) of the same.

It can be seen that the bypass capacitor has a bypass, decoupling, and energy storage.

2.3 FILTER CAPACITOR

Filtering, in general terms, is the filtering and selection of a waveform. A waveform is composed of one or more frequency components (known by the Fourier series expansion); filtering is to remove some of the frequency components, do not let them pass, retain some frequency components, and let them pass.[16] Filter capacitors in power supply networks are often called filter regulator capacitors, such as rectifier voltage output filtering, switching power supply output filtering, LDO regulator output filtering, and so on. Filter capacitors in signal networks are often referred to as filter selector capacitors, such as low-pass filtering, high-pass filtering, band-pass filtering, band-stop filtering, and so on.

The main function of the filter regulator capacitor in the rectifier circuit is to smooth the unidirectional pulsating DC voltage output from the rectifier diode to make the voltage more stable while storing energy to meet the real-time transient power demand of the back-end load. The main functions of the filtering and regulating capacitors in the switching power supply are as follows: (1) Smooth the pulsating DC voltage output from the inductor, filter out the ripple, and make the voltage more stable. (2) Provide a stable feedback loop for the regulator to suppress the feedback noise so that the regulator, according to the load, changes in response to the regulation in more real-time, more accurate, so that the output voltage is more stable and accurate. (3) Store energy to meet the transient power demand of the back-end load in real-time. The main function of the filtering and regulating capacitor in the LDO circuit is to smooth the output ripple of the regulator, filter out the AC noise, stabilize the voltage, and, at the same time, store the energy for the back-end circuit to meet the demand of the back-end load for power consumption with the sudden current. The internal dynamic adjustment of the LDO output according to the change of output voltage in real-time to meet the demand of the load for power consumption also generates a fine ripple internally, which is transmitted through the output capacitor to the back-end circuit. Small ripples will be generated internally, which will be smoothed and filtered out by the output capacitor to stabilize the voltage.

2.4 Pulse Energy Storage Capacitor

An energy storage capacitor is an energy storage device consisting of two parallel electrodes and a dielectric between the electrodes.[17, 18] When the capacitor is charged, the two electrodes are filled with equal and opposite charges, creating an external electric field. The dipole and free charges in the dielectric migrate locally in the electric field. This leads to polarization simultaneously, generating an internal electric field. The direction of the internal electric field is opposite to the external electric field and increases with charge accumulation. The charging is over once the internal electric field is aligned with the external electric field. When an external load is connected, the capacitor will act as a power source discharging against the process, enabling the utilization of electrical energy. Unlike conventional capacitors, dielectric capacitors have unique pulse energy storage characteristics due to the absence of chemical processes during charging and discharging. Dielectric capacitors offer lower cost, shorter charge/discharge times (<0.01 s), higher power density (>10^5 W/kg), lower losses, and higher operating voltages.[19] As a result, they are the first choice for high-voltage, low-cost, and mass-manufactured energy storage applications (Figure 5.2).

Capacitor energy storage is commonly used in small to medium-sized pulsed power systems. The key technologies for capacitive energy storage pulsed power systems are the following: energy conversion and storage, switching technology, further steepening of the power pulse, and power pulse loading and measurement technology. Pulse capacitors used as energy storage elements are different from general capacitors in that, in addition to the general requirements for capacitors, the main requirements are low inductance, long life, high outer energy density, allowance for high-current discharges (close to short-circuiting), and repetitive-frequency operation. There are generally three types of capacitors suitable for use as pulse capacitors. (1) Ceramic capacitor: suitable for small capacitance value and high-current discharge; (2) metallized film capacitor: developed in recent years with the advantages of high energy density and soft failure, etc., with the smallest volume under the same capacity, but weaker discharge current than metal foil capacitor; (3) metal foil capacitor (oil-immersed): the traditional pulse capacitor. The capacitor charging power supply is the source of energy in a pulsed power system, which converts the alternating current of the utility single-phase 220 V or three-phase 380 V into DC high-voltage electricity stored in a capacitor.[21]

Capacitors are not charged when the voltage at both ends is 0; after full charge voltage common small and medium-sized pulsed power system energy storage voltage is between 5 and 40 kV. For large-capacity capacitors, the started charging stage of the capacitor ends in an approximate short-circuit state; the voltage is 0, which

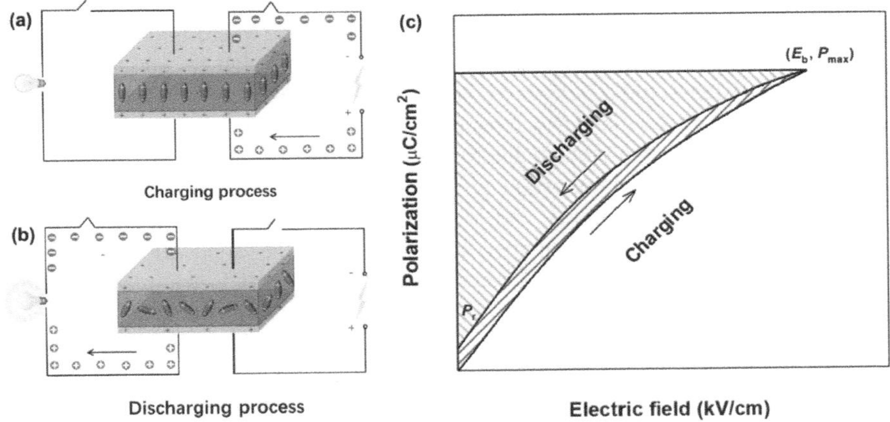

FIGURE 5.2 Energy storage principle for (a) charging process and (b) discharging process. (c) Schematic diagram of the energy storage performance evaluation based on the unipolar P–E loop of the dielectric capacitor.[20]

requires the output voltage of the capacitor charger to be adjusted from 0, allowing a long time of short-circuit operation.

Energy storage capacitors in the discharge voltage from tens of kilovolts instantly fall to 0, so the high-voltage charger should also have the output to cope with sudden short circuits. For certain high-frequency discharge applications, the energy storage inside the charger must be small enough so that frequent short circuits in the high-voltage output will have a reduced impact on the charger itself. The frequency of discharge can be increased.

Capacitor charging power supply in the charging process voltage is linearly rising, with conventional DC source charging having the problem of overshoot, so the voltage must be set at the time of stopping the output immediately in order to ensure that each test data is consistent with the charging repeatability is also a more important indicator.

A strong current pulse system in the discharge current can reach tens of kA or even hundreds of kA order of magnitude; such a strong discharge occasion generated by electromagnetic interference cannot be ignored, especially on the control system of the impact of the capacitor charger is best equipped with optical communication interface.

Due to the pulse characteristics of storage capacitors, they are widely used in both military and civilian pulse systems and are a core component of electric vehicle inverters, medical pacemakers, high-voltage DC transmission systems, and electromagnetic catapults on aircraft carriers.[5, 22]

3 MULTILAYER CERAMICS CAPACITORS

3.1 APPLICATIONS AND FUNCTIONS

Multilayer ceramic capacitors (MLCCs) are an important class of chip components due to their small size, low dielectric loss, high specific capacitance, and low price.

MLCCs are used in a large number of electronic products such as cell phones, scanners, and computers due to their small size, low dielectric loss, high specific capacitance, and low price. The rapid development of MLCC and its good application prospects have led to the development of related dielectric materials, whose application areas require high dielectric constants ε, good dielectric properties $\Delta C/C$, high insulation resistivity ρ and low dielectric loss tan δ.[23, 24]

MLCCs are the most important passive components for storing and releasing charge, providing high stability and low loss for resonant circuit applications and high volumetric efficiency for snubber, bypass, and coupling applications.[25] MLCCs are internally constructed as a staggered stack of ceramic dielectric material with a metal inner electrode and then encapsulated in a package with two end electrodes, as shown in Figure 5.3. The stacked structure is equivalent to the internal parallel connection of multiple bipolar plate capacitors, but the volume is smaller, and the layer thickness (equivalent to the distance between the two plates) is thinner. Compared to ceramic chip capacitors, it is easier to circuit surface-mounted technology (SMT) and has a high degree of integration.

MLCCs can be categorized into Class I and Class II capacitors.[26] Class I capacitors are electrically stable with little variation in temperature, voltage, and time and are suitable for high-frequency circuits where low loss and stability are required but

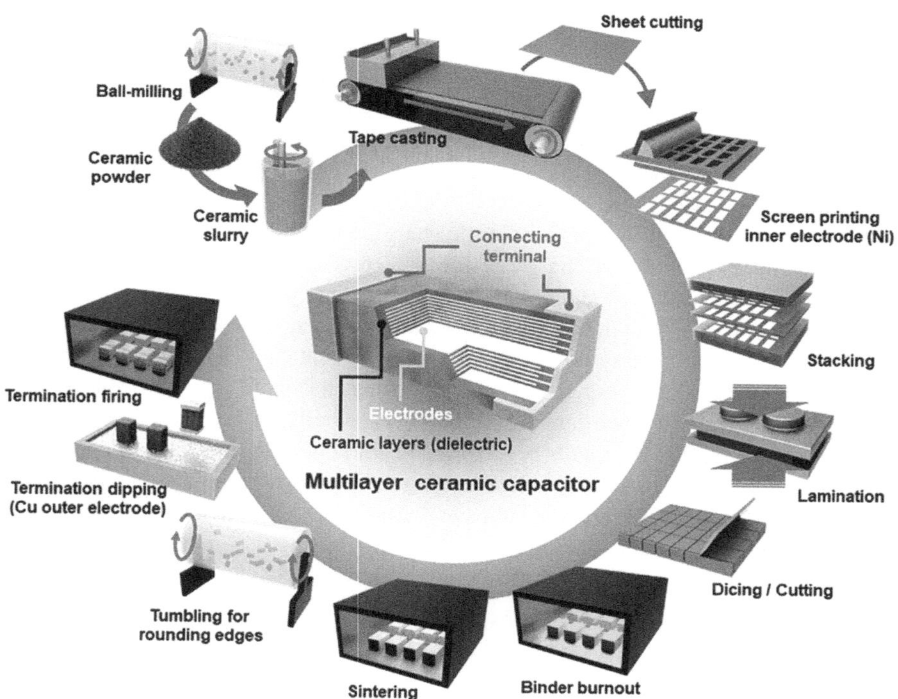

FIGURE 5.3   Schematics of the MLCC architecture and its fabrication process.[29]

generally smaller in capacity and are generally suitable for temperature-compensated, high-frequency, and filter circuits. Class I capacitors have a high dielectric constant and dielectric strength but are unsuitable for high-frequency circuits. Most Class I dielectrics are based on simple oxides such as rutile and chalcocite titanate.[27] Class I capacitors have high dielectric constants and are suitable for circuits with a wide range of capacities and low stability requirements, such as isolation, coupling, bypass, and frequency discrimination circuits.

The low voltage and high current of the current phase have led to the adoption of a decentralized power supply system for electronic equipment, in which several small Direct Current-Direct Current (DC-DC) converters are configured near the large-scale integration (LSI) or IC load from a bus converter, and several capacitors are connected externally. Output capacitors require high electrostatic capacity, especially for smoothing, so aluminum or button electrolytic capacitors are often used. Electrolytic capacitors have long been the preferred solution for applications requiring greater than 100 μF capacitance. However, the difficulty of miniaturizing electrolytic capacitors and their high equivalent series resistance can lead to excessive heat generation in the device or circuit and are often the first circuit components to fail.

Although MLCCs had excellent electrical characteristics in the early days, they could not be used in power supply circuits because of their small capacity. As the capacitance value of MLCCs continues to increase, it is becoming possible to replace electrolytic capacitors. In recent years, the development of MLCC (Multilayer Ceramic Capacitor) dielectric thin layer and multilayer technology has led to the industrialization of high-capacity MLCCs ranging from 10 to 100 μF or even higher. This advancement has positioned MLCCs as excellent substitutes for traditional electrolytic capacitors, offering the same high capacitance within the microelectronics industry. After replacing electrolytic capacitors, the low impedance characteristics of MLCCs can effectively reduce excessive heat generation caused by ripple currents, thus enhancing life and reliability, and the low backside of MLCCs also saves space on circuit boards, which is conducive to miniaturization of the overall modular circuits and high levels of integration. With the rapid development of 5G communications, consumer electronics, and new energy vehicles, the number of ultra-minimum high-capacity MLCCs has increased. Taking the most widely used cell phone as an example, a cell phone needs about 1,200 MLCCs to realize the functional operation of a cell phone; for new energy electric vehicles, their electronic control and automation systems need more than 10,000 MLCCs. The global MLCC market was valued at about $12.8 billion in 2018 and is expected to reach $24 billion in 2025.

In order to meet the ever-changing downstream end-market demand, the development trend of MLCC is mainly high-capacity, high-frequency, high-voltage, high-temperature, high-reliability, and miniaturization.[28]

1. High capacitance: to match the increasing functions of the terminal, the battery capacity growth, and the requirements of the MLCC toward the development of a large-capacity trend. Due to the increase in the number of terminal configuration functions, the battery capacity becomes large, stable, and fast charging large-capacity batteries, the need to configure

large-capacity, high-quality MLCCs. Part of the electronic circuit uses large-capacity specifications in order to reduce the number of MLCCs and, therefore, have a high demand for large-capacity. According to Japan's Murata (Murata) forecast, high-end smartphone electrostatic capacity is expected to grow to 4,000 µF in 2023, and medium smartphone electrostatic capacity is expected to grow to 2,000 µF in 2023. In short, the MLCC capacity to volume ratio will be with the downstream industry's requirements and improve year by year.

2. High-frequency: MLCC operating frequency has entered the millimeter-wave band range; in order to meet the high performance of electronic circuits and multi-functional requirements, LSI operating frequency is increasingly high, which also puts forward higher requirements for low-impedance power supply, the market for the ability to use low equivalent series resistance, low equivalent series inductance MLCCs in a wide range of frequencies ($10^6 \sim 10^9$ Hz) has become more urgent.

3. High-voltage: With the evolution of power supply device circuit design, LED lighting part of the demand is expected to rise, and 3~4 kV high-voltage capacitor demand will continue to increase. With the development of automotive electronics and electric vehicles in recent years, the demand for automotive MLCCs has soared.

4. High-temperature resistance and high reliability: Although car-specification MLCC does not have the size of the carving restrictions, its reliability requirements have become extremely important to the harsh environmental requirements, which puts forward higher requirements for the life of the MLCC high-reliability MLCC needs to be in high temperature, high humidity, and other extreme environments, stable operation. At the same time, they also need to be certified to AEC Q200 (Automotive Electronic Component Reliability Test Specification), which is a very demanding production standard. As a result, the high-reliability requirements for MLCCs will continue to rise in the future.

5. Miniaturization: The MLCC size model can be roughly divided into 1206, 0805, 0603, 0402, 0201, and so on. Such as 0805 model dimensions: length x width of 0.08 inch × 0.05 inch (2.03 × 1.27 mm). Smaller external dimensions mean the need to use smaller sizes of raw material powder, as well as more complex and sophisticated production processes. In the pursuit of miniaturization, it has become more difficult to maintain or even improve the volume efficiency of MLCCs.

China is the world's largest producer, seller, and consumer of MLCCs, but Japan, South Korea, and the United States still dominate the high-end market for MLCCs, and the development of the domestic industry is relatively slow. In recent years, Japan's advanced MLCC manufacturers adjusted production capacity, and in the context of the U.S.-China trade friction and the new crown epidemic, domestic downstream customers have begun to rely on the domestic MLCC supply chain, the space for domestic substitution is huge, China's MLCC R & D ushered in a golden period of development.

3.2 Fabrication Techniques of MLCC

From the present reports, the main dielectric systems used for MLCC are $BaTiO_3$-based, lead-based, $BaTiO_3$, and lead-based composite and tungsten bronzite dielectric ceramic materials. Although most of the dielectric materials used in MLCCs, whether at the experimental stage or commercially available, are $BaTiO_3$-based materials, this is related to the characteristics of $BaTiO_3$ itself, which is a typical perovskite structure with an internal O-atom forming an oxygen octahedral structure and a Ti atom located at the center of the oxygen octahedron, which has a large displacement space for Ti ions due to the large space of the oxygen octahedron. In particular, the temperature dependence of the dielectric constant near the Curie temperature TC (about 130°C) has a strong influence on the dielectric constant, and it is found that when the temperature is higher than the Curie temperature TC, the dielectric constant of $BaTiO_3$ can be drastically reduced. Pure $BaTiO_3$ has the disadvantages of a large temperature change rate, easy aging of the dielectric, high dielectric loss, and does not meet the application requirements of the product, so it can be used in the actual product by means of doping modification, microstructure adjustment, and process optimization to obtain a more desirable effect. Here, we first introduce the preparation process of MLCCs; the schematic is shown in Figure 5.3.[29]

1. Preparation of casting slurry

    The casting slurry with good fluidity is prepared by mixing the calcined ceramic powder with solvent, suspension dispersant, binder, and plasticizer at a certain ratio. The volume fraction of the ceramic powder in the slurry is often greater than 50% to ensure the qualified strength and relative density of the ceramic layer

2. Tape-casting process

    The mixed slurry is then transferred to the tape casting machine, where the slurry is scraped on the tape by a doctor blade. The thickness of the casted slurry and the resultant ceramic green body is controlled by adjusting the height of the blade. After volatilization of the excessive solvent in the slurry, a ceramic tape green body with good elasticity and flexibility can be obtained.

3. Preparation of inner electrodes

    The dried green body is cut into a designed shape, and a thin electrode layer is printed on one side of the tape by screen printing.

4. Lamination

    Tens of layers of green bodies printed with internal electrodes are stacked. The electrodes of adjacent green bodies are stacked in a misaligned manner to form positive and negative electrodes. Then, the stacked green bodies are subjected to warm isostatic pressing and cut into individual actuator green bodies.

5. Degreasing and co-firing

    The actuator green body is first heated slowly to 200°C–300°C to remove the remnant organics. Then, the heating temperature is raised to the ceramic sintering temperature for the co-firing of the ceramic layer and the inner electrodes.

6. External electrode printing, packaging, and testing

The sintered ceramic was chamfered to expose the inner electrode fully. Then, the electrode slurry was applied to the exposed internal electrodes to connect the internal electrode on the same side to form the external electrode. Low-temperature sintering of the ceramic is performed again to ensure the connection of the inner and outer electrodes. Finally, an electrical performance test is performed to eliminate defective products.

The ceramic layers prepared by tape casting often show high density, low porosity, and small grain size. These characteristics are all thought to be conducive to the capacitors achieving a higher breakdown of the electric field. In theory, the breakdown strength of MLCC increases exponentially with the decrease in single ceramic layer thickness. For example, for the Al_2O_3- modified $BaTiO_3$ dielectric layer-based capacitor, its breakdown field strength is significantly increased from 184 to 66 kV/cm when the thickness of the single dielectric layer reduces from 63 to 12 μm. The energy storage density of MLCC is also increased from 1.7 to 4 J/cm³.[30]

The selection of ceramic materials for MLCCs is one of the most critical factors in determining energy storage performances. With the development of dielectric materials and phase engineering, high-energy storage density and efficiency have been achieved for dielectric energy storage applications. For example, the MLCCs based on $0.85BaTiO_3$–$0.15Bi(Zn_{0.5}Ti_{0.5})O_3$ were reported to have an energy storage density of 2.8 J/cm³ at 33 MV/m.[31] By optimizing the composition, reducing the thickness of each layer to approximately 5 μm, and using the two-step sintering (TSS) method to avoid grain growth, Wang et al. successfully prepared MLCCs based on $0.87BaTiO_3$–$0.13Bi(Zn_{2/3}(Nb_{0.85}Ta_{0.15})_{1/3})O_3$ (BT–BZNT) with high-energy storage density (10.5 J/cm³) at 104.7 MV/cm.[32] Furthermore, BT–BZNT MLCC also shows good thermal stability, which only a relatively low variation (i.e., <±5%) can be observed over a temperature range of −75°C to 150°C at 400 kV/cm.

Although the BT–BZNT system obtains considerable energy storage density under the ultra-high electric field, its energy storage capacity is still slightly weaker than that of Bi-based multilayer capacitors. This indicates the prospective potential of Bi-based dielectric ceramics with high polarization in the application of ultra-high-energy storage density capacitors. Among all the MLCCs, $Bi_{0.5}Na_{0.5}TiO_3$–$Sr_{0.7}Bi_{0.2}TiO_3$ (BNT–SBT) binary system is an ideal energy storage dielectric. BNT possesses both relaxor and antiferroelectric characteristics, and the addition of relaxor ferroelectric SBT can enhance the relaxor features of the BNT–SBT. As a result, the P–E loop not only shows the near-zero remnant polarization of the antiferroelectric but also the slimness of the relaxor ferroelectric. Li et al.[33] fabricated MLCC using 0.55BNT–0.45SBT as the dielectric layers (single-layer thickness is 20 μm) and Pt as the internal electrode. Its breakdown field strength has been found with a significant increase to 720 kV/cm compared to its 200 μm thick bulk-like ceramics (200 kV/cm). The energy density and efficiency of MLCCs are 9.5 J/cm³ and 92%, respectively, measured at 720 kV/cm. The influences of temperature on the energy storage-related performances (e.g., energy density and efficiency, energy discharge, and cycle reliability) of MLCC have also been studied. It was found that the MLCCs exhibit satisfactory thermal stability in the temperature range of −60°C to 120°C, with a small variation (i.e. <10%) for both energy density and efficiency. In addition, the discharge behavior also shows a low-temperature variation (<15%) over the temperature range of 25°C–100°C. However, the capacitor performs better cycling reliability with cumulative numbers of up to 10^6 at 100°C compared with measurement at room temperature. The degradation of performance at room temperature is attributed to the mechanical stress from switching of the ferroelectric polar regions. At a higher temperature, the polar regions decrease, thus resulting in better fatigue behavior.

In MLCC, the electric breakdown of ferroelectric ceramics is often caused by strain-induced mechanical breakdown under the electric field. This strain effect has been found to be strongly linked with the grain orientation of the dielectric ceramics in MLCC. For example, by analyzing the <100>, <110> and <111>-textured single-layer 0.65BNT–0.35SBT capacitors using finite element simulation, Li et al. found that $h_{111}i$-oriented ceramic has the smallest local strain and stress concentration at the electrode end. Its elastic energy density is also much lower than <100> and <110> textured samples. As a result, the breakdown field strength of <111>-textured 0.65BNT–0.35SBT MLCC can reach up to 1,030 kV/cm, coming along with a large energy storage density of 21.5 J/cm^3.[34] This is the highest value in the reported lead-free MLCCs so far. Compared with non-textured MLCC (with a breakdown strength of 64 kV/cm), the breakdown strength of the <111> textured sample increased by 16 times (1,030 kV/cm), as shown in Figure 5.4. This is primarily attributed to the lowered strains in the textured sample, which has much more pores than the non-textured ceramics. These holes can reduce the possibility of microcrack initiation and crack-driven dielectric breakdown. Thus, the local electric field around the pores is reduced, and the growth rate of the partial discharge tree is also slowed. In addition, the difference between the maximum polarization of non-textured MLCC and <111>-textured MLCC is small, which helps to increase the energy storage density.

The core–shell structure has also proven to be effective in enhancing the performance of MLCC. Wang et al.[35] modified the core–shell structure of BF–BT grains

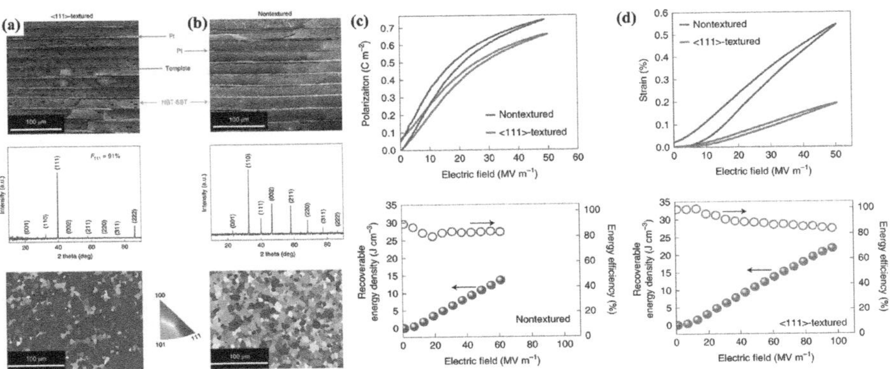

FIGURE 5.4 The microstructure, XRD, and EBSD of (a) textured MLCC and (b) non-textured MLCC; the (c) P–E loop, (d) S–E loop and energy storage properties of textured and non-textured MLCC.[34]

by adding $Nd(Zn_{0.5}Zr_{0.5})O_3$ to achieve an electrically homogeneous microstructure rather than chemically. They successfully prepared $0.62BiFeO_3$–$0.3BaTiO_3$–$0.08Nd(Zn_{0.5}Zr_{0.5})O_3$(0.62BF–0.3BT–0.08NZZ) ceramics with a relaxor feature for both core and shell parts. This chemically inhomogeneous structure can significantly increase the resistivity of the ceramic and thus improve the breakdown performance. Next, a Pt internal electrode fabricated MLCCs with 0.62BF–0.3BT–0.08NZZ. This capacitor is made of seven active ceramic layers. Each layer's thickness and active area are 16 mm and 33 mm_2, respectively. The as-prepared MLCC shows a high breakdown strength (700 kV/cm), a high P_{max} (35 $\mu C/cm^2$), and a low remnant polarization (1 $\mu C/cm^2$). As a result, a high-energy storage density of 10.5 J/cm^3 and an excellent efficiency of 87% have been achieved in this MLCC.

MLCC is a key electronic component used in almost all areas of electronics, and the number and variety of applications are steadily growing. Over the past few decades, MLCC manufacturing has focused on accelerating miniaturization, improving volumetric efficiency, enhancing environmental stability and reliability, and reducing costs, all of which have been made possible by improvements in material formulation and processing techniques. Most ceramic capacitors are based on $BaTiO_3$, which can be chemically or physically modified to exhibit the desired temperature-stable dielectric properties. Therefore, future research on MLCC dielectric ceramics should focus on the following aspects: improving the dielectric constant of the material and improving the dielectric temperature characteristics; finding suitable additive components and developing new material systems to increase the upper limit of the working temperature as much as possible while maintaining a relatively high dielectric constant; the performance of $BaTiO_3$-based dielectric materials changes greatly after the introduction of other components, but the microscopic mechanisms and theoretical studies around $BaTiO_3$ materials are not enough, so the basic theoretical studies are also an important research direction for this material.

3.3 Industry Research

The global MLCC industry has a high degree of industry concentration. Foreign MLCC manufacturers mainly include Japan Murat, TDK, Kyocera, and Taiyo Yuden. Domestic manufacturers mainly include Fenghua Advanced Technology, Torch Electronics, Chaozhou Three-circle, etc. However, in terms of technology and product supply varieties, China mainly produces low-end and medium-end MLCCs, and high-end MLCCs are concentrated in companies such as Murata and TDK in Japan. Lack of independent intellectual property rights and the development of high-performance ceramic key technologies result in high-end MLCC mainly dependent on imports, while high-performance ceramics are blocked. In the context of the U.S.-China trade friction and the new crown epidemic, domestic downstream customers began to rely on the domestic MLCC supply chain; domestic substitution space is huge, and China's MLCC research and development ushered in a golden period of development.

MLCC has a wide capacity range, good frequency characteristics, a wide range of operating voltage and temperature, ultra-small size, no polarity, and other excellent characteristics, such as noise bypass, power supply filtering, energy storage, differential, integral, oscillating circuits, etc. MLCC is the world's fastest development of one of the chip components, widely used in aerospace, aviation, ships, weapons, medical equipment, rail transportation, consumer electronics, automotive electronics, and other military, industrial, and consumer fields.

As the main application field of MLCC, consumer electronics has a huge consumption of MLCC. With the high-frequency updates and iteration of smartphones, product functions are constantly upgraded, and the electronic components behind them are becoming increasingly complex; the use of stand-alone MLCCs continues to increase. Taking Apple's cell phone as an example, data from the China Electronic Components Industry Association shows that the MLCC usage of a single iPhone 5S is about 400 pcs, iPhone 6 about 780 pcs, iPhone 7 about 850 pcs, iPhone 8 about 1,000 pcs, and iPhone X about 1,100 pcs, with high-end MLCC products accounting for a growing proportion. Along with the pursuit of convenience and thinness in smartphones, MLCCs are also developing in the direction of small size and high capacitance and are increasingly used in high-end models. Notebook computers, tablet PCs, and other consumer electronics products with high design precision will continue to increase the amount of MLCC.

A dual-carbon policy to promote the transformation of automotive intelligence automotive electronics drives the demand for automotive-grade MLCC. Automotive power systems, security systems, entertainment systems, and other systems need to use MLCC, and along with the development trend of intelligent traditional fuel vehicles, automotive electronic control circuits on the MLCC demand is increasing. According to the data of the China Electronic Components Industry Association, the new energy of automobiles makes the use of MLCC per vehicle increase from 1,000–3,000 pcs to 3,000–6,000 pcs, up to 10,000 pcs. Among them, the power system brings the largest increment of MLCC. While the number of MLCCs used in the power system of electric vehicles is 2,700–3,100, the number of MLCCs used in the power system of traditional fuel vehicles is 450–600. From the point of view of total demand, according to the Trend Force forecast, in 2022, the demand for MLCC in the automotive market will grow to 562 billion. Therefore, the continued increase in the penetration rate of new energy vehicles will further boost the booming development of the MLCC market.

From the demand side, China has become the world's largest MLCC market, with a demand share of 40%. According to the data from the China Electronic Components Industry Association, the global MLCC industry market size reached 101.7 billion yuan in 2020, while the market size of China's MLCC industry is about 46 billion yuan, accounting for about 45.23% of the global. The Chinese mainland MLCC manufacturers accounted for only 6% of the global share; mainland manufacturers' capacity supply is small. The huge imbalance between supply and demand in China's MLCC industry has resulted in the need to import a large number of MLCC products. According to China's General Administration of Customs data, China's MLCC import dependence is high. In 2020, China's MLCC imports were 3.08 trillion pcs, and exports were 1.63 trillion pcs. Imported products are mainly concentrated in the middle and high end. In 2020, the average unit price of MLCC imports was $26.35/million, and the export unit price was $23.60/million.

In summary, China's MLCC domestic substitution space is huge. The global competition in science and technology is becoming increasingly fierce; domestic MLCC manufacturers have ushered in a critical period of opportunity for domestic substitution. Fenghua Advanced Technology, Chaozhou Three-circle, torch electronics, etc., are actively increasing related production capacity; for example, Fenghua Advanced Technology 2020 has proposed to invest 7.505 billion yuan to build a new industrial park, which is a high-end capacitance-based project, with a monthly output of 45 billion MLCC capacitors. While the project is expected to reach production after the global ranking of Fenghua Advanced Technology to enter the top five, the overall competitiveness will be steadily improved. Vigorous research and development of new ceramic materials and advanced MLCC processes will surely realize the localization of high-performance ceramic capacitors and promote the development of global energy storage capacitors.

4 ORGANIC THIN FILM CAPACITORS

4.1 FILM CAPACITOR TYPES

As their name suggests, film capacitors employ a plastic film as a dielectric. The technical ancestor of this type of capacitor is the paper capacitor, invented in the second half of the 19th century. It consisted of paper impregnated with oil or paraffin, sandwiched between sheets of aluminum foil, and rolled into a round shape. A capacitor where the metal foil is replaced by a layer of metal directly created on the paper by means of vapor deposition is called an MP (metallized paper) capacitor. Based on this technique, film capacitors were developed in the 1930s. Compared to multilayer ceramic chip capacitors, it is difficult to make film capacitors small, but they offer high insulation resistance and high reliability. Film capacitors are found, for example, in electric home appliances, electronic circuits in cars, industrial equipment, and power electronics devices.

Depending on how the internal electrode is formed, film capacitors are divided into two main categories: foil electrode types and vapor deposition electrode (metallized film) types. Subcategories according to construction include wound types, laminated types, inductive and non-inductive types, etc.

4.1.1 Foil Electrode Type Film Capacitors

Wound-type film capacitors with internal electrodes are made of metal foil (aluminum, tin, copper, etc.) sandwiched between plastic film layers and rolled up. They come in inductive and non-inductive versions. Inductive types have lead wires attached to the internal electrodes before winding, while non-inductive types have lead wires or terminal electrodes connected to end faces. Compared to inductive types, non-inductive film capacitors have a lower inductance component and exhibit better high-frequency characteristics (Figure 5.5).

4.1.2 Vapor Deposition Electrode Type (Metallized Film Type) Film Capacitors

Instead of using foil as an electrode, this type of film capacitor uses a layer of metal (aluminum, zinc, etc.) deposited on the plastic film itself to form an internal electrode. Because the deposited film is very thin, the capacitor can be made smaller than the foil electrode type.

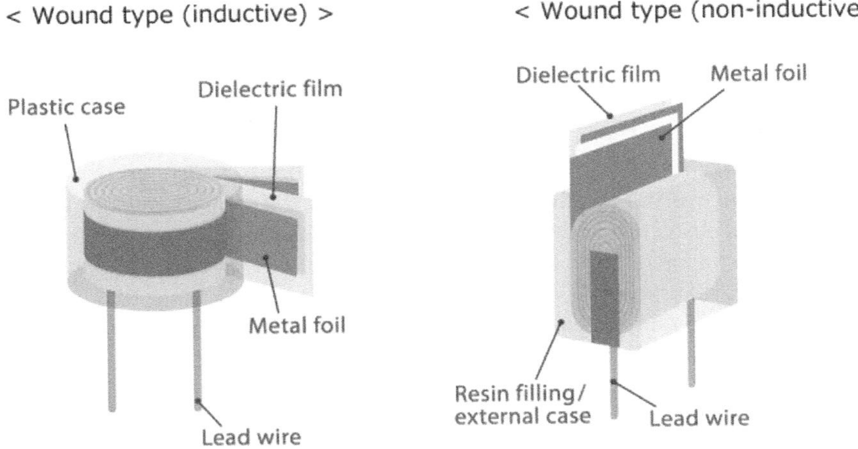

FIGURE 5.5 Structural schematic of foil electrode type film capacitors.

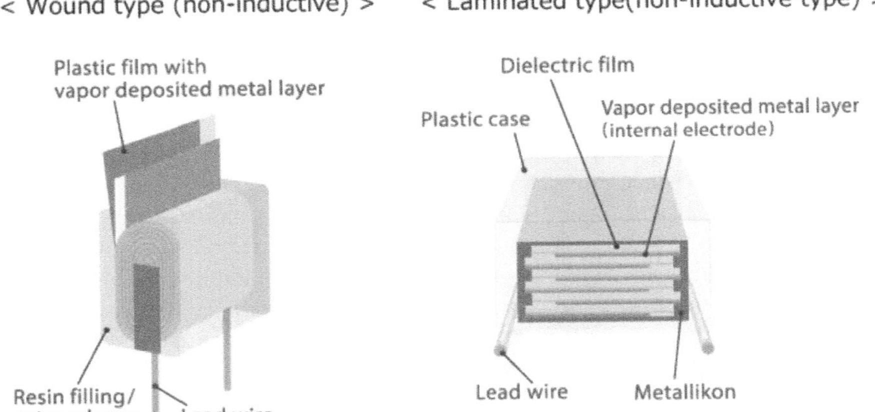

FIGURE 5.6 Structural schematic of vapor deposition electrode type (metallized film type) film capacitors.

Vapor deposition-type capacitors are non-inductive, where the electrode is connected to an end face. In terms of manufacturing methods, there are wound types and laminated types (Figure 5.6).

4.2 APPLICATIONS AND FUNCTIONS

Film capacitors are capacitors that utilize a plastic film as a dielectric. The technology originated from paper capacitors invented in the second half of the 19th century and was developed in the 1930s. They are characterized by high voltage resistance and reliability and are widely used in home appliances, automotive electronics, industrial equipment, and power electronics.[36]

Film capacitors are broadly classified into the foil electrode type and the evaporated electrode type (metallized film type), depending on the method of forming the internal electrodes. The evaporation electrode of the evaporation electrode type film capacitor has a self-healing function. When the film is punctured due to the application of an overvoltage, the evaporation film near the point of puncture is oxidized instantaneously, restoring its insulating state. In addition, safety features can be added to the vaporized film to further improve reliability. In contrast to the conventional whole-side evaporation method, evaporation is carried out by dividing into grids and connecting the different grids to each other with a narrow fusing section. In the event of an insulation breakdown that exceeds the limit of the film capacitor's self-healing function, the fusing section will fuse to prevent such a destructive insulation breakdown.

Depending on the structure, they are categorized into coiled, laminated, inductive, non-inductive, etc. Depending on the dielectric, capacitors can be divided into ceramic, aluminum, and film capacitors. A film capacitor is a capacitor that is constructed using metal foil as an electrode and plastic films, such as polyethylene, polypropylene, polystyrene, or polycarbonate, which overlap at both ends and are rolled into a cylinder. Depending on the type of plastic film, they are called polyethylene, polypropylene, polystyrene, and polycarbonate capacitors.[37]

Film capacitors exhibit different capabilities due to differences in film materials. Take the common polyester film and polypropylene film capacitors as an example. In terms of material properties, polyester film has a wide operating temperature range, high dielectric constant, high-capacity stability, positive temperature coefficient, high insulation resistance, good self-healing, and high-volume ratio. Polypropylene film capacitors, relative to polyester films, have higher dielectric strength, very low high-frequency loss, and high stability of capacitance. In addition, in high-voltage film capacitors, there are other advantages: the negative temperature coefficient is small, the insulation resistance is very high, and the dielectric absorption coefficient is very low.

In addition to the above general polypropylene film and polyester film, metallized film is the main development direction of organic film capacitors. Metallization refers to the use of the vacuum coating method in the organic film evaporated layer of aluminum or zinc and other metals as electrodes, with organic film as a medium, thus forming a capacitor. The advantage is that it has self-healing properties, which can enhance the working life and reliability of the device.

Compared with other capacitors, film capacitors have no polarity, high insulation impedance, high pulse resistance, excellent frequency characteristics (wide frequency response), and small dielectric losses. Based on these advantages, film capacitors are widely used in analog circuits of a certain power.[38]

    7. Small change in capacitance induced by temperature change

        The electrostatic capacitance of a capacitor changes with temperature. For example, in ceramic capacitors, the dielectric constant of the dielectric changes when the temperature changes, resulting in a change in electrostatic capacitance. In addition, for aluminum electrolytic capacitors, the electrolyte's conductivity and the electrodes' resistance change with temperature, and thus the electrostatic capacitance also changes. Ceramic capacitors and aluminum electrolytic capacitors exhibit a change in electrostatic capacitance of more than 10% with temperature variations. In contrast, film capacitors experience only a few percent change in electrostatic capacitance within the same temperature range.

8. High insulation resistance

Film capacitors use PP (polypropylene), PET (polyethylene terephthalate), PPS (polyphenylene sulfide), and PEN (polyethylene naphthalene dicarboxylate) as the dielectric and have features such as higher insulation resistance and a more desirable ability to maintain the saved electrics than ceramic capacitors and aluminum electrolytic capacitors. The insulation resistance of capacitors generally decreases with increasing temperature, but compared to other capacitors, the insulation resistance of film capacitors is less likely to decrease even when the temperature increases, thus maintaining performance. In addition, compared to other capacitors, film capacitors are characterized by low loss during electrical input or output. Among them, the film capacitor using PP dielectric has the distinct advantage of keeping the loss to a minimum even when the temperature changes.

9. No polarity

Some capacitors have polarity, while others do not. For example, aluminum electrolytic capacitors can only be used for direct current because of their polarity, but film capacitors do not have polarity and can be used for both direct current and alternating current.

10. Fewer failures and longer life

Plastic film as a dielectric for film capacitors is characterized by fewer failures and longer life than other capacitors due to its stable physical properties.

11. No change in bias characteristics

While the electrostatic capacitance of ceramic capacitors changes with the applied voltage, the electrostatic capacitance of film capacitors hardly changes with the applied voltage. Utilizing this characteristic, film capacitors have the advantage of less distortion and improved sound quality when used in audio circuits.

12. Excellent withstanding voltage and ripple current characteristics

Aluminum electrolytic capacitors have a withstand voltage of about 500 V, but film capacitors can be manufactured to withstand a high withstand voltage of nearly 4,000 V. Inverter power supplies are used to handle high voltages of 650 V for solar power generation systems, 48 to 750 V for HEVs, and 1,000 to 3,000 V for railroad cars. Film capacitors are indispensable for stabilizing the voltage of the inverter power supply (noise removal and smoothing). Film capacitors also have excellent ripple current resistance (allowable current) and are characterized by the fact that they are not susceptible to self-heating even when large currents flow through them.

Film capacitors have the disadvantages of high cost (the price is two to three times that of electrolytic capacitors) and large size. Currently, it is often used to balance its high cost by extending the device's life and reducing the capacitor's size by base film material improvement and metallization splitting technology. Base film is the main raw material of film capacitors and determines the performance of film capacitors. Toray of Japan, Mitsubishi Corporation, and DuPont of the United States are the world's high-quality base film suppliers.

Currently, the world's dielectric film materials used for film capacitors are mainly polyester film and polypropylene film, widely used in lighting, household appliances, industrial control, photovoltaic wind power, new energy vehicles, and other fields. Among them, polyester film capacitors are mainly used in lower frequency DC and pulsating circuits, mainly used in the production of DC capacitors, suitable for electronic products with a high degree of electronic integration; polypropylene film capacitors are mostly used in high-voltage DC and AC circuits or grids, mainly used in the production of AC capacitors, suitable for electronics, home appliances, communications, and electric power capacitors.

In household appliances, film capacitors are mainly used in electrode starting and other circuit control parts of air conditioners. Large household appliances such as refrigerators, washing machines, and air conditioners need to be driven by large motors; in the motors, especially in control circuits and frequency conversion circuits, different kinds of film capacitors play different roles, according to the different functions, and they can be divided into three main categories: (1) EMI suppression: In addition to the power supply side, they are also used in the input and output sides of frequency conversion circuits, both of which play a role in suppressing electromagnetic interference and preventing damage to components; (2) DC Link: Acts as a low-pass filter to suppress fast transient currents and smooth the output voltage to keep the DC bus voltage within the permissible range; (3) AC Film Capacitor: Used in the drive control circuit of induction motors, where a capacitor is needed to switch the motor to the desired direction by phase shift during the start-up of the induction motor. For appliances that consume a lot of power, such as refrigerators, washing machines, and air conditioners, if the frequency of the power supply can be adjusted according to the operating status of the appliance, the power consumption can be reduced, and the life of the appliance can be extended. An inverter consists of a rectifier, filter, inverter, brake, and drive units that control an AC motor by changing the motor's operating power supply frequency. The current in the inverter circuit is converted from AC to DC to AC, and film capacitors are required for EMI suppression at both the input and output points. Therefore, the demand for film capacitors in home appliances depends on the overall shipment of home appliances and the penetration rate of inverter appliances.

In new energy fields such as photovoltaic and wind power, capacitors are key devices in inverters/converters, mainly used for DC support (DC Link) and IGBT-supporting protection. In lighting, film capacitors are mainly used in ballasts for fluorescent lamps and high-intensity discharge lamps such as high-pressure mercury, high-pressure sodium, and metal halide lamps. Film capacitors are used in fluorescent lamps for two main functions: inside the starter for EMI suppression during induction and between the line and neutral for power factor correction.

The three core modules of an electric vehicle are the battery, motor, and motor controller. Among them, the motor controller, i.e., the inverter, is used to control the energy transmission between the power supply and the motor, and its core product is the IGBT power module. Due to the harsh internal working environment of the car, the engine compartment working temperature is high, and the start and stop will cause an impact on the device; the capacitor is required to have a strong resistance to high temperature and impact, and at the same time has a strong service life and reliability. Therefore, in the rapid development of new energy vehicles, the capacitors applied to power supply and motor control modules require higher requirements.

Film capacitors have excellent frequency characteristics, high ripple current tolerance, high voltage, high temperature, and shock resistance. They have a greater advantage in new energy vehicle applications than other capacitors. With the development of automotive electrification, the voltage and output power of the automotive circuit system have been substantially increased, and the key components require more stringent requirements for the voltage and shock resistance of electronic components. Changes in operating conditions put forward new requirements on the performance of capacitive components in the electronic system, thus driving the demand for film capacitors in the automotive industry.

Film capacitors are the main components in the drive circuit of electric vehicles, mainly playing the role of smoothing. The inverter will generate DC through the converter creating a small change in voltage and then through the IGBT switching elements into a rectangular wave similar to the alternating current; the resulting surge voltage is very large, hence the need to eliminate the use of smoothing capacitors. Early smoothing capacitors used aluminum electrolytic capacitors, but in order to improve efficiency, the maximum voltage of the motor drive was increased from 500 to 650 V, the voltage resistance of aluminum electrolytic capacitors was insufficient, and film capacitors became mainstream; for example, film capacitors have been used in the Toyota Prius since the second generation.

Film capacitors use organic film as an insulating medium. The organic film evaporates on the surface and forms the electrode with the metal layer. These capacitors are suitable for harsh conditions, offering stable capacity, low self-inductance, and long-lasting performance, making them ideal for high-power converter requirements.

4.3 Manufacturing Process and Equipment of Film Capacitors

Metal foil film capacitors can be applied directly on the polymer film with a thin metal foil, usually aluminum foil, which acts as electrodes. This process is relatively simple; the electrode wiring is convenient and can be applied to high-current occasions. Metallized film capacitors, by vacuum deposition process directly on the surface of the polymer film, form a very thin metal layer as the electrode. Due to the thin electrode thickness, it can be wound into a larger-capacity capacitor. However, due to the thin electrode thickness, it is only suitable for small current applications[39,40]. The following is an example of the preparation process of metallized polypropylene film capacitors, which are currently the most widely produced and used. Figure 5.7 shows the flow chart of the preparation process of metallized film capacitors. The main processes include biaxial stretching orientation, metallization (vaporization), cutting,

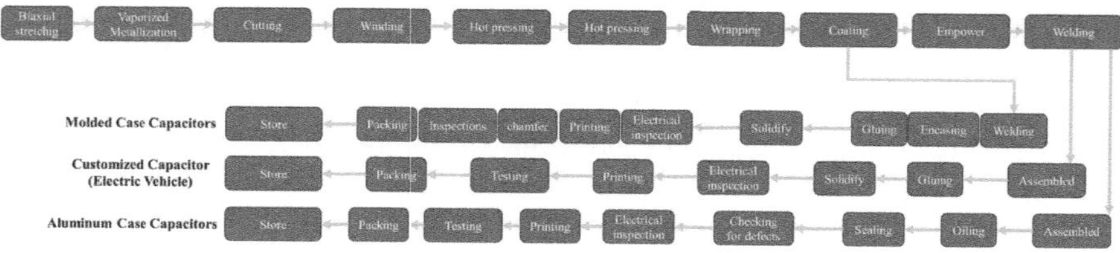

FIGURE 5.7 Fabrication process for metallized film capacitors.

winding, hot pressing, etc., followed by encapsulation, welding, etc., according to the requirements of different applications.

1. The biaxial orientation process involves co-extruding polymer particles to form a sheet and then stretching it in both longitudinal and transverse directions to obtain a film.
2. The metallization process is to vaporize, deposit, or coat metal electrode materials on the surface of the film.
3. Slitting process: Slitting the semi-finished film into the finished film according to the specified size.
4. Winding process: the finished film is rolled into a core. The core is the most basic and important element of the film capacitor.
5. Metal spraying process: Spray the two end surfaces of the core with a metal layer to facilitate the welding of the lead core.
6. Enabling process: After the core is sprayed with metal, the core is tested for charging and discharging. Test whether its capacity, loss, self-healing, and voltage withstand meet the design requirements.
7. Soldering assembly process: The core is soldered with leads and solder, and various wires and copper foils are combined for series and parallel connection.
8. Core and core group testing process: Conduct withstand voltage, capacity, and loss tests on the core and core groups.
9. Impregnation process: under high temperature and vacuum, the core and core group are impregnated to remove the air, water, and insulating oil in the core.
10. Assembly process: assembling the core into the shell, insulation treatment. Installation of leads, insulating film, insulating paper, shells, insulators, lead posts, etc.
11. Semi-finished product testing process: conduct voltage, capacity, loss, and insulation tests on semi-finished products.
12. Encapsulation and painting process: Encapsulation, sanding, cleaning, and painting of semi-finished products.
13. Finished product testing process: Voltage withstand, capacity, loss, insulation test for finished products. Life tests, accurate capacity tests, loss tests, and internal resistance and inductance tests are also carried out depending on the situation.

The following section will detail the biaxial orientation and metallization processes.

4.3.1 Biaxial Orientation Film Capacitors

The biaxial orientation (or biaxial stretching) process is an important process for industrial dielectric film capacitor manufacturing, which enables highly ordered alignment of polymer chains and is suitable for large scale, large area, continuous production of industrial capacitor films with uniform thickness, high structural quality isotropy, and superior capacitive performance. Conventional biaxially oriented equipment is a double-blown film equipment with a tensioning frame. The film thickness of the double-blown bubble equipment is usually around 15–250 microns, which is too thick for a capacitor, while the tenter frame can produce films less than 15 microns thick and is widely used for mass production (Figure 5.8).[41]

The first step of the biaxial orientation process is the extrusion section, where the polymer material is heated to its melting point and extruded through a die to produce flat sheets. The second step is the extruded flat sheets' continuous or simultaneous biaxial stretching. Two directions are used for stretching: one is machine direction (MD), which is the direction in which the flat sheet moves along the machine (length direction); the other is transverse direction (TD), which is perpendicular to MD (width direction). While simultaneous biaxial orientation is more desirable for improving the quality of stretch films, continuous biaxially oriented films are more popular in large-scale industrial production due to their relatively simple design and lower equipment costs.[41]

Biaxial stretching has two main advantages: (1) After biaxial stretching, the polymer film has a more uniform thickness distribution, which can prevent electric field distortion due to uneven thickness, thus meeting the demand for a high working electric field. The biaxially stretched polymer film has fewer defects and significantly improved breakdown performance. In addition, compared with other film preparation processes, biaxial stretching can obtain a thinner film thickness. Since the energy density per unit volume is inversely proportional to the film thickness, simultaneous improvement of dielectric permittivity and breakdown strength is

FIGURE 5.8 Picture of biaxial stretching orientation equipment.

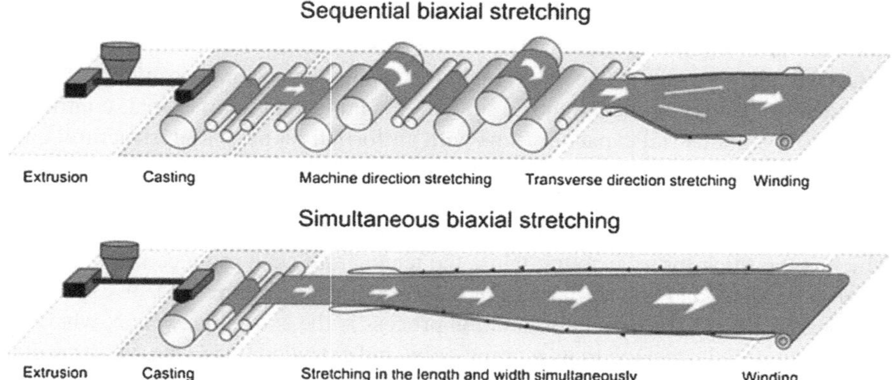

FIGURE 5.9   Typical tenter-frame biaxial stretching processes for the fabrication of polymer films.[43]

difficult in the short term. Therefore, it is a feasible and effective method to achieve high-energy density film capacitors by reducing the dielectric film thickness. The film thickness obtained by stretching is only 2~3 μm, and the thickness uniformity is good, with a variation of no more than 5% over a large area of film.[42] (2) Biaxial stretching facilitates the complete stretching and orientation of polymer molecular chains, which is very beneficial for improving the crystallinity of films and obtaining high breakdown strength (Figure 5.9).

Despite the excellent properties exhibited by the biaxially oriented film preparation process, which meets the industrial demand for large-scale film preparation, the complex process of preparing capacitors from dielectric polymers still leaves many problems for the conversion of laboratory results to industry. Compared with the industrial requirements for film size (over 100 m in length and 150 mm in width), the area of laboratory-prepared films is much smaller, and it is difficult for the laboratory-obtained films to meet the needs of large-area film preparation processes.[44] In addition, with the gradual increase in energy storage requirements, higher energy density per unit volume of capacitors is required, and the preparation of ultra-thin films (less than 5 μm) is essential. However, as the thickness decreases, the problem of thickness uniformity arises, and small deviations in film thickness can lead to severe electric field concentration and, thus, deterioration of capacitive performance. Therefore, the problem of performance degradation due to large-area film preparation from laboratory results to industrial scale requires the joint efforts of scientists and engineers from all related fields to obtain high-energy density film capacitors suitable for industrial production by rational optimization of the process.

4.3.2 Metallized Film Capacitors

The metallization process for metallized film capacitors generally employs evaporation technology, which is a method of preparing thin films by using high temperatures to evaporate metallic materials and deposit them on the surface of a substrate. Specifically, the metal material is first heated to its evaporation temperature to form an evaporation source, and then the gas around the evaporation source is extracted by a vacuum system to create a high vacuum environment. Next, a substrate is placed directly above the evaporation source so that the evaporated metal particles are deposited on the surface of the substrate to form a thin film.

Metallized aluminized films have the following characteristics compared to aluminum foil films:

1. Greatly reduce the amount of aluminum, saving energy and materials reducing costs. The thickness of the metal foil is mostly 7~8 μm, while the thickness of the aluminum layer of aluminized film is about 0.05 μm and its aluminum consumption is about 1/140~1/180 of aluminum foil, and the production speed can be as high as 450 m/min.
2. Excellent folding resistance and good toughness, few pinholes and cracks, no rubbing and cracking phenomenon, so the barrier to gas, water vapor, odor, light, etc. will set to improve.
3. The aluminized layer has good electrical conductivity and can eliminate static effects. Its sealing performance is good, especially when packaging powder products, and will not contaminate the sealing part to ensure the sealing performance of the packaging.

Due to the above characteristics, aluminized film has become a new type of composite film with excellent performance and economy and has replaced aluminum foil film in many aspects. With BOPP, PE, and another non-polar polymer film, the film surface must be corona-treated before vaporization so its surface tension reaches 38×10^{-5} N ~ 40×10^{-5} N. When coating, the roll film is placed in the vacuum chamber, and the vacuum chamber is closed to draw a vacuum. When the vacuum reaches a certain level, the evaporation source is heated up to 1,400°C ~ 1,500°C, and then the aluminum wire with a purity of 99.9% is continuously sent to the high-temperature area of the evaporation source. Adjustments are made in the unwinding, winding speed, wire feeding speed, and evaporation amount, and the cooling source is opened. The aluminum wire is continuously melted and evaporated in high-temperature zones to form a bright aluminum layer on the surface of the moving film, and the finished product is made after cooling.

Currently, BOPP metallization equipment is mainly used as a winding vacuum coating machine; the schematic diagram in Figure 5.10 shows the film from the roll-out side attached to the roll-through (cooling rollers) of the lower part and then into the vapor deposition chamber. The plated film is cooled and then rolled out by the rollers. The thinnest thickness of the film can be vaporized at 1 μm; the physical picture of the equipment is shown in Figure 5.11; the equipment combines mechanical, vacuum, refrigeration, and other technologies in one, with a very high degree of automation.

Vapor deposition technology is an important process method in the manufacturing process of film capacitors. Vapor deposition technology can prepare high-quality metal film to meet the needs of film capacitors in different fields. In the future, with the continuous development of science and technology, vapor deposition technology will continue to be improved and innovative to provide more possibilities for manufacturing film capacitors.

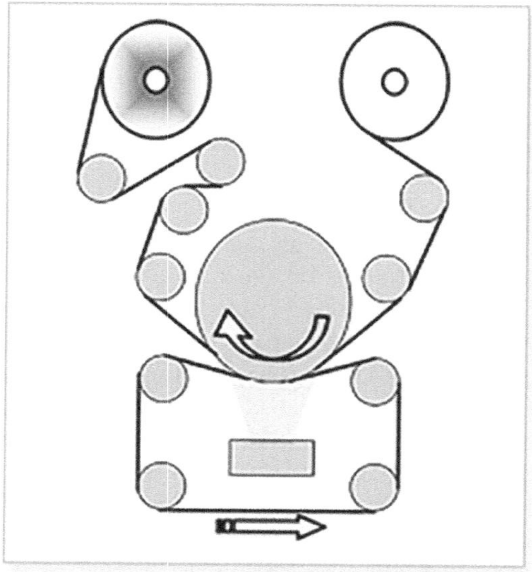

FIGURE 5.10   Schematic diagram of winding type vacuum coating machine.

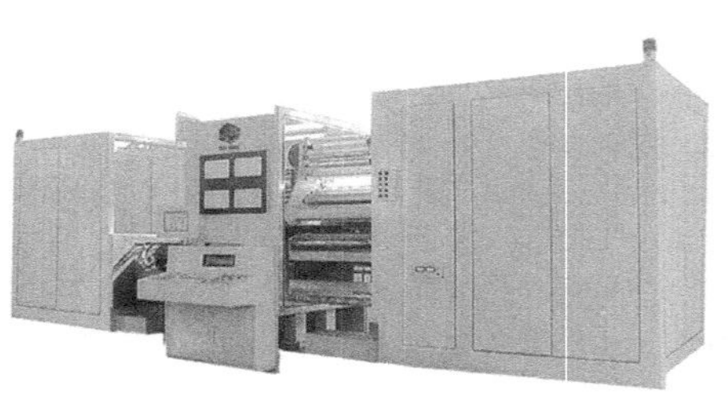 

FIGURE 5.11   Physical picture of the vacuum coating machine.

4.4 MANUFACTURING PROCESS FOR LAB-SCALE FILM

Limited by the intrinsic properties of materials and experimental equipment, not all polymers are suitable for preparation into thin films by the melt method; at present, polymer dielectric films in the laboratory are mainly prepared by solution casting (solvent evaporation) and spin coating methods. The polymer powder or particles are dissolved in a solvent, and the precursor solution is prepared by fully stirring and dissolving. The film is prepared on the substrate using a casting solution, spin coating, and spraying. Finally, dry and stabilized film is obtained by heat treatment. The solution casting process is suitable for the preparation of films with thicknesses of a few micrometers to tens of micrometers, while spin coating allows for the preparation of thinner films with thicknesses down to sub-micrometers. In addition, the thickness of the film can be precisely controlled by adjusting various parameters in the liquid-phase process, including squeegee height for the solution casting method, spin coating speed and time for the spin coating method, and solvent concentration. In addition, the quality of the film is also related to the nature of the solvent (viscosity, polarity, evaporation rate, etc.), the intrinsic properties of the polymer, and the post-treatment method.

The solution casting method is capable of preparing continuous polymer films over large areas in a continuous manner and is widely used in fields such as solar cells. Solution casting is a technique well-suited for large-scale coating, and it is also well-suited for producing thicker films from viscous solutions. It does not provide the nanoscale uniformity or extremely thin films that spin coating can provide.[45] A clean substrate is placed between rails of equal thickness and held in place, then the solution to be coated is dripped onto the side of the substrate near the squeegee, and the squeegee is moved at a constant speed to produce a thin, uniform, wet film. In this process, the film-forming time can be controlled by heating the substrate, and the thickness of the film can be controlled by a series of parameters, such as the size of the scraper, the concentration of the solution, the scraping speed, scraping temperature, and so on.

As one of the many film-forming methods, spin coating is widely used in the fields of microelectronics, biology, and organic functional films because of its ability to accurately control the film thickness and its many advantages, such as saving raw materials, low pollution, and environmental protection. The principle of spin coating is roughly divided into two stages: solvent evaporation and solute curing, as shown in Figure 5.12. The operation of spin coating is as follows: first, a drop of polymer solution is applied to the film-forming substrate fixed in the spin coater, and then the relevant coefficients of the spin coater are set, e.g., rotational speed, acceleration, and spin coating time. Then, the spin coater started to spin coat the film. The polymer solution is thrown out of the substrate due to gravity and centrifugal force, and the

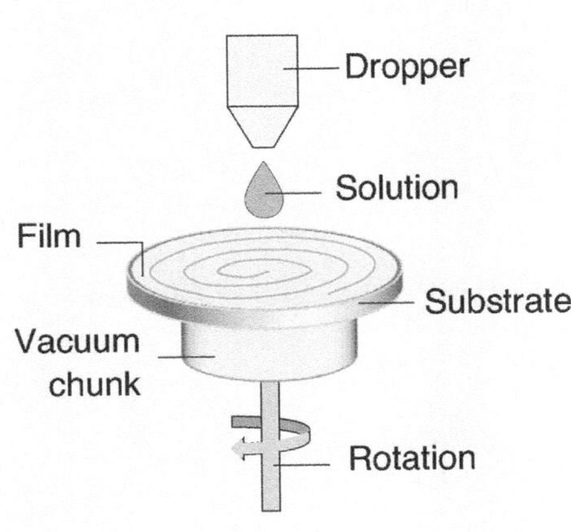

FIGURE 5.12 Schematic diagram of the spin coating method.[46]

remaining polymer solution continues to thin out on the substrate due to volatilization, and the viscosity increases due to the volatilization of the liquid so that the residual liquid is no longer able to move on the substrate, and the polymer molecule chains run through each other on the substrate, entanglement leads to the solidification of the solute, and a uniform polymer nanofilm is formed. A homogeneous polymer nanofilm is formed.[46]

The solution process is also suitable for the fabrication of polymer nanocomposite films. Fillers such as spherical nanoparticles, 1D nanorods and nanowires, and 2D nanosheets can be composited into the polymer matrix to increase the dielectric constant of the matrix. The homogeneous dispersion of fillers in the polymer matrix is crucial for the overall dielectric properties of the resulting composite films. Therefore, surface modification of the filler is very important in order to improve the interfacial interaction between the filler and the matrix. Due to the presence of solvents, liquid-phase film formation inevitably suffers from some internal defects, which can have some impact on the dielectric properties of the material, while the recycling and disposal of solvents also pose some difficulties. This is the reason why wet film formation has not yet been used for large-scale film production, but the simplicity and scalability of the liquid-phase method will be the focus of future research on commercialized films.

The dielectric polymer film still needs to go through a complex process, as shown in Figure 5.13, to finally obtain the finished film capacitor, whose subsequent steps mainly include surface metallization, film slitting, winding, end metallization, healing, end soldering, painting, end testing, etc.[47, 48] First, hundreds of nanometers of zinc-aluminum alloy need to be deposited on the stretched film surface using the physical vapor deposition (PVD) method as the surface electrode. Then the metallized film is slit to the target size by film slitting according to the actual demand, and next, the initial film capacitor is prepared by a high-speed winding process. Then the end face of the original film capacitor is capped and

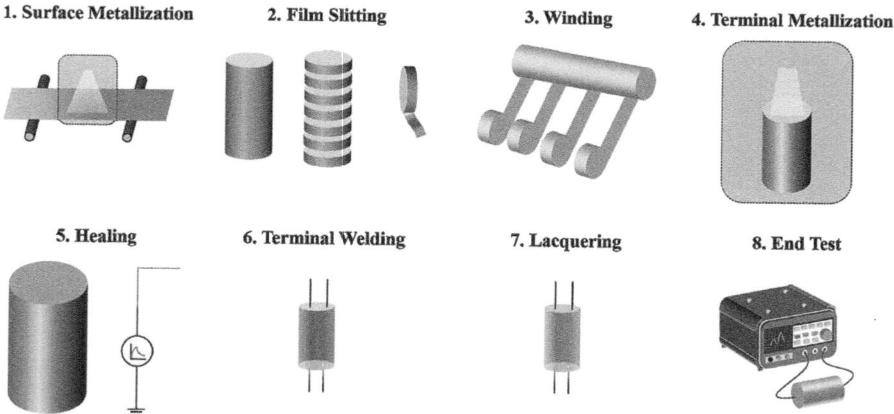

FIGURE 5.13   Typical manufacturing process flow for the fabrication of polymer film capacitors.[42]

metallized with zinc-aluminum alloy or gold. A healing process is then performed to eliminate film defects with surge voltages. The leads are then soldered to the terminal surfaces of the original film capacitors, followed by a lacquer treatment to protect the film capacitors from damage by external environments. Finally, in order to test the capacitors to meet qualification requirements, an end test of the capacitors is required.[49]

4.5　INDUSTRY RESEARCH

The United States, Germany, Japan, Taiwan, and mainland China are the main production area of film capacitors, of which the United States, Germany, Japanese manufacturers, and leading domestic manufacturers in the high-end film capacitor market occupies a major position; Taiwan and mainland manufacturers mainly occupy low-end film capacitor market. Among them, the more well-known supplier of domestic products, Xiamen Faratronic, promotes products which mainly focus on lighting, consumer electronics, and other fields. Its main competitors include Panasonic Electric, KEMET TDK, etc.

In the supply of film capacitors, there are more than 100 large film capacitor manufacturers in mainland China. It has a total annual production capacity of about 12 billion, of which 60% of the products are exported. The world's chip organic film capacitors production is in about 30 countries and regions, with about 50 production enterprises. The dielectric material used is also increasingly diversified; in the miniaturization of cutting-edge technology, Japan, Italy, and the United States are at the forefront of the world and are available to produce today's smallest size of 0603 products.

1. Panasonic – Leading Automotive Film Capacitors

   High market share of automotive film capacitors: According to Passive Devices Industry News on October 15, 2019, Panasonic's global market share of automotive film capacitors is as high as 70%. Self-developed technology and materials to enhance the performance of film capacitors: Panasonic uses a film with safety features in the electrodes using its original vapor deposition technology to ensure safety and adopts an externally developed material to achieve high moisture resistance to extend the life of film capacitors.

2. Nichicon – Expansion of film capacitors for inverters in new energy vehicles

   Focus on three product lines and four markets: Nichicon has three product lines, "Aluminum Electrolytic Capacitors," "Film Capacitors," and "Circuit Products," and focuses on energy, automotive and vehicle equipment, white goods, and industrial applications. The company focuses on four major markets: energy, automotive vehicle equipment, white goods and industrial inverters, and information and communication equipment (computers and tablets). Farad's film capacitors for inverters in new energy vehicles, with outstanding production and technology advantages: cooperation with leading film companies to develop film; establishment of film evaporation mass production system to enhance the price competitiveness of the products; small-sized and lightweight technology; unique evaporation format to achieve the safety of self-healing film capacitors; simulation technology.

3. Xiamen Faratronic – new energy vehicles + photovoltaic applications accounted for more than 50%, good earnings quality.

   In-depth layout in the field of new energy, rich customer resources: new energy vehicle industry customers include BYD, SAIC, Geely, Azure, Mercedes-Benz, BMW, Volkswagen, etc., and the company's new energy vehicle domestic market share in 19 years amounted to 40%, photovoltaic industry customers include Huawei, Sungrow, Power-one, Vestas, Sma, etc. Focus on film capacitor market, new energy revenue accounted for a high proportion: the company focuses on film capacitor market, 21 years revenue reached 2.811 billion yuan (+48.66% year-on-year), net profit of 831 million yuan (+49.51% year-on-year). We estimate that in 2021, the company's photovoltaic field revenue of about 871 million yuan (+53.62% year-on-year), accounting for 31% of revenue (+1 pct year-on-year), and new energy vehicle field revenue of about 759 million yuan (+122.99% year-on-year), accounting for 27% of revenue (+9 pcts year-on-year).

4. Nantong Jianghai Capacitor Co. Ltd. – joint Kemet production of film capacitors for automotive use

   The capacitor category is complete, and film capacitors are rapidly growing: aluminum electrolytic capacitors, film capacitors, and supercapacitors are three major product lines; film capacitors in consumer electronics and industrial equipment have entered the stage of high-volume applications, and in the new energy vehicles and military field began to deliver in batches. In 2021, the company achieved a revenue of 224 million yuan from film capacitors, with a gross profit margin of 20.62%. The company's goal for the next three years is to achieve an annual average growth of over 40% in revenue from film capacitors by producing them. Joint production of automotive film capacitors with Kemet: in 2018, the company's subsidiary Nantong New Jianghai Power Electronics and the leading film capacitor manufacturer Kemet each invested 50% to set up Nantong Haimei Electronics, a joint production of automotive customized film capacitors and axial aluminum electrolytic capacitors. HMI Electronics has won a number of film capacitor projects for electric drives of automobile brands and electric drives manufacturers.

5. Tongfeng Electronics–Film Capacitor Industry Chain Integration Layout

   Tongfeng Electronics is engaged in the research and development of film capacitors and film materials and owns the integrated industrial chain of film+metallized film+film capacitors. In 2021, the company's film capacitors achieved revenues of 400 million yuan, an increase of 9% year-on-year, with a gross profit margin of 20.60%, up 0.16% year-on-year, expanding the production of film capacitors for direct current transmission and automobiles. In June 2021, the company announced its intention to build a project to manufacture film capacitors for direct current grid transmission, and the project is expected to be completed shortly. In June 2021, the company announced that it would build a capacitor project for DC grid transmission, with an annual output of 8,000 capacitors for flexible DC transmission and 13,000 damping capacitors for extra-high voltage transmission, and the construction period of the project would last 24 months; in December 2021, the company announced that it would set up a joint venture with Sungmun Electronics Co., Ltd., a South Korean metallized film enterprise, with an annual output of 1,000,000 film capacitors for new energy automobile filtering, and the construction period of the project would last 24 months. (Source: Future Intelligence)

The global film capacitor market share in the world includes Japan's Nichicon, Germany's Wima, Italy's ICEL, the U.S.'s CDE, etc., who are the top film capacitor manufacturers. Among them, Wima's products are mainly used in high-quality audio, Nichion is mainly used in electronic products, and CDE is a professional manufacturer of film capacitors for inverters. Japan's Panasonic, EPCOS, Nichicon, and Japan's Gang Valley, the four heavenly kings of the film capacitor industry, remain the dominant force for quite a long time. In terms of production, Japan's Panasonic Electric, Germany's EPCOS, and the United States' Kemet are the world's leading film capacitor manufacturers. Domestically, Xiamen Faratronic is the leading film capacitor manufacturer, in addition to Anhui Tongfeng Electronics, Nantong Jianghai, etc., which has been a good development of enterprises in recent years. Although China's film capacitor industry, Xiamen Lafarge Electronics, ranks as the world's third largest production unit, when compared with developed countries, China's overall competitiveness in film capacitor enterprises is still weak.

With the development of new energy storage materials, and advanced energy storage devices around the world, China's manufacturers are also developing rapidly with the help of policy support and global capacity structure transfer. The future industrial development focus and innovation direction are as follows:

1. Higher operating voltage

   In order to further increase the range and improve the usage experience of new energy vehicles, the working voltage of the existing electrical architecture must be raised from 400 to 800 V. On the one hand, with the rise in the amount of single-vehicle power to make the car power battery power, if you still maintain the 400 V voltage specifications, the high power to increase the current will bring a series of power loss and thermal management issues; on the other hand, in order to achieve faster charging of new energy vehicles under lower power loss, also need to further increase the bus voltage. Therefore, the relevant electrical components also need to have the ability to work stably at higher voltages.

2. Smaller size

   Automotive continued electrification, digitalization, intelligent processes, and the relevant electrical system will be more complex and, therefore, require more compact design and implementation, which further puts forward the miniaturization requirements of passive components. In recent years, Panasonic film capacitors have shown a clear trend toward miniaturization in the process of volume change.

3. Higher reliability

During the driving process of new energy vehicles, especially in urban conditions, acceleration and braking are often required, and the capacitors in the electric drive system need to withstand the frequent impacts of instantaneous high-current conduction and shutdown, which further enhances the reliability requirements for capacitors and other related passive components.

5 CONCLUSION

This book provides an in-depth exploration of the energy storage theory associated with dielectric capacitors, delving into the research advancements across various dielectric materials. Commencing with an overview of the current global status of energy storage and emphasizing the advantages of dielectric capacitors in energy storage applications, the book proceeds to elucidate the definitions and calculation formulas for key energy storage parameters such as ε_r, E_b, U_d, and η.

The examination extends to the investigation of the impacts of multiple polarization mechanisms on εr and polarization properties, while also shedding light on the causes of breakdown and loss in dielectrics through an understanding of diverse breakdown mechanisms and conduction mechanisms. Chapters 2–4 categorize energy storage dielectrics into inorganic, organic, and composite types, systematically exploring the energy storage properties of various materials and strategies for enhancing their performance. Furthermore, the industrial preparation and application requirements of multilayer ceramic capacitors and polymer thin films are summarized.

The narrative in the chapters underscores a significant advancement in energy storage dielectrics, showcasing a simultaneous enhancement of ε_r and E_b. Despite notable progress, challenges persist in the realm of energy storage dielectrics, prompting a discussion on unresolved issues and charting future research priorities for dielectric materials, as detailed in the following sections:

1. Low ε_r or low E_b are the main problems that limit the energy storage applications of most dielectrics. Linear dielectrics have a high E_b and high η, but their ε_r is low. Normal ferroelectric dielectrics are the opposite. This limits the improvement of both U_d. Modification of ceramics by chemical modification, defect engineering, microstructure modulation, or modification of polymers by modification strategies such as blend, composite, and multilayer have proven to be effective methods to achieve energy storage density in dielectric materials. Meanwhile, relaxor ferroelectric and antiferroelectric dielectrics with narrow hysteresis loops exhibiting low remnant polarization are also expected to achieve a high ε_r and high η. However, the preparation and theory of relaxor ferroelectrics and antiferroelectrics are complicated. The way to prepare special ferroelectric polymers by simple methods may be a key direction for future research.

2. The current state-of-the-art studies generally treated their internal polarization and constant dielectric changes for energy storage dielectric materials as an approximately linear relationship. However, this dynamic process has complex polarization characteristics and requires more theoretical studies for guidance. Meanwhile, theories related to the internal ferroelectric phase transition, interface polarization mechanism, and multilayer electric field distribution still require a lot of effort from researchers in order to match the polarization curves of various linear, ferroelectric, and antiferroelectric dielectric materials through the mechanism, and to obtain multi-series systems with high-energy storage density, and high-energy storage efficiency.

3. When capacitors are used in high-temperature environments such as electric vehicles and oil and gas exploration, there are very strict requirements for the thermal stability of capacitors. Researchers have come up with feasible solutions based on studies on breakdown and conduction mechanisms. Both feasible approaches are the introduction of linear and paraelectric dielectrics to reduce losses, the preparation of multilayer structures, and the use of differences in dielectric properties between layers to impede charge and breakdown paths. In addition, there are two key directions for researchers to solve the high-temperature problem of polymers in the future. Adding a high thermal conductivity second phase or coating the dielectric surface with an insulating nanolayer to increase the Schottky barrier to inhibit charge injection to inhibit charge injection to improve capacitor performance.

4. It can be known from the previous introduction that inorganic ceramic dielectric materials have the characteristics of high dielectric constant and high-temperature resistance, but the disadvantage is the low breakdown field strength with higher loss. Organic polymer dielectric materials, on the contrary, have good temperature stability and high breakdown field strength. Therefore, many researchers have been committed to developing polymer-ceramic composites and two-phase composites for many years. However, due to the differences in intrinsic mechanical properties and the dielectric properties of the two phases, they are severely limited by the preparation methods, e.g., high-temperature sintering is required for inorganic materials, while stretching treatment is required for organic materials. It has become a research priority to develop a new process method that can synthesize the preparation of both materials, which is expected to take advantage of the respective properties of organic and inorganic dielectrics.

5. Although some dielectrics with ultra-high levels of energy storage properties have been successfully fabricated in the laboratory, they are not feasible for large-scale commercial production. Dielectrics applied to capacitors need to consider not only the energy storage properties of the material but also the mechanical and thermal stability, service life, and other comprehensive performance improvements to achieve the preparation of devices in industry.

In the future, the development of dielectric capacitors will be more rapid, and the energy storage requirements of dielectric materials will be more stringent. Based on computer simulations and analysis, it will become feasible for researchers to select suitable dielectric materials and design suitable structures to solve these problems through continuous experiments. The preparation of dielectric materials with better performance through various modification methods will contribute to the innovative development of the energy storage industry and effectively solve global energy problems. This requires the joint efforts and cooperation of staff and researchers in related fields.

REFERENCES

1. Ahmed, S. A.; Mohsin, M.; Ali, S. M. Z. Survey and technological analysis of laser and its defense applications, *Def. Technol.* **2021**, 17, 583–592. DOI: 10.1016/j.dt.2020.02.012.
2. Li, D. X.; Zeng, X. J.; Li, Z. P.; Shen, Z. Y.; Hao, H.; Luo, W. Q.; Wang, X. C.; Song, F. S.; Wang, Z. M.; Li, Y. M. Progress and perspectives in dielectric energy storage ceramics, *J. Adv. Ceram.* **2021**, 10, 675–703. DOI: 10.1007/s40145-021-0500-3.
3. Xu, Q. W.; Vafamand, N.; Chen, L. L.; Dragicevic, T.; Xie, L. H.; Blaabjerg, F. Review on advanced control technologies for bidirectional DC/DC converters in DC microgrids, *IEEE J. Emerging Sel. Top. Power Electron.* **2021**, 9, 1205–1221. DOI: 10.1109/Jestpe.2020.2978064.
4. Castro, M.; Kumar, B.; Feller, J. F.; Haddi, Z.; Amari, A.; Bouchikhi, B. Novel e-nose for the discrimination of volatile organic biomarkers with an array of carbon nanotubes (CNT) conductive polymer nanocomposites (CPC) sensors, *Sens. Actuators, B* **2011**, 159, 213–219. DOI: 10.1016/j.snb.2011.06.073.
5. Hao, X. A review on the dielectric materials for high energy-storage application, *J. Adv. Dielectr.* **2013**, 03, 1330001. DOI: 10.1142/S2010135X13300016.
6. Yang, L.; Kong, X.; Li, F.; Hao, H.; Cheng, Z.; Liu, H.; Li, J. F.; Zhang, S. Perovskite lead-free dielectrics for energy storage applications, *Prog. Mater. Sci.* **2019**, 102, 72–108. DOI: 10.1016/j.pmatsci.2018.12.005.
7. Veerapandiyan, V.; Benes, F.; Gindel, T.; Deluca, M. Strategies to improve the energy storage properties of perovskite lead-free relaxor ferroelectrics: A review, *Materials* **2020**, 13. DOI: 10.3390/ma13245742.
8. Palneedi, H.; Peddigari, M.; Hwang, G.-T.; Jeong, D.-Y.; Ryu, J. High-performance dielectric ceramic films for energy storage capacitors: Progress and outlook, *Adv. Funct. Mater.* **2018**, 28, 1803665. DOI: 10.1002/adfm.201803665.
9. Qi, H.; Xie, A. W.; Zuo, R. Z. Local structure engineered lead-free ferroic dielectrics for superior energy-storage capacitors: A review, *Energy Storage Mater.* **2022**, 45, 541–567. DOI: 10.1016/j.ensm.2021.11.043.
10. Poonam; Sharma, K.; Arora, A.; Tripathi, S. K. Review of supercapacitors: Materials and devices, *J. Energy Storage* **2019**, 21, 801–825. DOI: 10.1016/j.est.2019.01.010.
11. Wang, G.; Lu, Z. L.; Li, Y.; Li, L. H.; Ji, H. F.; Feteira, A.; Zhou, D.; Wang, D. W.; Zhang, S. J.; Reaney, I. M. Electroceramics for high-energy density capacitors: Current status and future perspectives, *Chem. Rev.* **2021**, 121, 6124–6172. DOI: 10.1021/acs.chemrev.0c01264.
12. Reis, R. L. D.; Neves, W. L. A.; Lopes, F. V.; Fernandes, D. Coupling capacitor voltage transformers models and impacts on electric power systems: A review, *IEEE Trans. Power Delivery* **2019**, 34, 1874–1884. DOI: 10.1109/Tpwrd.2019.2908390.
13. Zimmermann, T.; Keil, P.; Hofmann, M.; Horsche, M. F.; Pichlmaier, S.; Jossen, A. Review of system topologies for hybrid electrical energy storage systems, *J. Energy Storage* **2016**, 8, 78–90. DOI: 10.1016/j.est.2016.09.006.

14. Laadjal, K.; Cardoso, A. J. M. Multilayer ceramic capacitors: An overview of failure mechanisms, perspectives, and challenges, *Electronics-Switz* **2023**, 12. DOI: 10.3390/electronics12061297.

15. Rodríguez-Benítez, O.; Ponce-Silva, M.; Aqui-Tapia, J. A.; Rodríguez-Benítez, O. M.; Lozoya-Ponce, R. E.; Adamas-Pérez, H. Active power-decoupling methods for photovoltaic-connected applications: An overview, *Processes* **2023**, 11. DOI: 10.3390/pr11061808.

16. Allendorf, M. D.; Dong, R. H.; Feng, X. L.; Kaskel, S.; Matoga, D.; Stavila, V. Electronic devices using open framework materials, *Chem. Rev.* **2020**, 120, 8581–8640. DOI: 10.1021/acs.chemrev.0c00033.

17. Li, Q.; Tan, S.; Gong, H.; Lu, J.; Zhang, W.; Zhang, X.; Zhang, Z. Influence of dipole and intermolecular interaction on the tuning dielectric and energy storage properties of polystyrene-based polymers, *Phys. Chem. Chem. Phys.* **2021**, 23, 3856–3865. DOI: 10.1039/d0cp05233g.

18. Huang, X.; Sun, B.; Zhu, Y.; Li, S.; Jiang, P. High-k polymer nanocomposites with 1D filler for dielectric and energy storage applications, *Prog. Mater. Sci.* **2019**, 100, 187–225. DOI: 10.1016/j.pmatsci.2018.10.003.

19. Dong, J. F.; Deng, X. L.; Niu, Y. J.; Pan, Z. Z.; Wang, H. Research progress of polymer based dielectrics for high-temperature capacitor energy storage, *ACTA PHYS SIN-CH ED* **2020**, 69. DOI: 10.7498/aps.69.20201006.

20. Yuan, Q.; Chen, M.; Zhan, S.; Li, Y.; Lin, Y.; Yang, H. Ceramic-based dielectrics for electrostatic energy storage applications: Fundamental aspects, recent progress, and remaining challenges, *Chem. Eng. J.* **2022**, 446. DOI: 10.1016/j.cej.2022.136315.

21. Luo, B.; Wang, X.; Sun, H.; Li, L. Dielectric, ferroelectric, and thermodynamic properties of silicone oil modified PVDF films for energy storage application, *Appl. Phys. Lett.* **2016**, 108. DOI: 10.1063/1.4954174.

22. Johnson, R. W.; Evans, J. L.; Jacobsen, P.; Thompson, J. R.; Christopher, M. The changing automotive environment: High-temperature electronics, *IEEE Trans. Electron. Packag. Manuf.* **2004**, 27, 164–176. DOI: 10.1109/TEPM.2004.843109.

23. Sakabe, Y. Multilayer ceramic capacitors, *Curr. Opin. Solid State Mater. Sci.* **1997**, 2, 584–587.

24. Covaci, C.; Gontean, A. "Singing" multilayer ceramic capacitors and mitigation methods—A review, *Sensors-Basel* **2022**, 22, 3869. DOI: 10.3390/s22103869.

25. Zhao, P. Y.; Cai, Z. M.; Wu, L. W.; Zhu, C. Q.; Li, L. T.; Wang, X. H. Perspectives and challenges for lead-free energy-storage multilayer ceramic capacitors, *J. Adv. Ceram.* **2021**, 10, 1153–1193. DOI: 10.1007/s40145-021-0516-8.

26. Xu, J. H.; Liu, D. B.; Lee, C.; Feydi, P.; Chapuis, M.; Yu, J.; Billy, E.; Yan, Q. Y.; Gabriel, J. C. P. Efficient electrocatalyst nanoparticles from upcycled class II capacitors, *Nanomaterials-Basel* **2022**, 12. DOI: 10.3390/nano12152697.

27. Templeton, A.; Reed, N.; Hayes, H.; Davis, J.; Bultitude, J. Class I Multi-Layer Ceramic Capacitors (MLCCs) Performance as Wide Band Gap (WBG) Snubbers in Hard Switching Applications, 2023 Fourth International Symposium on 3d Power Electronics Integration and Manufacturing, 3d-Peim **2023**. DOI: 10.1109/3d-Peim55914.2023.10052607.

28. Zhang, H. B.; Wei, T.; Zhang, Q.; Ma, W. G.; Fan, P. Y.; Salamon, D.; Zhang, S. T.; Nan, B.; Tan, H.; Ye, Z. G. A review on the development of lead-free ferroelectric energy-storage ceramics and multilayer capacitors, *J. Mater. Chem. C* **2020**, 8, 16648–16667. DOI: 10.1039/d0tc04381h.

29. Hong, K.; Lee, T. H.; Suh, J. M.; Yoon, S. H.; Jang, H. W. Perspectives and challenges in multilayer ceramic capacitors for next generation electronics, *J. Mater. Chem. C* **2019**, 7, 9782–9802. DOI: 10.1039/c9tc02921d.

30. Wang, H.; Zeng, C.; Liu, B.; Wang, X. Energy storage properties of surface-modified BaTiO 3 ceramic films for multilayer capacitors applications, *J. Adv. Dielectr.* **2019**, 09, 1950027. DOI: 10.1142/S2010135X19500279.

31. Kumar, N.; Ionin, A.; Ansell, T.; Kwon, S.; Hackenberger, W.; Cann, D. Multilayer ceramic capacitors based on relaxor BaTiO3-Bi(Zn1/2Ti1/2)O3 for temperature stable and high energy density capacitor applications, *Appl. Phys. Lett.* **2015**, 106, 252901. DOI: 10.1063/1.4922947.

32. Zhao, P.; Wang, H.; Wu, L.; Chen, L.; Cai, Z.; Li, L.; Wang, X. High-performance relaxor ferroelectric materials for energy storage applications, *Adv. Energy Mater.* **2019**, 9, 1803048. DOI: 10.1002/aenm.201803048.

33. Li, J.; Li, F.; Xu, Z.; Zhang, S. Multilayer lead-free ceramic capacitors with ultrahigh energy density and efficiency, Advanced Materials **2018**, 30, 1802155. DOI: 10.1002/adma.201802155.

34. Li, J.; Shen, Z.; Chen, X.; Yang, S.; Zhou, W.; Wang, M.; Wang, L.; Kou, Q.; Liu, Y.; Li, Q.; Xu, Z.; Chang, Y.; Zhang, S.; Li, F. *Grain-orientation-engineered multilayer ceramic capacitors for energy storage applications, Nat Mater* **2020**, 19, 999–1005. DOI: 10.1038/s41563-020-0704-x.

35. Wang, G.; Li, J.; Zhang, X.; Fan, Z.; Yang, F.; Feteira, A.; Zhou, D.; Sinclair, D. C.; Ma, T.; Tan, X.; Wang, D.; Reaney, I. M. Ultrahigh energy storage density lead-free multilayers by controlled electrical homogeneity, *Energy Environ. Sci.* **2019**, 12, 582–588. DOI: 10.1039/c8ee03287d.

36. Tan, D. Q. Review of polymer-based nanodielectric exploration and film scale-up for advanced capacitors, *Adv. Funct. Mater.* **2020**, 30. DOI: 10.1002/adfm.201808567.

37. Wu, X. D.; Chen, X.; Zhang, Q. M.; Tan, D. Q. Advanced dielectric polymers for energy storage, *Energy Storage Mater.* **2022**, 44, 29–47. DOI: 10.1016/j.ensm.2021.10.010.

38. Feng, Q. K.; Zhong, S. L.; Pei, J. Y.; Zhao, Y.; Zhang, D. L.; Liu, D. F.; Zhang, Y. X.; Dang, Z. M. Recent progress and future prospects on all-organic polymer dielectrics for energy storage capacitors, *Chem. Rev.* **2022**, 122, 3820–3878. DOI: 10.1021/acs.chemrev.1c00793.

39. Chen, Q.; Shen, Y.; Zhang, S.; Zhang, Q. M. Polymer-based dielectrics with high energy storage density, *Annu. Rev. Mater. Res.* **2015**, 45, 433–458. DOI: 10.1146/annurev-matsci-070214-021017.

40. Valentine, N.; Azarian, M. H.; Pecht, M. Metallized film capacitors used for EMI filtering: A reliability review, *Microelectron. Reliab.* **2019**, 92, 123–135. DOI: 10.1016/j.microrel.2018.11.003.

41. Wang, D.; Dang, Z. M. Processing of polymeric dielectrics for high energy density capacitors, In *Dielectric Polymer Materials for High-Density Energy Storage*; Elsevier, **2018**, pp. 429–446.

42. Feng, Q. K.; Zhong, S. L.; Pei, J. Y.; Zhao, Y.; Zhang, D. L.; Liu, D. F.; Zhang, Y. X.; Dang, Z. M. Recent progress and future prospects on all-organic polymer dielectrics for energy storage capacitors, *Chem. Rev.* **2022**, 122, 3820–3878. DOI: 10.1021/acs.chemrev.1c00793.

43. Fan, B.; Zhou, M.; Zhang, C.; He, D.; Bai, J. Polymer-based materials for achieving high energy density film capacitors, *Prog. Polym. Sci.* **2019**, 97, 101143. DOI: 10.1016/j.progpolymsci.2019.06.003.

44. Tan, D. Q. Review of polymer-based nanodielectric exploration and film scale-up for advanced capacitors, *Adv. Funct. Mater.* **2020**, 30, 1808567. DOI: 10.1002/adfm.201808567.

45. Xie, L. F.; Wang, G. L.; Jiang, C.; Yu, F. P.; Zhao, X. Properties and applications of flexible poly(vinylidene fluoride)-based piezoelectric materials, *Crystals* **2021**, 11. DOI: 10.3390/cryst11060644.

46. He, Q. F.; Sun, K.; Shi, Z. C.; Liu, Y.; Fan, R. H. Polymer dielectrics for capacitive energy storage: From theories, materials to industrial capacitors, *Mater. Today* **2023**, 68, 298–333. DOI: 10.1016/j.mattod.2023.07.023.

47. Breil, J. Future trends for biaxially oriented films and orienting lines, In *Biaxial Stretching of Film*; Elsevier, **2011** pp. 240–273.

48. Ding, L.; Shao, L.; Bai, Y. Deciphering the mechanism of corona discharge treatment of BOPET film, *RSC Advances* **2014**, 4, 21782–21787. DOI: 10.1039/c4ra02289k.

49. Barbaro, G.; Galdi, M. R.; Di Maio, L.; Incarnato, L. Effect of BOPET film surface treatments on adhesion performance of biodegradable coatings for packaging applications, *Eur. Polym. J.* **2015**, 68, 80–89. DOI: 10.1016/j.eurpolymj.2015.04.027.

Index